Kenneth R. Lang

The Sun from Space

With 96 Figures
Including 46 Color Figures
and 32 Tables

Springer

Professor Kenneth R. Lang
Department of Physics and Astronomy
Robinson Hall
Tufts University
Medford, MA 02155, USA

Cover picture: X-ray View of the Solar Cycle. A montage of images from the Soft X-ray Telescope (SXT) aboard Yokhoh showing dramatic changes in the corona as the Sun's 11-year magnetic activity cycle changed from maximum to minimum. The 12 images are spaced at 120-day intervals from the time of the satellite's launch in August 1991, at the maximum phase of the 11-year sunspot cycle (*left*), to late 1995 near the minimum phase (*right*); see Fig. 7.12 on page 235. (Courtesy of Gregory L. Slater and Gary A. Linford, NASA, ISAS, the Lockheed-Martin Solar and Astrophysics Laboratory, the National Astronomical Observatory of Japan, and the University of Tokyo)

Library of Congress Cataloging-in-Publication Data.
Lang, Kenneth R. The sun from space/ Kenneth R. Lang. p. cm. – (Astronomy and astrophysics library). Includes bibliographical references and index. ISBN 3540669442 (alk. paper) 1. Sun. 2. Astronautics in astronomy. I. Title. II. Series. QB521.L25 2000 523.7–dc21 00-041038

ISSN 0941-7834
ISBN 3-540-66944-2 Springer-Verlag Berlin Heidelberg New York

Springer-Verlag Berlin Heidelberg New York
a member of BertelsmannSpringer Science+Business Media GmbH

Printed in Germany

Typesetting: Data conversion by LE-TeX, Leipzig
Cover design: *design & production* GmbH, Heidelberg

Printed on acid-free paper SPIN: 10717748 55/3141/ba - 5 4 3 2 1 0

This book is dedicated to everyone
who is curious, imaginative and intelligent,
thereby strengthening the human spirit.

Preface

Our familiar, but often inscrutable, star exhibits a variety of enigmatic phenomena that have continued to defy explanation. Our book begins with a brief account of these unsolved mysteries. Scientists could not, for example, understand how the Sun's intense magnetism is concentrated into dark sunspots that are as large as the Earth and thousands of times more magnetic. Nor did they know exactly how the magnetic fields are generated within the Sun, for no one could look inside it.

Another long-standing mystery is the million-degree solar atmosphere, or corona, that lies just above the cooler, visible solar disk, or photosphere. Heat should not emanate from a cold object to a hotter one anymore than water should flow up hill. Researchers have hunted for the elusive coronal heating mechanism for more than half a century.

The Sun's hot and stormy atmosphere is continuously expanding in all directions, creating a relentless solar wind that seems to blow forever. The exact sources of all the wind's components, and the mechanisms of its acceleration to supersonic velocities, also remained perplexing problems.

The relatively calm solar atmosphere can be violently disrupted by powerful explosions, filling the solar system with radio waves, X-rays, and gamma rays, and hurling charged particles out into space at nearly the speed of light. Other solar explosions, called Coronal Mass Ejections, throw billions of tons of coronal gases into interplanetary space, creating powerful gusts in the solar wind. Yet, we have only just begun to understand the detailed causes of the Sun's explosive outbursts, and no one has been able to predict exactly when they will occur.

In less than a decade, three pioneering spacecraft, named the SOlar and Heliospheric Observatory, or SOHO for short, Ulysses and Yohkoh, have transformed our perception of the Sun. They are also described in the introductory chapter, together with their principal scientific goals. This scientific troika has examined the Sun with exquisitely sensitive and precise instruments that have widened our range of perception, giving us the eyes to see the invisible and the hands to touch what cannot be felt. They have extended our gaze from the visible solar disk, down to the hidden core of the Sun and out in all directions through the Sun's tenuous, expanding atmosphere. SOHO, Ulysses and Yohkoh have together provided insights that are vastly more focused and detailed than those of previous solar missions, providing clues to many of the crucial, unsolved problems in our understanding of the Sun.

Scientific discoveries are usually not isolated "eureka" moments, occurring in a mental or experiential vacuum. They are built upon a foundation of previous investigations and extrapolated into the future. Our recent accomplishments in solar physics are the culmination of a long history of prior research, from the earliest optical and radio telescopes to rockets and then full-fledged spacecraft such as Skylab, Helios 1 and 2, and the Solar Maximum Mission.

The second chapter of our book describes the discovery of space, providing the historical framework needed to understand subsequent results. Here we also introduce ideas and terms that have become so common to our space-age vocabulary that even many practicing scientists have forgotten their origin and meaning.

As mentioned in Chap. 2, just half a century ago most people visualized out planet as a solitary sphere traveling in a cold, dark vacuum around the Sun. Inquisitive spacecraft have now shown that the space outside the Earth's atmosphere is not empty, and demonstrated that the Earth and other planets are immersed within an eternal, stormy wind. The solar wind is mainly composed of electrons and protons, set free from hydrogen atoms, the most abundant element in the Sun, but it also contains heavier ions and a magnetic field.

The second chapter also describes the discovery that intense magnetism pervades the Sun's atmosphere, creating an 11-year variation in the level of explosive solar activity, as well as the amount of cosmic rays reaching the Earth from the space outside our solar system. X-ray photographs from rockets and the Skylab space station demonstrated that the Sun's magnetic fields mold and constrain the million-degree gases found in the low solar atmosphere. Long-lasting coronal holes, without any detectable X-ray radiation in Skylab's photographs, were also compared with other spacecraft measurements to show that these places expel a high-speed solar wind.

Contextual and historical preludes are also woven into the texture of subsequent chapters, and usually presented in chronological order. Every chapter, except the first, contains a concluding time-line table that highlights key events and discoveries in the area under consideration. Complete references to seminal articles are given at the end of this book. The reader can consult these fundamental research papers for further details on important scientific results, from centuries ago to the present day.

Set-aside focus elements are also inserted throughout the text, that enhance and amplify the discussion with interesting details or fundamental physics. They will be read by the curious person or serious student, but do not interfere with the general flow of the text and can be bypassed by the general educated reader who wants to follow the main ideas.

The discovery of how the Sun shines by thermonuclear reactions within its hot, dense core is spelled out in the beginning of the third chapter of our book. It then describes how the unseen depths of the Sun are being probed using the new science of helioseismology. Today, we can peel back the outer layers of the Sun and glimpse inside it by observing widespread throbbing motions in the Sun's visible material. Helioseismologists decipher these oscillations, caused by sounds trapped within the Sun, to reveal its internal constitution. This procedure resembles Computed Axial Tomography (CAT) scans that derive views inside our bodies from numerous readings of X-rays that cross them from different directions.

The Sun's internal sound waves have been used as a thermometer, taking the temperature of the Sun's energy-generating core and showing that it agrees with model predictions. This strongly disfavors any astrophysical solution for the solar neutrino problem in which massive subterranean instruments always come up short, detecting only one-third to one-half the number of neutrinos that theory says they should detect. The ghostly neutrinos seem to have an identity crisis, transforming themselves on the way out of the Sun into a form we cannot detect and a flavor we cannot taste.

SOHO's helioseismology instruments have shown how the Sun rotates inside, using the Doppler effect in which motion changes the pitch of sound waves. Regions near the Sun's poles rotate with exceptionally slow speeds, while the equatorial regions spin rapidly. This differential rotation persists to about a third of the way inside the Sun, where the rotation becomes uniform from pole to pole. The Sun's magnetism is probably generated at the interface between the deep interior, that rotates at one speed, and the overlying gas that spins faster in the equatorial middle.

Internal flows have also been discovered by the SOHO helioseismologists. White-hot currents of gas move beneath the Sun we see with our eyes, streaming at a leisurely pace when compared to the rotation. They circulate near the equator, and between the equator and poles.

In the fourth chapter, we describe the tenuous, million-degree gas, called the corona, that lies outside the sharp apparent edge of the Sun. The visible solar disk is closer to the Sun's core than the million-degree corona, but is several hundred times cooler. This defies common sense, and violates the second law of thermodynamics, which holds that heat cannot be continuously transferred from a cooler body to a warmer one without doing work.

Attempts to solve the Sun's heating problem are also discussed in the fourth chapter. The paradox cannot be solved by sunlight, which passes right through the transparent corona, and spacecraft have shown that sound waves cannot get out of the Sun to provide the corona's heat. Moreover, when SOHO focuses in on the material just above the photosphere, it all seems to be falling down into the Sun, so nothing seems to be carrying the heat out into the overlying corona.

Yohkoh and SOHO images at X-ray and extreme ultraviolet wavelengths have provided some solutions to the coronal heating problem, in which the million-degree outer solar atmosphere overlies the Sun's cooler visible disk. They have shown that the hottest, densest material in the low corona is concentrated within thin, long magnetized loops that are in a state of continual agitation. Wherever the magnetism is strongest, the coronal gas is hottest. These magnetic loops often come together, releasing magnetic energy when they make contact. This provides a plausible explanation for heating the low corona.

The hot coronal gases are expanding out in all directions, filling the solar system with a ceaseless flow – called the solar wind – that contains electrons, protons and other ions, and magnetic fields. The fifth chapter provides a detailed discussion of this solar wind, together with recent investigations into its origin and acceleration. Early spacecraft measurements showed that the

Sun's wind blows hard and soft. That is, there are two kinds of wind, a fast one moving at about 750 thousand meters per second, and a slow one with about half that speed. Ulysses provided the first measurements of the solar wind all around the Sun, conclusively showing that much of the steady, high-speed wind squirts out of polar coronal holes. A capricious, gusty, slow wind emanates from the Sun's equatorial regions near the minimum in the 11-year solar activity cycle when the SOHO and Ulysses measurements were made. The high-speed wind is accelerated very close to the Sun, within just a few solar radii, and the slow component obtains full speed further out.

SOHO instruments have unexpectedly demonstrated that oxygen ions move faster than protons in coronal holes. Absorbing more power from magnetic waves might preferentially accelerate the heavier ions, as they gyrate about open magnetic fields. Ulysses has detected magnetic waves further out above the Sun's poles, where the waves apparently block incoming cosmic rays. Instruments aboard SOHO have also pinpointed the source of the high-speed wind; it is coming out of honeycomb-shaped magnetic fields at the base of coronal holes.

Our sixth chapter describes sudden, brief, and intense outbursts, called solar flares, that release magnetic energy equivalent to billions of nuclear bombs on Earth. The Sun's flares flood the solar system with intense radiation across the full electromagnetic spectrum from the longest radio wavelengths to the shortest X-rays and gamma rays. The radio bursts provide evidence for the ejection of very energetic particles into space, either as electron beams moving at nearly the speed of light or as the result of shock waves moving out at a more leisurely pace.

The Solar Maximum Mission (SMM) and Yohkoh spacecraft have shown that solar flares also hurl high-speed electrons and protons down into the Sun, colliding with the denser gas and emitting hard X-rays and gamma rays. Soft X-ray observations indicate that Earth-sized regions can be heated to about ten million degrees during the later stages of a solar flare, becoming about as hot as the center of the Sun.

Magnetic bubbles of surprising proportion, called Coronal Mass Ejections, or CMEs for short, are also discussed in the sixth chapter. They have been routinely detected with instruments on board several solar satellites, most recently from SOHO. The CMEs expand as they propagate outward from the Sun to rapidly rival it in size, and carry up to ten billion tons of coronal material away from the Sun. Their associated shocks accelerate and propel vast quantities of high-speed particles ahead of them.

The sixth chapter additionally describes how explosive solar activity can occur when magnetic fields come together and reconnect in the corona. Stored magnetic energy is released rapidly at the place where the magnetic fields touch. Here we also discuss how high-speed particles, released from solar flares or accelerated by CME shock waves, have been directly measured *in situ* by spacecraft in interplanetary space. Intense radiation, shock waves and magnetic fields are also detected by these spacecraft.

The seventh chapter describes how forceful coronal mass ejections can create intense magnetic storms on Earth, trigger intense auroras in the skies, damage or destroy Earth-orbiting satellites, and induce destructive power surges in long-distance transmission lines on Earth. These solar ejections travel to the Earth in a few days, so there is some warning time.

Intense radiation from a solar flare reaches the Earth's atmosphere in just eight minutes, moving at the speed of light – the fastest thing around. Flaring X-rays increase the ionization of our air, and disrupt long-distance radio communications. A satellite's orbit around the Earth can be disturbed by the enhanced drag of the expanded atmosphere.

The Earth's magnetic field shields us from most of the high-speed particles ejected by solar flares or accelerated by CME shock waves, but the energetic charged particles can endanger astronauts in space and destroy satellite electronics. Some of the particles move at nearly the speed of light, so there is not much time to seek protection from their effects. If we knew the solar magnetic changes preceding these violent events, then spacecraft could provide the necessary early warning. As an example, when the coronal magnetic fields get twisted into an S, or inverted S, shape, they are probably getting ready to release a mass ejection.

As our civilization becomes increasingly dependent on sophisticated systems in space, it becomes more vulnerable to this Sun-driven space weather, which is tuned to the rhythm of the Sun's 11-year magnetic activity cycle. It is of such vital importance that national centers employ space-weather forecasters, and continuously monitor the Sun from ground and space to warn of threatening solar events.

As also described in the seventh chapter, solar X-ray and ultraviolet radiation are extremely variable, changing in step with the 11-year cycle of magnetic activity. The fluctuating X-rays produce substantial alterations of the Earth's ionosphere, and the changing ultraviolet modulates our ozone layer. The varying magnetic activity also changes the Sun's total brightness, but to a lesser degree. During the past 130 years, the Earth's surface temperature has been associated with decade-long variations in solar activity, perhaps because of changing cloud cover related to the Sun's modulation of the amount of cosmic rays reaching Earth. Observations of the brightness variations of Sun-like stars indicate that they are capable of a wider range of variation in total radiation than has been observed for the Sun so far.

Radioactive isotopes found in tree rings and ice cores indicate that the Sun's activity has fallen to unusually low levels at least three times during the past one thousand years, each drop corresponding to a long cold spell of roughly a century in duration. Further back in time, during the past one million years, our climate has been dominated by the recurrent ice ages, each lasting about 100 thousand years. They are explained by three overlapping astronomical cycles which combine to alter the distribution and amount of sunlight falling on Earth.

Our book ends with a description of the Sun's distant past and remote future. The Sun is gradually increasing in brightness with age, by a startling 30 percent over the 4.5 billion years since it began to shine. This slow, inexorable brightening ought to have important long-term terrestrial consequences, but some global thermostat has kept the Earth's surface temperature relatively unchanged as the Sun grew brighter and hotter. A powerful atmospheric greenhouse might have warmed the young Earth, gradually weakening over time, or plants and animals might have beneficially controlled their environment for the last 3 billion years. The Sun's steady increase in luminous intensity is nevertheless expected to boil our oceans away in about 3 billion years; and the Sun will expand into a giant star in another 4 billion years. It will then engulf Mercury, melt the Earth's surface rocks, and turn the frozen moons of the distant giant planets into globes of liquid water.

I am very grateful to my expert colleagues who have read portions of this book, and commented on its accuracy, clarity and completeness, substantially improving the manuscript. They include Loren W. Acton, Markus J. Aschwanden, W. Ian Axford, Arnold O. Benz, Richard C. Canfield, Edward L. Chupp, Edward W. Cliver, George A. Doschek, A. Gordon Emslie, Bernhard Fleck, Peter V. Foukal, Claus Fröhlich, John W. Harvey, Hugh S. Hudson, Stephen W. Kahler, Mukul R. Kundu, John W. Leibacher, Michael E. Mann, Richard G. Marsden, Ronald L. Moore, Eugene N. Parker, Eric R. Priest, John C. Raymond, Andrew P. Skumanich, Charles P. Sonett, Barbara Thompson, Virginia L. Trimble, Bruce T. Tsurutani, Yi-Ming Wang, and David F. Webb.

Locating quality figures is perhaps the most time-consuming and frustrating aspect of producing a volume like this, so I am especially thankful for the support of ESA, ISAS and NASA for providing them. Individuals that were especially helpful in locating and providing specific images include Loren W. Acton, David Alexander, Cary Anderson, Frances Bagenel, Richard C. Canfield, Michael Changery, David Chenette, Fred Espenak, Bernhard Fleck, Eigil Friis-Christensen, Claus Fröhlich, Bruce Goldstein, Leon Golub, Steele Hill, Gordon Holman, Beth Jacob, Imelda Joson, Therese Kucera, Judith Lean, William C. Livingston, Michael E. Mann, Richard Marsden, Michael J. Reiner, Thomas Rimmele, Kazunari Shibata, Gregory Lee Slater, Barbara Thompson, Haimin Wang, and Joachim Woch.

In conclusion, this book uses the International System of Units (Système International, SI) for most quantities, but the reader should be warned that

Table P.1. Principal SI units and their conversion to corresponding c.g.s. units

Quantity	SI units	Conversion to c.g.s. units
Length	meter (m)	100 centimeters (cm)
	nanometer (nm) $= 10^{-9}$ m	10^{-7} cm = 10 Angstroms = 10 Å
Mass	kilogram (kg)	1,000 grams (g)
Velocity	meter per second (m s^{-1})	100 centimeters per second (cm s^{-1})
Energy*	Joule (J)	10,000,000 erg
Power	Watt (W)	10,000,000 erg s^{-1}
	= Joule per second (J s^{-1})	
Magnetic flux density	Tesla (T)	10,000 Gauss (G)

* The energy of high energy particles and X-ray radiation are often expressed in units of kilo-electron volts, or keV, where 1 keV $= 1.602 \times 10^{-16}$ Joule, or 1 MeV = 1,000 keV.

Table P.2. Solar quantities and fundamental constants*

Symbol	Name	Value
$L_\odot$	Luminosity of Sun	3.854×10^{26} Joule s^{-1}
$M_\odot$	Mass of Sun	1.989×10^{30} kilograms
$R_\odot$	Radius of Sun	6.955×10^{8} meters
AU	Mean Distance of Sun	1.496×10^{11} meters
$T_{e\odot}$	Effective Temperature of Photosphere	5,780 degrees Kelvin
c	Velocity of Light	2.9979×10^{8} m s^{-1}
G	Gravitational Constant	6.6726×10^{-11} N m^{2} kg^{-2}
k	Boltzmann's Constant	1.38066×10^{-23} J K^{-1}
h	Planck's Constant	6.6261×10^{-34} J s
a	Radiation Density Constant	7.5659×10^{-16} J m^{-3} K^{-4}
m_e	Mass of Electron	9.1094×10^{-31} kg
e	Charge of Electron	1.6022×10^{-19} C
m_H	Mass of Hydrogen Atom	1.673534×10^{-27} kg
m_p	Mass of Proton	1.672623×10^{-27} kg
ε_0	Permittivity of Free Space	$10^{-9}/(36\pi) = 8.854 \times 10^{-12}$ F m^{-1}
μ_0	Permeability of Free Space	$4\pi \times 10^{-7} = 12.566 \times 10^{-7}$ N A^{-1}

* Adapted from Lang (1991). One Joule per second $= 1\,J\,s^{-1}$ is equal to one watt $= 1$ W. The unit symbols are J for Joule, s for second, kg for kilogram, m for meter, K for degree Kelvin, C for Coulomb, N for Newton and A for Ampere.

centimeter-gram-second (c.g.s.) units are employed in nearly all of the seminal papers referenced in this book. Moreover, the c.g.s. units are still extensively used by many solar astronomers. Table P.1 provides unit abbreviations and conversions between units. Some other common units are the nanometer (nm) with 1 nm $= 10^{-9}$ meters, the Angstrom unit of wavelength, where 1 Angstrom $= 1$ Å $= 10^{-10}$ meters $= 10^{-8}$ centimeters, the nanoTesla (nT) unit of magnetic flux density, where 1 nT $= 10^{-9}$ Tesla $= 10^{-5}$ Gauss, the electron volt (eV) unit of energy, with 1 eV $= 1.6 \times 10^{-19}$ Joule, and the ton measurement of mass, where 1 ton $= 10^{3}$ kilograms $= 10^{6}$ grams. The accompanying Table P.2 provides numerical values for solar quantities and fundamental constants.

Medford, February 2000 *Kenneth R. Lang*

Contents

1. Instruments for a Revolution ... 1
Overview ... 1
1.1 Solar Mysteries ... 1
1.2 The Sun Does Not Set for SOHO ... 3
1.3 Ulysses Moves into Unexplored Territory ... 8
1.4 Yohkoh Detects Unrest on an Awesome Scale ... 12

2. Discovering Space ... 17
Overview ... 17
2.1 Space is not Empty ... 17
2.2 Touching the Unseen ... 28
2.3 Cosmic Rays ... 31
2.4 Pervasive Solar Magnetism ... 35
2.5 Invisible Radiation from the Sun ... 44
Key Events in the Discovery of Space ... 53

3. Exploring Unseen Depths of the Sun ... 59
Overview ... 59
3.1 What Makes the Sun Shine? ... 59
3.2 Taking the Sun's Pulse ... 72
3.3 Looking Within the Sun ... 78
3.4 How the Sun Rotates Inside ... 83
3.5 Internal Flows ... 86
3.6 Sunquakes and Active Regions ... 91
Key Events in the Development of Helioseismology ... 93

4. Solving the Sun's Heating Problem ... 95
Overview ... 95
4.1 Mysterious Heat ... 95
4.2 Pumping up the Corona in Active Regions ... 99
4.3 Tuning in and Dropping out ... 105
4.4 Magnetic Connections in the Quiet Corona ... 110
Key Events in Coronal Heating ... 120

5. Winds Across the Void ... 123
Overview ... 123
5.1 The Two Solar Winds ... 123
5.2 From Whence do the Winds Blow? ... 129

5.3 Some Like It Hot ... 138
5.4 Magnetized Waves, Exotic Particles and the Edge of the Solar System ... 145
Key Events in Studies of the Solar Wind ... 150

6. Our Violent Sun ... 155
Overview ... 155
6.1 Brightening in the Chromosphere ... 156
6.2 Solar Radio Bursts ... 157
6.3 X-ray Flares ... 166
6.4 Gamma Rays from Solar Flares ... 178
6.5 Magnetic Bubbles ... 182
6.6 Making Solar Explosions Happen ... 191
6.7 Interplanetary Protons, Electrons and Shock Waves ... 200
Key Events in Understanding Explosive Solar Activity ... 205

7. The Sun–Earth Connection ... 211
Overview ... 211
7.1 Earth's Magnetic Storms and the Aurora Lights ... 212
7.2 Danger Blowing in the Wind ... 221
7.3 Solar Ultraviolet and X-rays Transform Our Atmosphere ... 234
7.4 Varying Solar Activity and Climate Change ... 239
7.5 Climate Change over Millions and Billions of Years ... 253
Key Events in the Discovery of Solar-Terrestrial Interactions ... 262

Appendix ... 269
A. SOHO Principal Investigators and Their Institutions ... 269
B. Ulysses Principal Investigators and Their Institutions ... 270

Internet Addresses ... 271
I. Fundamental Data ... 271
II. Information About SOHO, Ulysses and Yohkoh ... 271
III. Information About Other Solar Missions ... 272
IV. Information About NASA's Sun-Earth Connection ... 272
V. Information About Helioseismology ... 272
VI. Other Solar Topics ... 273
VII. Information About Cosmic and Heliospheric Physics ... 273
VIII. Information About Solar Activity and Space Weather ... 273

References ... 275

Author Index ... 337

Subject Index ... 343

1. Instruments for a Revolution

Overview

Three pioneering spacecraft, the SOlar and Heliospheric Observatory, or SOHO for short, Ulysses and Yohkoh, have discovered clues to a variety of enigmatic problems in our understanding of the Sun. They have transformed human perception of the star on which our lives depend, examining it with exquisitely sensitive and precise instruments and perceiving worlds that are otherwise invisible. Major new instruments aboard these spacecraft have traced the flow of energy and matter from down inside our star to the Earth and beyond, providing insights that are vastly more focused and detailed than those of previous solar missions. When combined with ground-based observations, and with each other, the SOHO, Ulysses and Yohkoh spacecraft have created a renaissance in solar science.

1.1 Solar Mysteries

From afar, the Sun does not look very complex. To the casual observer, it is just a smooth, uniform ball of gas. Close inspection, however, shows that the star is in constant turmoil. The seemingly calm Sun is a churning, quivering and explosive body, driven by intense, variable magnetism. This new perception of the volatile, ever-changing Sun has also resulted in several fundamental, unsolved mysteries.

Scientists could not understand how the Sun generates its magnetic fields, which are responsible for most solar activity. Nor did they know why some of this intense magnetism is concentrated into so-called sunspots, dark islands on the Sun's surface that are as large as the Earth and thousands of times more magnetic. Furthermore, physicists could not explain why the Sun's magnetic activity varies dramatically, waning and intensifying again every 11 years or so. The answers to these problems have been hidden deep down inside the Sun where its powerful magnetism is generated.

Further out, the normally invisible solar atmosphere, or corona, presents one of the most puzzling paradoxes of modern science. It is unexpectedly hot, reaching temperatures of a few million degrees Kelvin, which are several hundred times hotter than the Sun's underlying visible disk. Heat simply should not flow outward from a cooler to a hotter region. It violates the second law of thermodynamics and all common sense as well. After more than a half century of speculations, scientists still could not explain precisely how the million-degree corona is heated.

The Sun's stormy atmosphere is forever expanding in all directions, filling the solar system with a ceaseless flow – called the solar wind – that contains electrons, ions and magnetic fields. The hot corona creates an outward pressure that overcomes the Sun's gravitational attraction, enabling this perpetual flow. Earlier spacecraft obtained important information about the solar wind, but the exact sources of all the wind's components and the mechanisms for accelerating it to high velocities remained as other long-standing mysteries.

The Sun is a vigorous and violent place of writhing gases, moving magnetic fields and powerful explosions. Without warning, the relatively calm solar atmosphere can be torn asunder by sudden outbursts of awesome scale, releasing stored magnetic energy equivalent to billions of nuclear bombs and raising the temperature of Earth-sized regions to tens of millions of degrees. The resultant explosions, called solar flares, emit intense X-ray and gamma-ray radiation, as well as energetic charged particles moving at nearly the speed of light. Other ejections from the Sun hurl billions of tons of coronal gases into interplanetary space, punctuating the steady outward flow of the relentless solar wind. Yet, no one has been able to predict when the Sun's explosive outbursts will occur.

Understanding how the Sun behaves is of crucial importance to all of us on Earth. It affects our everyday lives. Space near the Earth is energized and threatened by the tempestuous Sun. In tens of minutes, intense explosions hurl out energetic particles that can endanger astronauts and destroy satellite electronics. Forceful solar ejections can compress the Earth's magnetic field, damage or destroy Earth-orbiting satellites, produce intense geomagnetic storms and create power surges that can blackout entire cities. As our civilization becomes increasingly dependent on sophisticated systems in space, it becomes more vulnerable to this Sun-driven space weather that we cannot now forecast with any certainty.

So there were many perplexing enigmas that resulted from our new understanding of the Sun – the internal generating mechanism for its all-important magnetic field, the heating of its million-degree corona, the origin and driving force of the solar wind, and the triggering and energy source of its unpredictable explosions. While these mysteries were quite well established empirically, very little was understood about the underlying physical causes. To clarify and help solve many of these outstanding puzzles – and to better predict the Sun's impact on our planet – three new solar spacecraft were launched into orbit about the Sun.

As it turned out, this scientific armada, SOHO, Ulysses and Yohkoh, accomplished most of the things that scientists said it would. Their extraordinary results have been enhanced by combining them with each other, and with observations from the ground at optical and radio wavelengths. We will next give an account of each satellite, and provide the background needed to understand their results. The main part of this book will then explain their fascinating discoveries.

1.2 The Sun Does Not Set for SOHO

A great observatory of the Sun, the SOlar and Heliospheric Observatory, or SOHO for short, has stared the Sun down with an unblinking eye for years. This 1.33 ton (1,350 kilogram), one billion dollar spacecraft was launched from Cape Canaveral Air Station, Florida aboard a two stage Atlas/Centaur rocket on 2 December 1995 and reached its permanent position on 14 February 1996.

SOHO is located sunward at about 1.5 billion (1.5×10^9) meters out in space, or at about one percent of the way to the Sun. At this place, the spacecraft is balanced between the pull of the Earth's gravity and the Sun's gravity. Such a position is known in astronomy as the inner Lagrangian, or L_1, point after the French mathematician Joseph Louis Lagrange, who first calculated its location near the end of the nineteenth century. From this strategic vantage point, SOHO can monitor the Sun with a continuous, uninterrupted view 24 hours a day, every day, all year round.

The spacecraft has been keeping watch over the Sun through all its tempestuous seasons – from a lull in its 11-year magnetic activity cycle at the time of launch to an activity maximum near the turn of the century and millennium. Except, that is, for a three-month interlude when controllers lost contact with the spacecraft (Focus 1.1).

If SOHO was positioned exactly at the Lagrangian point, NASA's tracking telescopes would look directly at the Sun. The intense solar noise would then severely limit satellite communications with Earth and the bright glare might even burn terrestrial telescope receivers up. As a result, SOHO flies as close to the Lagrangian point as possible, in an elliptical orbit about it with a radius of about 600 million meters. When viewed from Earth, the spacecraft forms a closed curve around the Sun, like a halo, so its trajectory is often called a halo orbit. As it loops about the Lagrangian point, SOHO also orbits the Sun in step with Earth, experiencing perpetual day and continuously gazing at the Sun with an unobstructed view.

Focus 1.1

SOHO Lost in Space and Recovered from Oblivion

After more than two years of uninterrupted views of the Sun, and completing its primary mission with unqualified success, SOHO's eyes were abruptly closed. During routine maintenance maneuvers and calibrations on 25 June 1998, the spacecraft spun out of control and engineers could not re-establish radio contact with it.

A group of experts, with the ponderous title "The SOHO Mission Interruption Joint ESA/NASA Investigation Board", was assembled to find out what went wrong. Their post-mortem found no fault with the spacecraft itself; a sequence of operational mistakes added up to its loss. A combination of human error and faulty computer command software, that had not been previously used or adequately tested, disabled some of SOHO's stabilizing gyroscopes. Continuous firings of its jet thrusters failed to bring the spacecraft into balance, and instead sent it spinning faster and faster.

For nearly a month, the crippled spacecraft failed to respond to signals sent daily. Scientists feared that SOHO might be drifting away from its expected orbit, gone forever and never to be heard from again. However, the wayward satellite was found nearly a month after it was lost by transmitting a powerful radar pulse to it from the world's largest radio telescope located in Arecibo, Puerto Rico. The faint radio echo indicated that SOHO was located in the right, predicted part of space, turning relatively slowly at the rate of one revolution every 53 seconds. This gave everyone renewed hope and optimism that radio contact would eventually be established and the spacecraft ultimately revived.

SOHO's solar panels were then turned edge-on toward the Sun, so its electrical power could not be renewed. Its internal energy had drained away, and the spacecraft was unable to receive or send communications. Power might nevertheless be regained as the panels slowly turned to a more favorable alignment with the Sun during SOHO's annual orbit around the star. Antennas in NASA's Deep Space Network therefore continued to send the satellite wake up messages, asking it to call home.

After six weeks of silence, a feeble and intermittent response was received from the dormant spacecraft, like the faint, erratic heartbeat of a patient in a coma or the worn-out, distressed cries of a tired, lost child. The elated European engineers, who built the spacecraft, knew that SOHO was alive and immediately began regaining control of it.

It was a long, slow recovery. The onboard batteries had to be recharged, and the inner workings had to be warmed up after an enforced period of deep freeze. Some of the propellant in its tanks had been frozen solid, and the pipes that carry fuel to the craft's jet thrusters also had to be thawed out. Fortunately, the fuel, named hydrazine, does not expand when it freezes, so the fuel pipes did not crack open as sometimes happens when a building's water pipes freeze during the loss of heat in a severe winter storm. Altogether, it took three months from the initial loss of radio contact to full recovery.

As luck would have it, SOHO's tribulations were not yet over, since its gyroscopes acted up just a few months after recovery. The satellite had to constantly fire its onboard jets to keep it balanced and pointed toward the Sun, and this was rapidly exhausting its fuel supply. Ingenious engineers fixed the problem by instructing the spacecraft to bypass the gyroscopes and use stars to determine its position, somewhat like the ancient mariners who navigated by the stars. Since then, all of SOHO's sensitive instruments have been restored to full health, showing no signs of damage from their unexpected ordeal, and the mission is expected to continue until 2003, covering the maximum in solar activity that will peak in 2001.

The fantastic rescue from deep space was an intense, unprecedented drama, and a remarkable achievement by a heroic international team working together in the face of overwhelming obstacles. In the end, it was human ingenuity, hope and perseverance that brought SOHO back to life.

All previous solar observatories have either been on the Earth or in orbit around our planet. Terrestrial telescopes are limited by inclement weather conditions and atmospheric distortion of the Sun's signal, and of course they cannot observe the Sun at night. Although the weather problem has been removed in orbit around the Earth, the observations are still periodically interrupted when an Earth-orbiting spacecraft enters our planet's shadow. So, SOHO is the first solar observatory to look at the Sun nonstop.

SOHO is a joint project of the European Space Agency (ESA) and the United States (US) National Aeronautics and Space Administration (NASA). The spacecraft was designed and built in Europe. NASA launched SOHO and operates the satellite from the Experimenters' Operations Facility at the Goddard Space Flight Center in Greenbelt, Maryland. SOHO studies the Sun using a dozen instruments, including telescopes; European scientists provided nine of them and US scientists a further three. The twelve instruments were developed by international consortia involving scientific institutes in 15 countries. The spacecraft is included within the International Solar-Terrestrial Physics (ISTP) program that includes more than 20 satellites, coupled with ground-based observations and modeling centers, that allow scientists to study the Sun, the Earth and the space between them in unprecedented detail.

SOHO's instruments have examined the Sun from its deep interior, through its million-degree atmosphere and ceaseless wind, to our home planet, Earth. Three devices probe the Sun's internal structure and dynamics; six measure the solar atmosphere; and three keep track of the star's far-reaching winds. The acronyms and measurements of these instruments are given in Table 1.1; the Principal Investigators and their institutions are provided in Appendix A.

The hidden interior of the Sun is illuminated by its in-and-out, heaving motions. These oscillations arise from sounds that move inside the Sun. On striking the Sun's apparent edge and rebounding back inside, the sound waves cause the visible gas to move up and down. The oscillating motions these sounds create are imperceptible to the naked eye, but SOHO instruments routinely pick them out.

The Sun's atmosphere can be probed by observing ultraviolet, extreme ultraviolet or X-ray radiation (Fig. 1.1). This is because hot material emits most of its energy at these wavelengths. Also, the underlying visible surface of the Sun is too cool to emit intense radiation at these wavelengths, so it appears dark under the hot gas.

To map out structures across the solar disk, ranging in temperature from 6,000 to two million degrees Kelvin, the SOHO instruments tune into ultraviolet or extreme ultraviolet spectral features emitted by different ions at various definite wavelengths. Atoms in a hotter gas lose more electrons through collisions, and so they become more highly ionized. The wavelength-specific radiation of different ions therefore serves as a kind of thermometer. SOHO telescopes transform the emission from a given ion into high-resolution images of the structures formed at the relevant temperature. We can also infer the speed of the material moving in these regions from the Doppler wavelength changes of the spectral features that SOHO records.

Table 1.1. SOHO's instruments arranged alphabetically by acronym within three areas of investigation

Instrument	Measurement
Helioseismology Instruments	
GOLF	The Global Oscillations at Low Frequencies device records the velocity of global oscillations within the Sun.
MDI/SOI	The Michelson Doppler Imager/Solar Oscillations Investigation measures the velocity of oscillations, produced by sounds trapped inside the Sun, and obtains high resolution magnetograms.
VIRGO	The Variability of solar IRradiance and Gravity Oscillations instrument measures fluctuations in the Sun's brightness, as well as its precise energy output.
Coronal Instruments	
CDS	The Coronal Diagnostics Spectrometer records the temperature and density of gases in the corona.
EIT	The Extreme ultraviolet Imaging Telescope provides full-disk images of the chromosphere and the corona.
SUMER	The Solar Ultraviolet Measurements of Emitted Radiation instrument gives data about the temperatures, densities and velocities of various gases in the chromosphere and corona.
LASCO	The Large Angle Spectrometric COronagraph provides images that reveal the corona's activity, mass, momentum and energy.
UVCS	The UltraViolet Coronagraph Spectrometer measures the temperatures and velocities of hydrogen atoms, oxygen ions and other ions in the corona.
SWAN	The Solar Wind ANisotropies device monitors latitudinal and temporal variations in the solar wind.
Solar Wind "In-Situ" Instruments	
CELIAS	The Charge, ELement and Isotope Analysis System quantifies the mass, charge, composition and energy distribution of particles in the solar wind.
COSTEP	The COmprehensive SupraThermal and Energetic Particle analyzer determines the energy distribution of protons, helium ions and electrons.
ERNE	The Energetic and Relativistic Nuclei and Electron experiment measures the energy distribution and isotopic composition of protons, other ions and electrons.

Other SOHO instruments directly sample and analyze pieces of the Sun *in situ*, or in their place. The properties of energetic solar electrons, protons and heavier ions are measured as they sweep past the satellite. By determining the abundance of these elements, we can reach conclusions about conditions in the Sun's atmosphere where its winds originate. SOHO also detects very energetic particles generated during violent explosions on the Sun.

SOHO's instruments uniquely examine many different levels in the solar atmosphere simultaneously, including for the first time the locations where the million-degree gases are heated and the Sun's winds are accelerated.

Fig. 1.1. The Ultraviolet Sun. This composite image, taken by two SOHO instruments and joined at the black circle, reveals the ultraviolet light of the Sun's atmosphere from the base of the corona to millions of kilometers above the visible solar disk. The region outside the black circle, obtained by UVCS, shines in the ultraviolet light emitted by oxygen ions flowing away from the Sun to form the solar wind. The inner image, obtained by EIT, shows the ultraviolet light emitted by iron ions at a temperature near two million degrees Kelvin. Dark areas, called coronal holes, are found at both poles of the Sun (*top* and *bottom*) and across the disk of the Sun; they are the places where the highest-speed solar wind originates. UVCS has discovered that the oxygen ions flowing out of coronal holes have extremely high energies corresponding to temperatures of over 200 million degrees Kelvin, and accelerate to supersonic outflow velocities within 2.5 solar radii of the Sun's center. The structure of the corona is controlled by the Sun's magnetic field which forms the bright active regions on the solar disk and the ray-like structures extending from the coronal holes. [Courtesy of the SOHO UVCS consortium (*outer region*) and the SOHO EIT consortium (*inner region*). SOHO is a project of international cooperation between ESA and NASA]

Moreover, many of the instruments complement each other and are operated together, establishing connections between various phenomena at different places in the volume of the Sun and in the interplanetary medium. Their combined data can link events in the Sun's atmosphere and solar wind to changes taking place within the solar interior. In effect, SOHO can combine data from its instruments to look all the way from deep inside the Sun to our planet. It provides a completely new perspective of how agitation inside the Sun, transmitted through the solar atmosphere, directly affects us on Earth.

Large radio telescopes around the world, which form NASA's Deep Space Network, are used to track the satellite and retrieve its data. Each day over a thousand images are routed to the Experimenters' Operation Facility. Here solar physicists from around the world work together, watching the Sun night and day from a room without windows. Many of the unique images and movies they receive move nearly instantaneously to the SOHO home page (http://sohowww.nascom.nasa.gov) on the World Wide Web.

The SOHO mission has three principal scientific goals: to measure the structure and dynamics of the solar interior; to gain an understanding of the heating mechanisms of the Sun's million-degree atmosphere, or solar corona; and to determine where the solar wind originates and how it is accelerated.

SOHO has accomplished many of these objectives. Internal rotation and currents have been measured, discovering a previously unknown solar meteorology and providing clues to the origin of solar magnetism. Ever-changing magnetic forces have been monitored as they interact to help heat the solar atmosphere and shape, mold and constrain it. The fastest solar winds emanate from a magnetic network at the bottom of holes in the Sun's Polar Regions. These winds are accelerated very close to the Sun where heavier ions move at faster speeds and magnetic waves are found. The slower wind puffs out of the equatorial regions and takes a longer time to get up to speed. SOHO has also scrutinized gigantic magnetic bubbles that are formed during global restructuring of the Sun's magnetic fields and can be hurled toward the Earth with potentially dangerous consequences. All of these accomplishments are discussed in subsequent chapters of this book.

1.3 Ulysses Moves into Unexplored Territory

Gases expand away from the Sun in all directions, carving out a bubble in the space around our star. This vast region, known as the heliosphere, is occupied by the Sun's atmosphere and dominated by the outflow of its winds. Yet, until quite recently, most of our knowledge of the immense heliosphere came from spacecraft in – or very close to – the ecliptic, the plane of the Earth's orbital motion about the Sun. Because the solar equator nearly coincides with the ecliptic, scientific probes have been unable to directly sample any regions more than 16 degrees north or south solar latitude, confining their perspective to a narrow, two-dimensional, "flat-land" slice of the heliosphere. (The solar latitude is the angular distance from the plane of the Sun's equator.) Then in the 1990s a single craft, Ulysses, ventured out of this thin zone into uncharted interplanetary space, moving over the poles of the Sun and probing the third dimension of the heliosphere for the first time.

Like SOHO, the Ulysses mission is a joint undertaking of the European Space Agency (ESA) and the United States (US) National Aeronautics and Space Administration (NASA). The spacecraft and its operations team have been provided by ESA, the launch of the spacecraft, radio tracking, and data management operations are provided by NASA. Scientific experiments have been provided by investigation teams both in Europe and the United States.

The spacecraft is named Ulysses in honor of the mythical Greek adventurer. Ulysses' exploits are recounted in Homer's *Iliad*, and his long wanderings before he reached home in the *Odyssey*. In the 26th Canto of Dante Alighieri's *Inferno* Ulysses recalls his restless desire

> To venture the uncharted distances;

Exhorting his friends to embark on one last occasion

> Of feeling life and the new experience
> Of the uninhabited world behind the Sun ...
> To follow after knowledge and excellence.

Like the legendary explorer, the Ulysses spacecraft is travelling into previously unexplored regions.

Interplanetary spacecraft usually travel within the ecliptic. This is because they often rendezvous with another planet whose orbit lies near that plane, and also because their launch vehicles obtain a natural boost by traveling in the same direction as the Earth's spin and in the plane of the Earth's orbit around the Sun. Moreover, the thrust and high speed needed to send a probe directly on a trajectory over the poles of the Sun is beyond the capability of today's most powerful rockets. To move out of the ecliptic, a spacecraft must lose the momentum it receives from its launch platform, Earth, which travels around the Sun with a mean orbital speed of 29,780 meters per second.

Ulysses had to first move outward away from the Sun, not toward it, in order to voyage across the top of our star. By traveling out to Jupiter, the 367-kilogram spacecraft could use the planet's powerful gravity to cancel excess momentum acquired from the Earth and to accelerate, propel and reorient the spacecraft in slingshot fashion into an inclined orbit that sent it under the Sun.

Launched by NASA's Space Shuttle *Discovery* on 6 October 1990, Ulysses encountered Jupiter on 8 February 1992, which hurled the spacecraft into a Sun-centered elliptical orbit, with a period of 6.2 years. Its distance from the Sun varies from 5.4 AU, near Jupiter's orbit, to 1.3 AU at its closest approach to the Sun; the distance over the Sun's poles ranges from 2.0 AU (north) to 2.3 AU (south). The astronomical unit, 1.0 AU, is the mean distance from the Sun to the Earth, or about 150 billion (1.5×10^{11}) meters. Thus, at the time of polar passage, Ulysses was not close to the Sun; it was more than twice as far from the Sun as the Earth.

After traveling about 3 million million (3×10^{12}) meters for nearly four years since leaving Earth, brave stalwart Ulysses reached the summit of its trajectory beneath the South Pole of the Sun on 13 September 1994. It arrived there at a maximum southern latitude of 80.2 degrees and 344 billion (3.44×10^{11}) meters (2.3 AU) from the Sun. In response to the Sun's gravitational pull, Ulysses arched up toward the solar equator, and crossed over the solar North Pole on 31 July 1995.

Ulysses carries four European and five American instruments that make measurements of fundamental parameters as a function of distance from the Sun and solar latitude. They include instruments to measure the solar wind speed, a pair of magnetometers that measure magnetic fields, and several in-

Table 1.2. Ulysses' instruments arranged alphabetically by acronym*

Instrument	Measurement
COSPIN	The COsmic ray and Solar Particle INstrument records ions with energies of 0.3 to 600 MeV per nucleon and electrons with energies of 4 to 2,000 MeV.
DUST	Measures interplanetary dust particles with mass between 10^{-19} and 10^{-10} kilograms.
EPAC/GAS	The Energetic PArticle Composition and interstellar neutral GAS instrument records energetic ion composition for energies of 80 keV to 15 MeV per nucleon and neutral helium atoms.
GRB	The Gamma Ray Burst instrument measures solar flare X-rays and cosmic gamma-ray bursts with energies of 15 to 150 keV.
GWE	The Gravitational Wave Experiment records Doppler shifts in the satellite radio signal that might be due to gravitational waves.
HISCALE	The Heliosphere Instrument for Spectra, Composition, and Anisotropy at Low Energies instrument measures ions with energies of 50 keV to 5 MeV and electrons with energies of 30 to 200 keV.
SCE	The Solar Corona Experiment uses the radio signals from the spacecraft to measure the density, velocity and turbulence spectra in the solar corona and solar wind.
SWICS	The Solar Wind Ion Composition Spectrometer records elemental and ionic-charge composition, temperature and the mean speed of solar wind ions for speeds of 145 thousand (H^+) to 1.35 million (Fe^{+8}) meters per second.
SWOOPS	The Solar Wind Observations Over the Poles of the Sun instrument measures ions from 237 eV to 35 keV per charge and electrons from 1 to 860 eV.
URAP	The Unified Radio And Plasma wave experiment records plasma waves at frequencies from 0 to 60 kHz, remotely senses traveling solar radio bursts exciting plasma frequencies from 1 to 940 kHz, and measures the electron density.
VHM/FGM	A pair of magnetometers, the Vector Helium Magnetometer and the scalar Flux Gate Magnetometer, that measure the magnetic field strength and direction in the heliosphere and the Jovian magnetosphere from 0.01 to 44,000 nanoTeslas or from 10^{-7} to 0.44 Gauss.

* Particle energies are often expressed in electron volts, or eV, where $1\,\text{eV} = 1.602 \times 10^{-19}$ Joule $= 1.602 \times 10^{-12}$ erg. High energy particles can have energies expressed in kilo-electron volts, or keV, and Mega-electron volts, or MeV, where $1\,\text{keV} = 10^3\,\text{eV} = 1.602 \times 10^{-16}$ Joule and $1\,\text{MeV} = 10^6\,\text{eV} = 1.602 \times 10^{-13}$ Joule. The magnetic field strength is given in units of Tesla, where 1 Tesla = 10,000 Gauss or 1 nanoTesla = 10^{-9} Tesla = 10^{-5} Gauss.

struments that study electrons, protons and heavier ions that come from both the quiescent and active Sun and from interstellar space (cosmic rays). The spacecraft radio is used to study the Sun's outer atmosphere and to search for gravitational waves in interplanetary space, in addition to its communication functions. The acronyms and measurements of all of the instruments on Ulyssesare given in Table 1.2, while the Principal Investigators and their institutions are tabulated in Appendix B.

The data from Ulysses are gathered by NASA's Deep Space Network, and spacecraft operations and data analysis are performed at the Jet Propulsion Laboratory (JPL) in Pasadena, California by a joint ESA/JPL team. Fundamental information about the various experiments can be found on the Ulysses home page (http://ulysses.jpl.nasa.gov/) or (http://helio.estec.esa.nl/ulysses/).

The primary scientific objective of the Ulysses spacecraft is to explore and define the heliosphere in three dimensions, characterizing it as a function of solar latitude. The solar wind speed and magnetic field strength have been measured from pole to pole, compositional differences have been established for the expanding solar atmosphere, and the flux of cosmic rays has been measured as a function of solar latitude. Interplanetary dust, solar energetic particles, plasma waves and solar radio bursts have also been measured. All of these accomplishments are discussed in subsequent chapters of this book. Not included here, are Ulysses' study of cosmic gamma ray bursts and its investigations of the Jovian magnetosphere obtained during the Jupiter flyby.

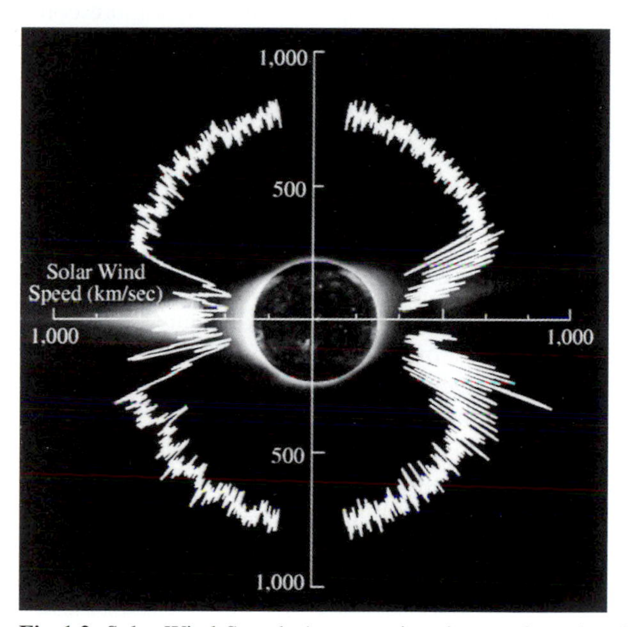

Fig. 1.2. Solar Wind Speed. A composite picture of an ultraviolet image of the solar disk and a white-light image of the solar corona form the backdrop for a radial plot of solar-wind speed versus solar latitude. The wind escapes into interplanetary space much faster from the Sun's Polar Regions (*top* and *bottom*) than from near its equator (*left center* and *right center*). These wind speed measurements were made by the Ulysses spacecraft between 1991 and 1996, near a minimum in the 11-year solar activity cycle when the corona had a relatively simple structure. Dark coronal holes are found at the poles, while pronounced streamers ring the Sun's equatorial regions. At least some of the fast solar wind originates from the coronal holes, and the slow wind emanates from the equatorial streamer regions. There are substantial fluctuations in the wind speed near the equator (*right center*) indicating that the slow wind is gusty; you see fewer fluctuations on the left side because the spacecraft was then moving rapidly over the equatorial regions. The wind speed data are from the Ulysses SWOOPS instrument, the solar disk has been imaged by the SOHO EIT and the coronal streamer data are from the SOHO LASCO C2 and Mauna Loa MK3 coronagraphs. (Courtesy of Peter Riley. Both SOHO and Ulysses are projects of international cooperation between ESA and NASA)

Ulysses has probed the heliosphere over the full range of solar latitudes, above and below the Sun's equator and across both poles of the Sun. It has thereby provided the first detailed three-dimensional analysis of the interaction of a star with its surroundings. These initial high-latitude observations were all taken during the quiescent (minimum) part of the 11-year cycle of solar activity, when the heliosphere had a relatively simple, well-organized structure (Fig. 1.2).

The hardy spacecraft has completed its first trip around the Sun, and begun its voyage back out to the orbit of Jupiter, where Ulysses will turn once again toward the Sun, returning to its vicinity at the turn of the millennium during the maximum part of the Sun's activity cycle. It will pass beneath the Sun's South Pole on 27 November 2000 (80.2 degrees south), and over its North Pole on 13 October 2001 (80.2 degrees north). Ulysses will then probe the effects of increasing solar activity when the heliosphere has a more complex shape, and is punctuated by frequent powerful solar outbursts. It will be complemented by related observations of solar activity with the SOHO and Yohkoh spacecraft.

1.4 Yohkoh Detects Unrest on an Awesome Scale

Scientists in Japan, the land of the rising Sun, have been particularly interested in sudden, powerful explosions called solar flares. These dynamic, short-lived events involve the sudden release of enormous amounts of energy, both as intense radiation in all regions of the electromagnetic spectrum and as accelerated particles whose speeds approach the velocity of light. Such flares are best studied during the maximum in the 11-year solar activity cycle, when they occur most frequently.

Following its successful Hinotori, or fire-bird, solar flare mission, launched on 21 February 1981 near a solar maximum, the Institute of Space and Astronautical Science (ISAS) in Japan organized a new mission, initially called Solar-A. It was designed to investigate solar flares with improved angular and energy resolution during the next maximum in solar activity; quiescent, non-flaring structures and pre-flare activity would also be scrutinized.

The 390-kilogram spacecraft was launched by ISAS from the Kagoshima Space Center on 30 August 1991, into a 96-minute, nearly-circular Earth orbit. After launch, the mission was renamed, Yohkoh, which means "sunbeam" in English.

Yohkoh detects high-energy radiation from the Sun. This radiation resembles light waves, except with shorter wavelengths and greater energy. Scientists have provided names for the high-energy radiation depending on its wavelength or energy; they are soft X-rays, hard X-rays and gamma rays. A soft X-ray is one of relatively long wavelength and low energy; this form of electromagnetic radiation has energies of 1 to 10 keV and wavelengths between 10^{-9} and 10^{-10} meters. Hard X-rays have shorter wavelengths, between 10^{-10} and 10^{-11} meters, and higher energy, between 10 keV

Table 1.3. Yohkoh's instruments arranged alphabetically by acronym*

Instrument	Measurement
BCS	The Bragg Crystal Spectrometer measures X-ray spectral lines of highly ionized iron, Fe XXV and Fe XXVI, calcium, Ca XIX, and sulfur, S XV, between 0.18 and 0.51 nanometers with a time resolution as short as 0.125 seconds and across the full solar disk.
HXT	The Hard X-ray Telescope images flare radiation with energy from 20 to 80 keV in four channels, with an angular resolution of about 7 arcseconds, a time resolution of 0.5 seconds and a field of view that includes the entire visible disk.
SXT**	The Soft X-ray Telescope achieves 4 arcsecond spatial resolution and 2 second temporal resolution, detecting radiation with energy between 1 and 4 keV (1.2 to 0.3 nanometers) with a wide dynamic range (up to 200,000). It renders images taken at a uniform rate across the full solar disk for both the faint, quiescent and intense flaring X-ray structures with temperatures above 2 to 3 million degrees.
WBS	The Wide Band Spectrometer measures X-rays and gamma-rays from 3 keV to 30 MeV during solar flares, and is also sensitive to neutrons emitted during flares; it observes the full solar disk with a time resolution as short as 0.125 seconds.

* Energetic radiation is often expressed in terms of its energy content, in units of kilo-electron volts, or keV, rather than the more conventional units of wavelength. For conversions, $1\,\text{keV} = 10^3\,\text{eV} = 1.602 \times 10^{-16}$ Joule and the photon energy, E, associated with a wavelength λ in nanometers is $E = 1.986 \times 10^{-16}/\lambda$ Joule. The wavelength units of Angstroms are often used, where one Angstrom $= 1\,\text{Å} = 10^{-10}$ meter $= 0.1$ nanometer.
** The SXT was prepared by the Lockheed-Martin Solar and Astrophysics Research Laboratory, the National Astronomical Observatory of Japan and the University of Tokyo with the support of NASA and ISAS.

and 100 keV. Gamma rays are even shorter and more energetic. The electron volt, or eV, is a unit of energy, equivalent to 1.602×10^{-19} Joule, so $1\,\text{keV} = 1{,}000\,\text{eV} = 1.602 \times 10^{-16}$ Joule.

The spacecraft carries four co-aligned instruments. They are a high-resolution soft X-ray Bragg Crystal Spectrometer (BCS) provided by the United Kingdom, a Hard X-ray Telescope (HXT) provided by ISAS, a Soft X-ray Telescope (SXT) that is a collaborative effort between ISAS and NASA, and a hard X-ray and gamma-ray Wide Band Spectrometer (WBS) provided by ISAS. The measurements of all four instruments are provided in Table 1.3. No previous solar spacecraft has had available such a potent combination of hard and soft X-ray imaging and spectrometry instruments operating over such a wide energy range.

The Yohkoh telescopes have full-Sun fields of view with unparalleled angular and spatial resolution. The Hard X-ray Telescope, or HXT, images hard X-rays emitted by high-speed electrons accelerated in impulsive flares. Both the flaring and quiescent, or non-flaring, Sun are detected at soft X-ray wavelengths with a rapid, uniform rate using the Soft X-ray Telescope, or SXT for short. It routinely images high-temperature gas, above 2 to 3 million degrees Kelvin, across the Sun (Fig. 1.3). The SXT is pointed at the Sun with an ac-

Fig. 1.3. The X-ray Sun. Million-degree gas is constrained within ubiquitous magnetic loops, giving rise to bright emission from active regions in this X-ray image of the Sun taken on 8 May 1992 with the Soft X-ray Telescope (SXT) onboard the Yohkoh mission. Relatively faint magnetic loops connect active regions to distant areas on the Sun, or emerge within quiet regions away from active ones. The extended corona rings the Sun, and dark coronal holes are found at its poles (*top* and *bottom*). This image has been corrected for instrumental effects and processed to enhance solar features. (Yohkoh is a project of international cooperation between the Japanese Institute of Space and Astronautical Science (ISAS) and NASA. The SXT was prepared by the Lockheed-Martin Solar and Astrophysics Laboratory, the National Astronomical Observatory of Japan, and the University of Tokyo with the support of NASA and ISAS)

curacy and steadiness that permits images at invisible X-ray wavelengths that are as sharp and clear as pictures made in visible wavelengths from telescopes on the ground.

Yohkoh is an excellent example of international cooperation. Even with the challenge of a formidable language barrier, the Yohkoh team of scientists from Japan, America and England has worked together and shared resources in a collegial and trusting atmosphere, working shoulder-to-shoulder operating the spacecraft, analyzing data and publishing results in an extremely austere staffing environment.

Yohkoh's principal telemetry and operation control center is provided by ISAS near Tokyo, Japan, while NASA obtains telemetry data capture during passes over its Deep Space Network ground stations in Goldstone, California, Canberra, Australia, and Madrid, Spain. The operation of the Yohkoh mission is the responsibility of the Yohkoh science team. Data from all four instruments are freely shared within this team, and ultimately by the full scientific community. For example, solar images from the soft X-ray telescope, which produces most of the data that make Yohkoh so widely known, are available at the instrument home page (http://www.lmsal.com/SXT/) on the World Wide Web; its images may also be obtained from the Solar Data Analysis Center at http://umbra.nascom.nasa.gov/.

The primary scientific objective of Yohkoh is to obtain high-resolution spatial and temporal information about high-energy flare phenomena, permitting detailed scrutiny of where and how flare energy is released and particle acceleration takes place. The location and geometry of the flaring sources have been related to the topology of the solar magnetic field in order to determine how stored magnetic energy powers the explosive flares. Oppositely directed magnetic fields can merge together, releasing the necessary energy at the place where they touch. Contorted, twisted magnetic structures may also play a role in triggering powerful explosions on the Sun.

The soft X-ray telescope has been additionally used to image the Sun's dynamic million-degree atmosphere, with both high time resolution and high spatial resolution, when a flare is not in progress. This data has been used to study how the hot gases evolve, interact and change their magnetic structure on both small and global scales, resulting in the discovery of new flaring phenomena and providing insights to the heating and expansion of the high-temperature material. Subsequent chapters of this book discuss all of these accomplishments in greater detail.

Yohkoh has now gathered information for about a decade after its launch, keeping careful watch over the Sun for almost a complete 11-year cycle of magnetic activity. Its unique measurements of the high energy and high temperature aspects of the Sun are being continued during the activity maximum near the turn of the century. This time the scientific return of Yohkoh is being greatly enhanced by simultaneous observations with the SOHO and Ulysses missions that were not available during the previous peak in solar activity.

Armed with the most sophisticated detectors and telescopes ever focused on the Sun and space, this scientific armada of missions, named SOHO, Ulysses and Yohkoh, has succeeded beyond our wildest expectations. To fully interpret their results, we must first understand the landscape that they have either moved within or viewed from afar. We therefore next describe the early exploration of space, providing the historical context from which our modern discoveries came. This will establish the framework needed for envisaging the new results, while introducing basic concepts and fundamental terms.

2. Discovering Space

Overview

Half a century ago, most people visualized out planet as a solitary sphere traveling in a cold, dark vacuum around the Sun. But we now know that the space outside the Earth's atmosphere is not empty; it is filled with pieces of the Sun. These solar corpuscles were first suggested by terrestrial auroras and geomagnetic storms. In fact, the Earth and other planets are immersed within the solar atmosphere, called the corona, so we are actually living inside the outer part of the Sun.

The chemical ingredients of the Sun are fingerprinted by spectral features in sunlight, indicating that the lightest element, hydrogen, is the most abundant substance in our star. Similar features in the corona's radiation indicate the presence of heavy elements ionized at a temperature of a million degrees, and that the solar atmosphere must therefore consist mainly of electrons and protons torn out of the abundant hydrogen atoms during high-temperature collisions.

The hot, charged atmosphere expands and flows away from the Sun, forming a perpetual solar wind that was inferred from comet tails, suggested by theoretical considerations, and fully confirmed by direct measurements with spacecraft. The relentless wind moves at supersonic speeds and consists mainly of electrons, protons, and magnetic fields, but with an exceedingly low density and even lesser amounts of heavier ions.

Space also contains the subatomic remnants of other stars, called cosmic rays, that rain down on the Earth in all directions, traveling at nearly the velocity of light from outside our solar system.

Intense magnetism pervades the Sun's atmosphere and extends into space. The solar cycle of magnetic activity causes an 11-year variation in the number and positions of sunspots, the shape of the corona, the level of explosive solar activity, and even the amount of cosmic rays reaching the Earth.

Early photographs of the Sun's X-rays, taken from rockets and the Skylab space station, indicated that magnetized coronal loops shape, mold and constrain some of the million-degree corona close to the Sun, and that places with no detectable X-rays, called coronal holes, are the source of a high-speed solar wind.

2.1 Space is not Empty

The first suggestive evidence for the fullness of space came not from looking at the Sun or cosmos, but instead at our home planet where auroras light up the Earth's polar regions and quivering compass needles record variations in the terrestrial magnetic field. Both phenomena suggested that the Sun sends matter as well as light into space.

The *aurora borealis* and *aurora australis*, or northern and southern lights, illuminate the Arctic and Antarctic skies, where curtains of multi-colored light dance and shimmer across the night sky far above the highest clouds (Fig. 2.1). The Earth is a huge dipolar magnet, with north and south magnetic poles located near the geographic ones – that is why a compass points roughly north or south. But compass needles fluctuate, jiggle about, and do not always point in exactly the same direction. Both auroras and varying terrestrial magnetism were attributed long ago to invisible, electrically charged particles, called solar corpuscles, sent out into space from our daytime star, the Sun.

Prophetic insights were provided by Edmund Halley in his account of the great aurora on 17 March 1716; the English astronomer attributed the aurora light to "magnetical effluvia" circulating poleward in the Earth's dipolar magnetic field. In 1733, the French scientist Jean Jacques d'Ortous de Mairan asserted that the aurora is a cosmic phenomenon, arising from the entry of solar gas into the Earth's atmosphere. He also suggested a possible connection between the "frequency, the cessation and the return of sunspots and the manifestation of the aurora borealis". Nearly one and a half centuries later, Elias Loomis, Professor of Natural Philosophy and Astronomy at Yale College, demonstrated a correlation between great auroras and the times of maximum in the number of sunspots, which varies with a period of about 11 years.

Fig. 2.1. Aurora Borealis. Spectacular curtains of red and green light are found in these photographs of the fluorescent Northern Lights, or Aurora Borealis, taken by Forrest Baldwin in Alaska. (Courtesy of Kathi and Forrest Baldwin, Palmer, Alaska)

At the turn of the century, the Norwegian physicist Kristian Birkeland argued that the Earth's magnetism focuses incoming electrons to the Polar Regions where they produce auroras. This is because magnetic fields create an invisible barrier to charged particles, causing them to move along their magnetic conduits rather than across them. Birkeland demonstrated his aurora theory in laboratory experiments by sending cathode rays, or electrons, toward a magnetized sphere, called a terella, with a dipolar magnetic field, using phosphorescent paint to show where the electrons struck it. The resulting light was emitted near the magnetic poles, with glowing shapes that reproduced many of the observed features of the auroras.

As Birkeland expressed it in 1913:

> It seems to be a natural consequence of our points of view that the whole of space is filled with electrons and flying ions of all kinds. We have assumed that each stellar system in evolutions throws off electric corpuscles into space. It does not seem unreasonable therefore to think that a greater part of the material masses in the universe is found, not in the solar systems or nebulae, but in 'empty' space.

Intense, global disturbances in the Earth's magnetic field, called geomagnetic storms, are synchronized over the years with solar activity (Sect. 7.1). Charged particles from the Sun might therefore also be buffeting, compressing and distorting the Earth's magnetism, causing at least some of the geomagnetic storms that make compass needles fluctuate.

During a nine-month northern polar expedition, in 1902 and 1903, Birkeland monitored auroras and geomagnetic activity, noting that there are always detectable variations in the geomagnetic field at high latitudes and that a faint aurora borealis is also usually present. So, electrically charged particles seemed to be always flowing from the Sun to the Earth, indirectly accounting for both the northern lights and geomagnetic storms, but with varying brilliance and intensity that seemed to be controlled by the Sun.

Birkeland thought in terms of electron beams and currents of electrons from the Sun. Such ideas were challenged first by Arthur Schuster in 1911 and then in 1919 by Frederick A. Lindemann, Professor of Physics at Oxford and science advisor to Winston Churchill. They both showed that a beam of solar electrons would disperse, and effectively blow itself apart, because of the repulsive force of like charges, long before reaching the Earth. Lindemann noted, however, that a stream of solar electrons and protons in equal numbers could retain its shape and travel to the Earth to initiate magnetic disturbances there. In 1930–31 Sydney Chapman and Vincent Ferraro described how sudden changes in the geomagnetic field can result when clouds of solar electrons and protons collide with the Earth's magnetic field. These ideas are taken up again in Sect. 7.1 that contains related modern explanations for geomagnetic storms and terrestrial auroras.

How could the apparently calm, serene and unchanging Sun be sending material into space? Our star is enveloped by an invisible atmosphere that extends all the way to the Earth and beyond. So, the sharp outer rim of our Sun is illusory, and we actually live inside the outer parts of it.

The visible disk of the Sun is called the photosphere, which simply means the sphere in which sunlight originates – from the Greek *photos* for light. We consider the photosphere to be the surface of the Sun, but it is not really a surface. Being entirely gaseous, the Sun has no solid surface and no permanent features.

The photosphere merely marks the level beyond which the solar gases become tenuous enough to be transparent. The atmosphere just above the round, visible disk of the Sun is far less substantial than a whisper, and more rarefied than the best vacuum on Earth. It is so tenuous that we see right through it, just as we see through the Earth's clear air. This diaphanous atmosphere of the Sun includes, from its deepest part outward, the photosphere, chromosphere and corona.

When the photosphere's light is spread out into its different colors, or wavelengths, it is cut by several dark gaps, now called absorption lines, that were first noticed by William Hyde Wollaston in 1802 and investigated in far greater detail by Joseph von Fraunhofer in 1814–15 (Table 2.1). The dark spectral features are called absorption lines because they each look like a line in a display of light intensity at different wavelengths, called a spectrum, and

Table 2.1. Prominent absorption lines in photosphere sunlight*

Wavelength (nanometers)	Fraunhofer Letter	Element Symbol and Name
393.368	K	Ionized Calcium, Ca II
396.849	H	Ionized Calcium, Ca II
410.175	h	Hydrogen, H_δ, delta transition
422.674	g	Neutral Calcium, Ca I
431.0 ± 1.0	G	CH molecule
434.048		Hydrogen, H_γ, gamma transition
438.356	d	Neutral Iron, Fe I
486.134	F	Hydrogen, H_β, beta transition
516.733	b_4	Neutral Magnesium, Mg I
517.270	b_2	Neutral Magnesium, Mg I
518.362	b_1	Neutral Magnesium, Mg I
526.955	E	Neutral Iron, Fe I
588.997	D_2	Neutral Sodium, Na I
589.594	D_1	Neutral Sodium, Na I
656.281	C	Hydrogen, H_α, alpha transition
686.719	B	Molecular Oxygen, O_2, in our air
759.370	A	Molecular Oxygen, O_2, in our air

* Adapted from Lang (1999). The wavelengths are in nanometer units, where 1 nanometer $= 10^{-9}$ meters. Astronomers have often used the Angstrom unit of wavelength, where 1 Angstrom $= 1$ Å $= 0.1$ nanometers. The letters were used by Joseph von Fraunhofer around 1814 to designate the spectral lines before they were chemically identified, but the subscripts denote components that were not resolved by Fraunhofer. A Roman numeral I after an element symbol denotes a neutral, or unionized atom, with no electrons missing, whereas the Roman numeral II denotes a singly ionized atom with one electron missing. The lines A and B are produced by molecular oxygen in the terrestrial atmosphere

because they are produced when atoms in a cool, tenuous gas absorb the radiation of hot, dense underlying material.

The absorption lines in the solar spectrum have been used to identify the chemical ingredients of the Sun. Since each element, and only that element, produces a unique set of dark lines, their specific wavelengths can be used to fingerprint the atom or ion from which they originated.

By comparing the solar absorption lines with those emitted by elements vaporized in the laboratory, Gustav Kirchhoff was able to identify many of the elements in the solar spectrum. Working with Robert Bunsen, inventor of the Bunsen burner, Kirchhoff showed that the Sun contains sodium, calcium and iron (Table 2.1). This suggested that the Sun is made out of the same material as the Earth, but that is only partly true.

Detailed investigations of the Sun's absorption-line intensities, by Albrecht Unsöld in 1928, suggested that the Sun is mainly composed of the lightest element, hydrogen, which is terrestrially rare. In contrast, the Earth is primarily made out of heavy elements that are relatively uncommon in the Sun. Hydrogen is, for example, about one million times more abundant than iron in the Sun, but iron is one of the main constituents of the Earth which cannot even retain hydrogen gas in its atmosphere.

Just above the photosphere lies a thin layer, about 2.5 million meters thick, called the chromosphere, from *chromos*, the Greek word for color. Still higher, above the chromosphere is the *corona*, from the Latin word for crown. The linkage between the chromosphere and the corona occurs in a very thin transition region, less than 100 thousand meters thick. We now know that the outermost tenuous layer of the solar atmosphere, the corona, extends to the planets and beyond.

The corona becomes visible to the unaided eye for only a few minutes when the Sun's bright disk is blocked out, or eclipsed, by the Moon. During such a total solar eclipse, the corona is seen at the limb, or apparent edge, of the Sun, against the blackened sky as a faint halo of white light, or all the visible colors combined (Fig. 2.2). The thin chromosphere also becomes visible during a solar eclipse, but just for a few seconds, revealing narrow, pink or rose-colored features at the solar limb.

Because of their very low densities, the chromosphere and corona do not create absorption lines. Instead of absorbing radiation, the chromosphere and corona emit it. When such a tenuous gas is heated to incandescence, the energized atoms or ions produce emission lines that shine precisely at the same wavelengths as the dark absorption lines produced by the same substance. So, with the photosphere eclipsed, the dark absorption lines in the solar spectrum are replaced by bright emission lines.

The spectrum of the chromosphere is dominated by the bright red emission line of hydrogen, dubbed hydrogen alpha, at 656.3 nanometers. The unusual intensity of the chromosphere's hydrogen line confirmed the great abundance of hydrogen in the solar atmosphere, for greater amounts of a substance tend to produce a brighter spectral line. The observed luminosity of the Sun additionally requires that the entire star must be predominantly composed of hydrogen.

Fig. 2.2. Gossamer Corona. The Sun's corona as photographed during the total solar eclipses of 26 February 1998 (*top*) and 11 August 1999 (*bottom*), observing from Oranjestad, Aruba and Lake Hazar, Turkey, respectively. To extract this much coronal detail, several individual images, made with different exposure times, were combined and processed electronically in a computer. The resultant composite images show the solar corona approximately as it appears to the human eye during totality. Note the fine rays and helmet streamers that extend far from the Sun and correspond to a wide range of brightness. (Courtesy of Fred Espenak)

Helium is so rare on the Earth that it was first discovered from its appearance as an unidentified yellow line in the solar chromosphere spectrum, during the eclipse of 18 August 1868, and outside of eclipse as well. It was probably not until the following year that Norman Lockyer convinced himself that the yellow line could not be identified with any known terrestrial element, and coined the term "helium" after the Greek Sun god, *Helios*. Helium was not found on Earth until 1895, when William Ramsay discovered it as a gaseous emission from radioactive material. Yet, helium is the second most abundant element in the Sun. Ramsay received the Nobel Prize in Chemistry in 1904 for his discovery of inert gaseous elements in the air and his determination of their place in the periodic system.

The solar corona presents one of the most puzzling enigmas of solar physics. It is hundreds of times hotter than the underlying photosphere. The corona's searing heat was suggested by the identification of emission lines, first observed during solar eclipses more than a century ago. The exceptionally hot corona was fully confirmed by observations of the Sun's radio emission about half a century later.

During the solar eclipse of 7 August 1869, both Charles Young and William Harkness first found that the spectrum of the solar corona is characterized by a conspicuous green emission line. Within half a century, eclipse observers had detected at least ten coronal emission lines and the number doubled in a few decades more, but not one of them had been convincingly explained.

Since none of the coronal features had been observed to come from terrestrial substances, astronomers concluded that the solar corona consisted of some mysterious ingredient, which they named "coronium". Belief in the new element lingered for many years, until it became obvious that there was no place for it in the atomic periodic table, and it was therefore not an unknown element, but a known substance in an unusual state.

The solution to the coronium puzzle was provided by Walter Grotrian, of Potsdam, and the Swedish spectroscopist, Bengt Edlén, in 1939 and 1941 respectively, who attributed the coronal emission lines to elements known on Earth, but in an astonishingly high degree of ionization and at an unexpectedly hot temperature. The spectral lines were attributed to relatively rare transitions in a very tenuous gas (Table 2.2). They are called forbidden emission lines because collisions can keep them from happening in even the best vacuum on Earth.

The identified features were emitted by atoms deprived of 9 to 14 electrons (Table 2.2). The coronal particles would have to be moving very fast, with temperatures of millions of degrees, to have enough energy to rip off so many electrons during atomic collisions.

In 1946 Australian scientists, led by Joseph L. Pawsey, used war-surplus radar (radio detection and ranging) equipment to monitor the Sun's radio emission at 1.5 meters wavelength, showing that its intensity, though highly variable, almost never fell below a threshold. Pawsey's colleague, David F. Martyn, noticed that the source of the steady radio component must be the corona, since radiation at meter wavelengths would be reflected by this outer

Table 2.2. Strong coronal forbidden emission lines*

Wavelength (nanometers)	Ion	Name	Wavelength (nanometers)	Ion
338.8	Fe XIII		670.2	Ni XV
423.2	Ni XII		789.2	Fe XI
530.3	Fe XIV	Green Line	802.4	Ni XV
569.4	Ca XV	Yellow Line	1,074.7	Fe XIII
637.4	Fe X	Red Line	1,079.8	Fe XIII

* Adapted from Edlén (1941). The symbols Ca, Fe and Ni denote, respectively, Calcium, Iron and Nickel. Subtract one from the Roman numeral to obtain the number of missing electrons. Thus, the ion Fe XIII is an iron atom missing 12 electrons; Fe I is a neutral, unionized atom of iron. The wavelength is in units of nanometers, or 10^{-9} meters. Astronomers have often used the Angstrom unit of wavelength, where 1 Angstrom $= 1$ Å $= 0.1$ nanometers $= 10^{-10}$ meters.

part of the Sun and one could not use long radio waves to look past it. Moreover, the observed intensity of the radio emission from such a tenuous gas corresponded to a temperature of approximately a million degrees. The existence of a million-degree corona, first suggested by the corona's emission lines, was thereby confirmed from its radio emission, in papers published in 1946 by Martyn, Pawsey and independently by the Russian physicist Vitaly L. Ginzburg.

The corona is so intensely hot that its abundant hydrogen is torn into numerous electrons and protons; each hydrogen atom consists of one electron and one proton. The solar atmosphere therefore consists mainly of electrons and protons, with smaller amounts of heavier ions created from the less abundant elements in the Sun. These electrons, set free from former hydrogen atoms, scatter the photosphere sunlight that strikes them, providing the pearl-white coronal radiation seen during a total solar eclipse.

Although the corona's electrons, protons and other ions are electrically charged, the corona is electrically neutral and has no net charge. In such an ionized gas, often called a plasma, the total negative charge of all the electrons is equal to the combined positive charge of the ions.

Somehow the corona is heated to a few million degrees Kelvin just above the photosphere with a temperature of 5,780 degrees Kelvin, which is completely unexpected. It is something like watching a glass of water boiling on your cool kitchen counter. Heat normally moves from a hotter to a colder region, not the other way around. Simply put, there is too much heat in the corona. In Sect. 4.1 we discuss early attempts to explain this heating paradox, and in Sect. 4.4 modern results are described that are closing in on an explanation of it.

But for now the main point, as far as the space near Earth is concerned, is that the corona is so hot that it can't stay still. At a million degrees, the sizzling heat is too hot to be entirely constrained by either the Sun's inward gravitational pull or its magnetic forces. An overflow corona is forever expanding in all directions, filling the solar system with a great eternal wind that blows from the Sun.

The existence of an electrified stream of solar corpuscles, continuously flowing away in all directions from the Sun, was inferred by the German astrophysicist Ludwig Biermann between 1951 and 1957 from the observed motions of comet tails. The comets appear almost anywhere in the sky, with tails that always point away from the Sun. Moreover, the ions in a comet tail move with velocities many times higher than could be caused by the weak pressure of sunlight. Biermann proposed that a perpetual flow of electrically charged particles pours out of the Sun at all times and in all directions, colliding with the cometary ions and imparting momentum to them. This would accelerate the comet ions and push them radially away from the Sun in straight tails.

Focus 2.1

Escape From the Sun, Earth and Moon

When the kinetic energy of motion of an object or a particle of mass, m, moving at velocity, V, is just equal to the gravitational potential energy exerted on it by a larger mass, M, we have the relation:

$$\text{Kinetic Energy} = \frac{mV^2}{2} = \frac{GmM}{D} = \text{Gravitational Potential Energy}\,,$$

where the Newtonian gravitational potential is $G = 6.67 \times 10^{-11}\,\mathrm{N\,m^2\,kg^{-2}}$, and D is the distance between the centers of the two masses. When we solve for the velocity, we obtain

$$V_{\text{escape}} = \sqrt{\frac{2GM}{D}}\,,$$

where the $\surd$ sign denotes the square root of the following term and the subscript escape has been added to show that the object or particle must be moving faster than V_{escape} to leave a larger mass, M. This expression is independent of the smaller mass, m.

At the solar photosphere, where D becomes the solar radius of 696 million (6.96×10^8) meters, the equation gives

$$V_{\text{escape}}(\text{Sun's photosphere}) = 617 \text{ thousand meters per second}\,,$$

where the Sun's mass is 1.989×10^{30} kilograms.

By way of comparison, the escape velocity at the Earth's surface is:

$$V_{\text{escape}}(\text{Earth's surface}) = 11{,}200 \text{ meters per second}\,,$$

where the mean radius of the Earth is 6.378 million (6.378×10^6) meters and the mass of the Earth is 5.9742×10^{24} kilograms. The escape velocity from the surface of the Moon is:

$$V_{\text{escape}}(\text{Moon's surface}) = 2{,}370 \text{ meters per second}\,.$$

The Moon's mass is 7.348×10^{22} kilograms and its mean radius is 1.738×10^{6} meters.

A rocket must move faster than 11,200 meters per second if is to move from the Earth into interplanetary space, and if it travels at a slower speed the rocket will crash back down into the Earth. A lunar craft only needs to be propelled at about one fifth of this speed to leave the Moon. There is no atmosphere on the Moon because it has a very low escape velocity, and molecules can therefore easily leave it.

An atom, ion or molecule moves about because it is hot. Its kinetic temperature, T, is defined in terms of the thermal velocity, V_{thermal}, given by the expression equating the thermal energy to the kinetic energy of motion,

$$\text{Thermal Energy} = \frac{3}{2}kT = \frac{1}{2}mV_{\text{thermal}}^2 = \text{Kinetic Energy}$$

or solving for the thermal velocity:

$$V_{\text{thermal}} = \sqrt{\frac{3kT}{m}},$$

where Boltzmann's constant $k = 1.38 \times 10^{-23}$ Joule second, the temperature of the particle is denoted by T, and its mass by m. We see right away that at a given temperature, lighter particles move at faster speeds. Colder particles of a given mass travel at slower speed. Anything will cease to move when it reaches absolute zero on the Kelvin scale of temperature.

Everything in the Universe moves, and there is nothing completely at rest. You might say that motion seems to define existence. When you stop moving it is all over.

For the lightest known element, hydrogen, our expression gives

$$V_{\text{thermal}}(\text{hydrogen atoms}) = 157\,T^{1/2} \text{ meters per second},$$

where the mass of a hydrogen atom is $m = 1.674 \times 10^{-27}$ kilograms. So, hydrogen atoms move at about 12.0 thousand and 15.7 thousand meters per second in the photosphere and the chromosphere, where the respective temperatures are 5,780 and 10,000 degrees Kelvin. Since these velocities are way below the Sun's escape velocity, hydrogen and any other heavier element must be retained in the low solar atmosphere. Even at the corona's temperature of 2 million degrees, the thermal velocity of a hydrogen atom is 222 thousand meters per second. To leave the Sun, hot coronal hydrogen has to be given an extra push out to a distance of a few solar radii, where the solar gravity and escape velocity have become smaller. The same conclusion applies to protons that have essentially the same mass as a hydrogen atom. Since a free electron is 1,836 times lighter than a proton, it has a thermal velocity that is 42.8 times faster at a given temperature.

The comet tails serve as probes of the solar particles streaming away from the Sun; it is something like putting a wet finger in the wind to show it is there. Biermann estimated a speed of 500 thousand to one million meters per second from moving irregularities and the directions of comet tails. As we shall see, his estimated number density of up to 1,000 million electrons or ions per cubic meter, near the Earth's orbit, turned out to be a factor of 100 too high because he assumed that the comet tail effects are from particle collisions alone. Most of the comet interaction is via the interplanetary magnetic field carried in the Sun's wind. The density estimate may have also been in error because of a faulty identification of particles responsible for the zodiacal light with electrons instead of cosmic dust.

By 1957 the English geophysicist Sydney Chapman had presented mathematical arguments that seemed to show that Biermann was wrong. Chapman noticed that the free electrons in the million-degree corona make it a very good thermal conductor, even better than a metal. The electrons in the corona would therefore carry its intense heat far into space, somewhat like an iron bar that is heated at one end and therefore becomes hot all over. This meant that a static corona must spread out to the Earth's orbit and beyond. According to Chapman, this extended, non-expanding corona would block any outward corpuscular stream from the Sun. Biermann disagreed, arguing that the solar corpuscles would sweep any stationary gas out of the solar system.

These conflicting ideas, of a static corona and a stream of charged particles, were reconciled by Eugene N. Parker of the University of Chicago. In 1958, he added dynamic terms to Chapman's equations, showing how a relentless flow might work, and dubbing it the solar wind. A very hot coronal gas can create an outward pressure that becomes greater than the inward pull of the Sun's gravity at increasing distances from the Sun, where the gas accelerates and expands radially at high, supersonic speeds in all directions. Parker also argued that the wind would pull the Sun's magnetic field into surrounding interplanetary space, obtaining a spiral shape due the combination of radial flow and solar rotation, and that Earth-bound cosmic rays would be modulated by the interplanetary magnetic field.

Parker's theoretical conclusions were very controversial, and were initially received with a great deal of skepticism. The Sun might be sporadically ejecting material from localized regions, but it was difficult to envisage a continual ejection over the entire corona. Critics also wondered how the solar atmosphere could be hot enough to sustain such a powerful wind.

Even with a temperature of one million degrees, the thermal energy of protons is several times less than the Sun's gravitational pull on them at the bottom of the corona. In scientific terms, the proton's thermal velocity is less than the escape velocity of the Sun (Focus 2.1). The hot, million-degree protons are not moving fast enough, on average, to overcome the gravity in the low corona, next to the visible Sun, and move out into the solar wind. So, some additional source of energy seemed to be required to help the corona break away from the Sun's powerful gravitational grasp.

Nevertheless, in less than a decade American and Soviet scientists confirmed the solar wind's existence with direct observations of the interplanetary medium, verifying many aspects of Parker's model.

2.2 Touching the Unseen

Our civilization was forever changed with the launch of *Prosteyshiy Sputnik*, the simplest satellite, by the Soviet Union on 4 October 1957. It began a space age that has led to our daily use of artificial satellites for communications, navigation and weather forecasting. To many people in the United States, *Sputnik* confirmed the reality of a missile gap, and also verified the threat that the Soviet Union posed to world peace. So it is not surprising that many of the space-age improvements in our life grew out of military applications. They include reconnaissance satellites that monitor enemy activity, navigation satellites that accurately target missiles launched from ships or airplanes, and intercontinental ballistic missiles that can either carry nuclear warheads to distant countries or toss satellites into space. Yet, there has always been a strong scientific component to the American and Russian space programs from the very beginning.

Some of the earliest spacecraft reached out to touch, feel and identify the invisible constituents of space. For instance, America's first Satellite 1958α, was launched into orbit on 1 February 1958. This satellite is better known today as Explorer 1. It included James A. Van Allen's instruments that detected energetic charged particles trapped by the Earth's magnetic field into donut-shaped regions that girdle our planet's equator (Sect. 7.2, Focus 7.2, and Fig. 7.6). Other spacecraft soon characterized the material components of the solar wind outside the Earth's magnetic domain.

The first direct measurements of the solar wind's corpuscular, or particle, content were made by a group of Soviet scientists led by Konstantin I. Gringauz, using four ion traps aboard the Lunik 2 spacecraft launched to the Moon on 12 September 1959. Each trap contained external, charged grids that acted as gates to exclude low energy ions and to keep energetic electrons out. Only high-speed ions could pass through the trap door. All four ion traps detected the energetic ions, leading Gringauz and his colleagues to report in 1960 that "the corpuscular emission of the Sun ... has thus been observed for the first time in the interplanetary space outside the magnetic field of the Earth".

In the following year, Gringauz reported that the maximum current in all four ion traps corresponded to a solar wind flux of 2 million million (2×10^{12}) ions (presumably protons) per square meter per second. This is in rough accord with all subsequent measurements. No evidence for a stationary component of the interplanetary gas was found. Moreover, since the ion traps were strategically placed around the rotating spacecraft, the group could show that the wind flows from the Sun and not toward it.

Solar wind measurements by American experimenters were first carried out by a Massachusetts Institute of Technology group with instruments aboard Explorer 10 in 1961, but unfortunately the spacecraft did not travel completely beyond the Earth's magnetic field to pristine, "undisturbed" interplanetary space.

All reasonable doubt concerning the existence of the solar wind was removed by measurements made on board Mariner 2, launched on 27 August 1962. Marcia Neugebauer and Conway W. Snyder of the Jet Propulsion Laboratory used more than one hundred days of Mariner 2 data, obtained as the spacecraft traveled to Venus, to show that charged particles are continuously emanating from the Sun, for at least as long as Mariner observed them. The velocity of the solar wind was accurately determined, with an average speed of 500 thousand meters per second, in rough accord with Biermann's and Parker's predictions.

The solar wind flux determined by Neugebauer and Snyder was in good agreement with the values measured with the ion traps on Lunik 2. The average wind ion density was shown to be 5 million (5×10^6) protons per cubic meter near the distance of the Earth from the Sun. We now know that such a low density close to the Earth's orbit is a natural consequence of the wind's expansion into an ever greater volume, but that variable wind components can gust with higher densities.

The Mariner data unexpectedly indicated that the solar wind has a slow and a fast component. The slow one moves at a speed of 300 thousand to 400 thousand meters per second; the fast one travels at twice that speed. The low-velocity wind was identified with the perpetual expansion of the million-degree corona. The high-velocity component swept past the spacecraft every 27 days, suggesting long-lived, localized sources on the rotating Sun. Moreover, peaks in geomagnetic activity, also repeating every 27 days, were correlated with the arrival of these high-speed streams at the Earth, indicating a direct connection between a region on the Sun and geomagnetic disturbances. As viewed from the Earth, the solar equator rotates once every 27 days.

Interplanetary space probes have been making *in situ* (Latin for "in original place", or literally "in the same place") measurements for decades, both within space near the Earth and further out in the Earth's orbital plane. Unlike any wind on Earth, the solar wind is a rarefied plasma or mixture of electrons, protons and heavier ions, and magnetic fields streaming radially outward in all directions from the Sun at supersonic speeds of hundreds of thousands of meters per second.

This perpetual solar gale brushes past the planets and engulfs them, carrying the Sun's corona out into interstellar space at the rate of almost a million tons (10^6 tons $= 10^9$ kilograms) every second. As the corona disperses, it must be replaced by gases welling up from below to feed the wind. Exactly where this material comes from is an important subject of contemporary space research, subsequently discussed in greater detail (Sect. 5.2).

Although the Sun is continuously blowing itself away, the outflow can continue for billions of years without significantly reducing the Sun's mass. Every second, the solar wind blows away a billion (10^9) kilograms. That sounds like a lot of mass loss, but it is four times less than the amount consumed every second during the thermonuclear reactions that make the Sun shine. To supply the Sun's present luminosity, hydrogen must be converted into helium, within the Sun's energy-generating core, with a mass loss of about 4 million tons every second. It is carried away by the Sun's radiation, whose energy and momentum flux vastly exceeds those of the solar wind (Table 2.3). A more significant concern is the depletion of hydrogen; the Sun will run out of hydrogen in its core in about 7 billion years, when our star will expand into a giant star (Sect. 7.5). By that time, the Sun will have lost only about 0.005 percent of its mass by the solar wind at the present rate.

The hot corona extends all the way to the Earth, where it has only cooled to a little more than one hundred thousand degrees Kelvin. Even though the electrons near the Earth are awfully hot, they are so scarce and widely separated that an astronaut or satellite will not burn up when venturing into interplanetary space. Although the velocity is high, the density of the gossamer corona is so low that if we could go into space and put our hands on it, we would not be able to feel it.

The reason that space looks empty is that these subatomic pieces of the Sun are very small and moving incredibly fast, and there really are not very many of them when compared even to our transparent atmosphere. By the time it reaches the Earth's orbit, the solar wind is diluted to about 5 million electrons and 5 million protons per cubic meter, a very rarefied gas (Table 2.4). By way of comparison, there are 25 million, billion, billion (2.5×10^{25}) molecules in every cubic meter of our air at sea level.

Still, at a mean speed of about 400 thousand meters per second, the flux of solar wind particles is far greater than anything else out there in space. Between one and ten million million (10^{12} to 10^{13}) particles in the solar wind cross every square meter of space each second (Table 2.4). That flux far surpasses the flux of more energetic cosmic rays that enter our atmosphere (Sect. 2.3, Table 2.5).

Table 2.3. Relative importance of solar wind and radiation*

	Solar Wind	Sun´s Radiation
Mass Loss Rate	10^9 kg s^{-1}	4×10^9 kg s^{-1}
Total Mass Loss (in 5 billion years)	0.00005 $M_\odot$	0.0002 $M_\odot$
Energy Flux (1 AU)	0.00016 Joule m^{-2} s^{-1}	1,400 Joule m^{-2} s^{-1}
Momentum Flux (1 AU)	8×10^{-10} kg m^{-1} s^{-1}	5×10^{-6} kg m^{-1} s^{-1}
Despin time	10^{16} years	10^{12} years

* Adapted from Axford (1985). $M_\odot$ denotes the Sun's mass of 2×10^{30} kilograms. The mean distance of the Earth from the Sun is one astronomical unit, or 1 AU for short, where 1 AU $= 1.496 \times 10^{11}$ meters.

Table 2.4. Mean values of solar-wind parameters at the Earth's orbit*

Parameter	Mean Value
Particle Density, N	$N \approx 10$ million particles per cubic meter (5 million electrons and 5 million protons)
Velocity, V	$V \approx 400$ thousand meters per second and $V \approx 800$ thousand meters per second
Flux, F	$F \approx 10^{12}$ to 10^{13} particles per square meter per second.
Temperature, T	$T \approx 120$ thousand degrees Kelvin (protons) to 140 thousand degrees Kelvin (electrons)
Particle Thermal Energy, kT	$kT \approx 2 \times 10^{-18}$ Joule ≈ 12 eV
Proton Kinetic Energy, $0.5 m_{\rm p} V^2$	$0.5 m_{\rm p} V^2 \approx 10^{-16}$ Joule $\approx 1{,}000$ eV $= 1$ keV
Particle Thermal Energy Density	$NkT \approx 10^{-11}$ Joule m^{-3}
Proton Kinetic Energy Density	$0.25 N m_{\rm p} V^2 \approx 10^{-9}$ Joule m^{-3}
Magnetic Field Strength, H	$H \approx 6 \times 10^{-9}$ Tesla $= 6$ nanoTesla $= 6 \times 10^{-5}$ Gauss

* These solar-wind parameters are at the mean distance of the Earth from the Sun, or at one astronomical unit, 1 AU, where 1 AU $= 1.496 \times 10^{11}$ meters; the Sun's radius, $R_{\odot}$, is $R_{\odot} = 6.96 \times 10^{8}$ meters. Boltzmann's constant $k = 1.38 \times 10^{-23}$ Joule K^{-1} relates temperature and thermal energy. The proton mass $m_{\rm p} = 1.67 \times 10^{-27}$ kilograms. The flux of cosmic ray protons arriving at the Earth, discussed in the next section, is about one ten billionth (10^{-10}) that of the solar wind at the Earth's orbit. The most abundant cosmic ray protons near Earth have an energy of 10^{-10} Joule but a local energy density of about 10^{-13} Joule m^{-3}, or about one ten thousandth (10^{-4}) the kinetic energy density of protons in the solar wind.

2.3 Cosmic Rays

Although the Sun's winds dominate the space within our solar system, cosmic rays form an important additional ingredient. These extraordinarily energetic charged particles enter the Earth's atmosphere from all directions in outer space. Many of them were probably accelerated to their awesome energies by the explosions of dying stars.

Cosmic rays were discovered in 1912, when the Austrian physicist Victor Franz Hess, an ardent amateur balloonist, measured the amount of ionization at different heights within our atmosphere. It was already known that radioactive rocks at the Earth's surface were emitting energetic "rays" that ionize the atmosphere near the ground, but it was expected that the ionizing substance would be completely absorbed after passing though sufficient quantities of the air.

Although the measured ionization at first decreased with altitude, as would be expected from atmospheric absorption of rays emitted by radioactive rocks, the ionization rate measured by Hess from balloons increased at even higher altitudes to levels exceeding that at the ground. This meant that some penetrating source of ionization came from beyond the Earth. By flying his balloon at night and during a solar eclipse, when the high-altitude signals persisted, Hess showed that they could not come from the Sun, but from some other source outside the solar system. In 1936 he was awarded the Nobel Prize in Physics for his discovery of these cosmic rays.

Further balloon observations by Werner Kolhörster of Berlin showed that the ionization continued to increase with height. By 1926 the American physicist Robert A. Millikan had used high-altitude balloon measurements to confirm that the "radiation" comes from beyond the terrestrial atmosphere, and incidentally gave it the present name of cosmic rays. Millikan was the first president of the California Institute of Technology, and the first American to win the Nobel Prize in Physics – in 1923 for his work on the elementary charge of electricity and the photoelectric effect.

Cosmic rays were initially believed to be high-energy radiation, in the form of gamma rays, but global measurements showed that they are charged particles deflected by the Earth's magnetic field toward its magnetic poles (Focus 2.2). In 1927 and 1928, the Dutch physicist Jacob Clay published the results of cosmic ray measurements during ocean voyages between Genoa and the Dutch colony of Java. He found a lower cosmic ray intensity near the Earth's equator, suggesting that the cosmic rays consisted, in part at least, of charged particles. Clay's results were confirmed and extended between 1930 and 1933 by Arthur H. Compton of the University of Chicago. He conclusively demonstrated that the cosmic-ray intensity increases with latitude, and also made measurements at mountain altitudes, where the latitude increase was even stronger.

The latitude increase showed that cosmic rays are electrically charged, but both negative and positive charges would show a similar effect. The sign of the charge was inferred in the late 1930s from the hemispheric distribution of the cutoff energy below which no vertically arriving particles are found. Lower-energy particles of positive charge will be observed if they arrive from the west; negatively-charged cosmic rays of lower energy will be found in the east. Measurements of this east-west effect showed that the most abundant cosmic rays are positively charged, and most likely protons, for more cosmic rays were found when the detector was directed to the west rather than the east.

By the late 1940s, instruments carried by high-altitude balloons established that the cosmic rays consist of hydrogen nuclei (protons), helium nuclei (alpha particles) and heavier nuclei (Table 2.5). Cosmic ray electrons were not discovered until 1961, mainly because they are far less abundant than the cosmic ray protons at a given energy. The presence of very energetic electrons in interstellar space had nevertheless been inferred from the radio emissions of our Galaxy.

Since the cosmic rays enter the atmosphere with energies greater than could be produced by particle accelerators on Earth, physicists used them as colossal atom destroyers, examining their subatomic debris after colliding with atoms in our air. These studies of fundamental particles led to several other Nobel Prizes in Physics – to Charles T. R. Wilson in 1927 for

Table 2.5. Average fluxes of primary cosmic rays at the top of the atmosphere*

Type of Nucleus	Flux (particles $m^{-2} s^{-1}$)
Hydrogen (protons)	640
Helium (alpha particles)	94
Carbon, Nitrogen, Oxygen	6

* Adapted from Friedlander (1989). The flux is in units of nuclei per square meter per second for particles with energies greater than 1.5 billion (1.5×10^9) electron volts per nucleon, arriving at the top of the atmosphere from directions within 30 degrees of the vertical.

his cloud chamber that makes the tracks of the electrically charged particles visible by condensation of vapor; to Carl D. Anderson in 1936 for his detection of the positron, the anti-particle of the electron, amongst the byproducts of cosmic rays colliding in the air; to Patrick M. S. Blackett in 1948 for his development of the Wilson cloud chamber method, and his discoveries therewith in the fields of nuclear physics and cosmic radiation; and to Cecil F. Powell in 1950 for his discovery of a fundamental particle, called the meson, from high-altitude studies of cosmic rays using a photographic method.

Although few in number, cosmic rays contain phenomenal amounts of energy. That energy is usually measured in units of electron volts, abbreviated eV; for conversion use $1\,\text{eV} = 1.602 \times 10^{-19}$ Joule. The greatest flux of cosmic rays arriving at Earth occurs at about 1 GeV, or at a billion (10^9) electron volts of energy. This is about a million times more than the kinetic energy of a typical proton in the solar wind. At this energy, a cosmic ray proton must be traveling at 88 percent of the speed of light (Table 2.6) while one in the Sun's wind moves about 500 times slower.

Despite their awesome energy, there are far fewer cosmic rays than solar wind particles in the vicinity of the Earth. The cosmic ray protons at the top of our atmosphere have a flux of 640 protons per square meter per second (Table 2.5). At the Earth's orbit, protons in the solar wind have a flux that is about 10 billion times (10^{10}) times larger. Even at their peak flux, the local energy density of cosmic rays is about one million electron volts per cubic meter, or about one ten thousandth (10^{-4}) the kinetic energy density of solar wind protons.

Focus 2.2

Charged Particles Gyrate Around Magnetic Fields

A charged particle cannot move straight across a magnetic field, but instead gyrates around it. If the particle approaches the magnetic field straight on, in the perpendicular direction, a magnetic force pulls it into a circular motion about the magnetic field line. Since the particle can move freely in the direction of the magnetic field, it spirals around it with a helical trajectory.

The size of the circular motion, called the radius of gyration and designated by R_g, depends on the velocity, $V_\perp$, of the particle in the perpendicular direction, the magnetic field strength, B, and the mass, m, and charge, Ze, of the particle. That gyration radius is described by the equation:

$$R_g = [m/(Ze)][V_\perp/B] ,$$

provided that the velocity is not close to the velocity of light, c. The period, P, of the circular orbit is $P = 2\pi R_g/V_\perp = 2\pi m/(eZB)$, and the frequency, ν_g, is $\nu_g = eZB/(2\pi m_e) = 2.8 \times 10^{10} B$ Hz for an electron when B is in Tesla.

This expression for the gyration radius provides a short-hand method of explaining a lot of intuitively logical aspects of the particle motion. It says that faster particles will gyrate in larger circles, and that a stronger magnetic field tightens the gyration into smaller coils. The attractive magnetic force increases with the charge, also resulting in a tighter gyration.

For an electron with mass $m_e = 9.1094 \times 10^{-31}$ kilograms and charge $e = 1.602 \times 10^{-19}$ Coulomb, with $Z = 1.0$, the corresponding gyration radius is:

$$R_g(\text{electron}) = 5.7 \times 10^{-12}[V_\perp/B] \text{ meters} ,$$

where $V_\perp$ is in meters per second and B is in Tesla. Since the proton has the same charge, but a mass, m_p, that is larger by the ratio $m_p/m_e = 1{,}836$, the proton's gyration radius for the same velocity is:

$$R_p(\text{proton}) = 1.05 \times 10^{-8}[V_\perp/B] \text{ meters} .$$

At high particle velocities approaching that of light, the radius equation is multiplied by a Lorentz factor $\gamma = [1 - (V/c)^2]^{-1/2}$, which becomes unimportant at low velocities when $\gamma = 1$. For cosmic ray protons of high velocity and large energies, $E = \gamma m_p c^2$, our equation becomes:

$$R_g = 3.3E/B \text{ meters} ,$$

when the energy E is in units of GeV, $1\,\text{GeV} = 1.602 \times 10^{-10}$ Joule, and the magnetic field strength is in Tesla. Thus, the path of a 1 GeV cosmic ray proton will be coiled into a gyration radius of about 10^{10} meters in the interstellar medium where $B \approx 10^{-10}$ Tesla $= 10^{-6}$ Gauss, which is about 15 times smaller than the mean distance between the Earth and the Sun. When that 1 GeV proton encounters the terrestrial magnetic field, whose strength is roughly 10^{-4} Tesla or 1 Gauss, its radius of gyration is $R_g \approx 10{,}000$ meters, or hundreds of times smaller than the Earth whose mean radius is 6.37 million meters. That means that the proton will gyrate about the magnetic field, and move toward Earth's magnetic poles.

Table 2.6. Particle speeds at different particle energies, expressed as fractions of the speed of light, c *

Particle Kinetic Energy (keV)	Electron Speed (times c)	Proton Speed (times c)
1 keV	$0.063c$	$0.0015c$
1,000 keV = 1 MeV	$0.94c$	$0.046c$
100,000 keV = 100 MeV	$0.999987c$	$0.43c$
1,000,000,000 keV = 1 GeV	$0.99999987c$	$0.88c$

* An energy of one kilo-electron volt is $1\,\text{keV} = 1.6022 \times 10^{-16}$ Joule, and the speed of light $c = 2.99792458 \times 10^{8}$ meters per second.

Cosmic rays do not all have the same energy, and there are fewer of them with higher energy. The number of particles with kinetic energy E is proportional to $E^{-\alpha}$, where the index $\alpha = 2.5$ to 2.7 for energies from a billion to a million million (10^9 to 10^{12}) electron volts. At higher energies there are relatively few particles. At lower energies the magnetized solar wind acts as a valve for the more abundant lower-energy cosmic rays, controlling the amount entering the solar system.

2.4 Pervasive Solar Magnetism

To most of us, the Sun looks like a perfect, white-hot globe, smooth and without a blemish. However, detailed scrutiny indicates that our star is not perfectly smooth, just as the texture of a beautiful face increases when viewed close up. Magnetism protrudes to darken the skin of the Sun in Earth-sized spots detected on the visual disk or photosphere. Our understanding of what causes these dark, ephemeral sunspots is relatively recent, at least in comparison to how long they have been known. The earliest Chinese records of sunspots, seen with the unaided eye, date back 3,000 years.

In 1908 George Ellery Hale first established the existence of intense, concentrated magnetic fields in sunspots, using subtle wavelength shifts in spectral lines detected at his solar observatory on Mt. Wilson, California. Such a shift, or line splitting, is called the Zeeman effect after Pieter Zeeman who observed it with magnetic fields on the Earth in 1896; he received the Nobel Prize in Physics in 1902 for his researches into the influence of magnetism upon radiation.

The size of the wavelength shifts measured by Hale indicated that sunspots contain magnetic fields as strong as 0.3 Tesla, or 3,000 Gauss (Focus 2.3). That is about ten thousand times the strength of the terrestrial magnetic field which orients our compasses. The intense sunspot magnetism acts as both a valve and a refrigerator, choking off the outward flow of heat and energy from the solar interior and keeping the sunspots cooler and darker than their surroundings.

Magnetic fields are described by lines of force, like those joining the north and south poles of the Earth or the opposite poles of a bar magnet. When the lines are close together, the force of the field is strong; when they are far apart, the force is weak. The direction of the lines of force, and the orientation of the magnetic fields, can be inferred from the polarization of the spectral lines. Magnetic field lines pointing out of the Sun have positive magnetic polarity, while inward-directed fields have negative polarity.

Sunspots tend to travel in pairs of opposite polarity, roughly aligned in the east-west direction on the Sun. The twinned sunspots are joined by invisible magnetic loops that rise above the photosphere into the corona, like an arching bridge connecting two magnetic islands. The magnetic field lines emerge from the Sun at a spot with one polarity and re-enter it at another one of opposite polarity.

The magnetized atmosphere in, around and above bipolar sunspot groups is called a solar active region. Active regions are places of concentrated, enhanced magnetic fields, large enough and strong enough to stand out from the magnetically weaker less intense areas. These disturbed regions also emit intense X-rays and are prone to awesome explosions. They mark one location of intense unrest on the Sun.

Unlike the Earth, the most intense solar magnetism does not consist of just one simple dipole. The Sun is spotted all over, like a young child with the measles, and contains numerous pairs of opposite magnetic polarity. The invisible magnetized bridges that join them have become known as coronal loops. Thus, powerful magnetism, spawned deep inside the Sun, threads its way through the solar atmosphere, creating ubiquitous coronal loops with magnetic lines of force.

The Sun's magnetism is forever changing and is never still, just like everything else in the Universe. Sunspots, and the magnetic loops that join them, are temporary. They come and go with lifetimes ranging from hours to months. Moreover, the total number of sunspots varies periodically, from a maximum to a minimum and back to a maximum, in about 11 years (Fig. 2.3 and Table 2.7). This periodic variation was discovered in the early 1840s by Samuel Heinrich Schwabe, an amateur astronomer of Dessau, Germany. At the maximum in the sunspot cycle, we may find 100 or more spots on the visible hemisphere of the Sun at one time; at sunspot minimum very few of them are seen, and for periods as long as a month none can be found.

The number of active regions, with their bipolar sunspots and coronal loops, varies in step with the sunspot cycle, peaking at sunspot maximum. They then dominate the structure of the inner corona. At sunspot minimum, the active regions are largely absent and the strength of the ultraviolet and X-ray emission of the corona is greatly reduced.

Brief, powerful bursts of energy, called solar flares, can be suddenly and unpredictably ignited in solar active regions. These incredible explosions become more frequent and violent when the number of sunspots is greatest; several solar flares can be observed on a busy day near the maximum of the sunspot cycle. They are not caused by sunspots, but are instead powered by

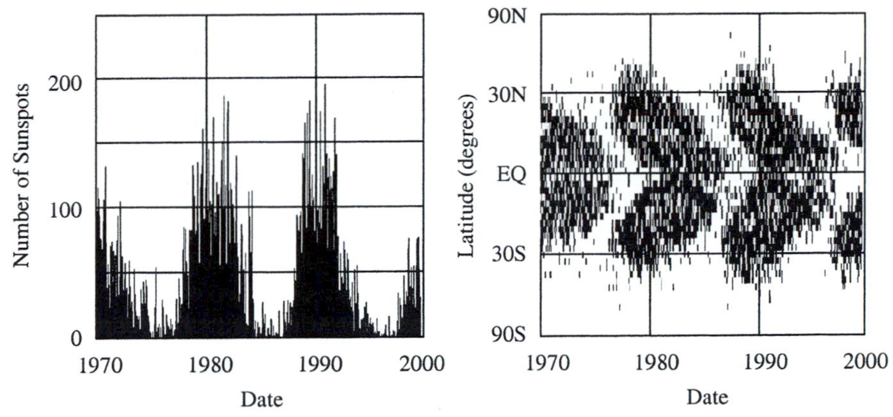

Fig. 2.3. Activity Cycle. The 11-year solar cycle of magnetic activity plotted from 1970 to 2000. Both the numbers of sunspots (*left*) and the positions of sunspots (*right*) wax and wane in cycles that peak every 11 years. Similar 11-year cycles have been plotted for more than a century. At the beginning of each cycle, the first sunspots appear at about 30 degrees latitude and then migrate to 0 degrees latitude, or the solar equator, at the cycle's end. This plot of the changing positions of sunspots resembles the wings of a butterfly, and has therefore been called the butterfly diagram. The cycles overlap with spots from a new cycle appearing at high latitudes when the spots from the old cycle persist in the equatorial regions. The solar latitude is the angular distance from the plane of the Sun's equator, which is very close to the plane of the Earth's orbit about the Sun, called the ecliptic. (Courtesy of David Hathaway, NASA/MSFC)

magnetic energy stored higher in the corona. Since most forms of solar activity are magnetic in origin, the sunspot cycle is also called the solar cycle of magnetic activity.

The places that sunspots emerge and disappear also vary over the slow 11-year sunspot cycle (Fig. 2.3). At the beginning of the cycle, when the number of sunspots just starts to grow, sunspots break out in belts of activity at middle latitudes, parallel to the equator in both the northern and southern hemisphere. The two belts move to lower latitudes, or toward the equator, as the cycle progresses and the number of sunspots increases to a maximum. The zones of sunspot activity then keep on moving closer and closer to the equator, but they never reach it. The sunspots fizzle out and gradually disappear at sunspot minimum, just before coming together at the equator.

Table 2.7. Dates of the minimum and maximum in solar activity since 1960*

Cycle Number	Date of Minimum	Date of Maximum
Cycle 20	1964.9	1968.9
Cycle 21	1976.5	1979.9
Cycle 22	1986.8	1989.6
Cycle 23	1996.9	2000.5

* Courtesy of David Hathaway.

Focus 2.3

The Zeeman Effect

When an atom is placed in a magnetic field, it acts like a tiny compass, adjusting the energy levels of its electrons. If the atomic compass is aligned in the direction of the magnetic field, the electron's energy increases; if it is aligned in the opposite direction, the energy decreases. Since each energy change coincides with a change in the wavelength or frequency emitted by that electron, a spectral line emitted at a single wavelength by a randomly oriented collection of atoms becomes a group of three lines of slightly different wavelengths in the presence of a magnetic field. This magnetic transformation has been named the Zeeman effect, after Pieter Zeeman, who first noticed it in the terrestrial laboratory.

The size of an atom's internal adjustments, and the extent of its spectral division, increases with the strength of the magnetic field. We can understand this by considering the motion of a free electron in the presence of a magnetic field. The electron will circle about the magnetic field with a radius, R_g, described in Focus 2.2, and with a period, P, given by:

$$P = 2\pi R_g / V_\perp = 2\pi m_e/(eB) \, .$$

At the velocity $V_\perp$, the electron goes once around the circumference $2\pi R$ in the period P. The frequency, ν_g, of this motion is

$$\nu_g = 1/P = eB/(2\pi m_e) = 2.8 \times 10^{10} B \text{ Hz} \, ,$$

where B is the magnetic field strength in Tesla. Here e and m_e respectively denote the charge and mass of the electron.

When an atom is placed in a magnetic field, a very similar thing happens to its electrons and the spectral lines they emit. A line that radiates at a wavelength λ_0 without a magnetic field, becomes split into two or three components depending on the orientation of the magnetic field. For the three component split, the shift, $\Delta\lambda$, in wavelength of the two outer components from the central wavelength is given by:

$$\Delta\lambda = \lambda^2 \Delta\nu / c = 47 \lambda^2 B \text{ meters} \, ,$$

where the shift in frequency is $\Delta\nu = \nu_g/2 = 1.4 \times 10^{10}$ B Hz, the velocity of light is $c = 2.9794 \times 10^8$ meters per second, both $\Delta\lambda$ and λ are in meters, and B is in Tesla.

The separation is thus proportional to the magnetic field strength B. In 1908 George Ellery Hale made measurements of this Zeeman splitting in sunspots, showing that they have magnetic field strengths of about 0.3 Tesla or 3,000 Gauss.

As old spots linger near the equator, new ones break out about a third of the way toward the poles (Fig. 2.3). But the magnetic polarities of the new spots are reversed with north becoming south and *vice versa* – as if the Sun had turned itself inside out. The magnetic field at the solar poles approximates that of a dipole; this field also reverses polarity every 11 years.

Magnetographs are now used to chart magnetic fields running in and out of the Sun's photosphere. These instruments consist of an array of tiny detectors that measure the Zeeman effect at different locations on the visible solar disk. Two images are produced, one in each sense of circular polarization, and the difference produces a magnetogram (Fig. 2.4), with strong magnetic fields displayed as bright or dark regions depending on their direction.

The Zeeman effect measures the longitudinal component of the magnetic field, or the component which is directed toward or away from us. The circular polarization of the radiation provides the in or out direction. In addition to the longitudinal components, some modern instruments, called vector magnetographs, also measure the component of the magnetic field directed across

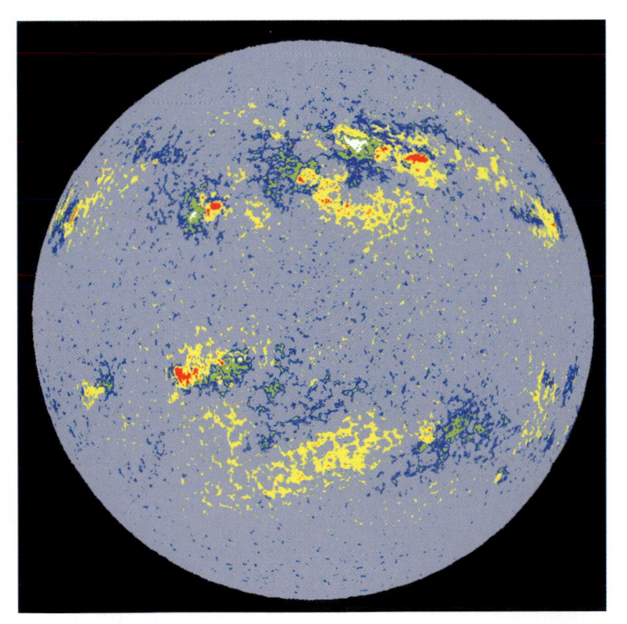

Fig. 2.4. Magnetogram. This map shows the strength and polarity of solar magnetic fields on 12 February 1989, close to sunspot maximum. Magnetic polarity is color coded: yellow represents positive or north polarity, with red the strongest fields which are around sunspots; blue is negative or south polarity, with green the strongest. In the Northern Hemisphere (*top half*) positive fields lead, in the Southern Hemisphere (*bottom half*) the polarities are exactly reversed and the negative fields lead. The Sun rotates east to west so that leading parts of active regions are to the right. This magnetogram was taken in the light of iron, Fe, at 868.8 nanometers. (Courtesy of William C. Livingston, National Solar Observatory, National Optical Astronomy Observatories)

our line-of-sight, or the transverse component of the magnetic field using linear polarization data. The field's strength and direction are then detertmined.

Focus 2.4

Magnetic and Gas Pressure

A magnetic field tends to restrain a collection of electrons and protons, called a plasma, while the plasma exerts a pressure that opposes this. The pressure, P_B, produced by a magnetic field transverse to its direction is given by:

$$P_B = B^2/(2\mu_0) ,$$

for a magnetic field of strength B in Tesla, and the permeability of free space $\mu_0 = 4\pi \times 10^{-7} NA^{-2}$. As expected, a stronger magnetic field applies a greater restraining pressure.

A hot plasma generates a gas pressure, P_{G}, owing to the motions of its particles. It is described by the expression

$$P_{\mathrm{G}} = NkT ,$$

where N is the particle number density, $k = 1.38 \times 10^{-23}\,\mathrm{J\,K^{-1}}$ is Boltzmann's constant, and T is the temperature. Hotter particles move faster and create greater pressure to oppose the magnetic field, and a denser plasma also results in greater pressure.

The two kinds of pressure compete for control of the solar atmosphere. In the low solar corona, strong magnetic fields in active regions hold the hot, dense electrified gas within coronal loops. The magnetic and gas pressure become equal for a magnetic field, B, given by:

$$B = [(2\mu_0 k)NT]^{1/2} = [3.46 \times 10^{-29} NT]^{1/2}\ \mathrm{Tesla} .$$

If a coronal loop contains a hot, dense plasma, with $N = 10^{17}$ electrons per cubic meter and $T = 10^6\,\mathrm{K}$, the magnetic field must be stronger than $B = 0.002$ Tesla $= 20$ Gauss. The magnetic field strengths of coronal loops in active regions are therefore strong enough to hold this gas in, at least within the low corona near sunspots.

In contrast, coronal holes have relatively weak magnetic fields, and the normally constraining magnetic forces relax and open up, allowing the unencumbered outward flow of charged particles into interplanetary space. Far from the Sun, the magnetic fields of coronal loops also become too weak to constrain the outward pressure of the hot gas, and the loops expand or break open to allow electrons and protons to escape, contributing to the solar wind and carrying the magnetic fields away. Within the solar wind, the gas pressure of the electrons and protons is roughly equal to the magnetic pressure of the interplanetary magnetic field.

Magnetograms provide the ground-zero reference for magnetic extrapolations into the overlying corona. They also indicate that at sunspot minimum there is still plenty of magnetism, especially at the solar poles, even though there are no large spots on the Sun. Much smaller magnetized regions continuously bubble and well up all over the solar surface throughout the 11-year solar cycle. Moreover, the solar magnetic fields are clumped together into intense bundles, that cover only a few percent of the Sun's surface, both at sunspot minimum and maximum.

About 100 years ago Frank H. Bigelow used the shape of the corona, detected during total eclipses of the Sun, to make some very prescient speculations about solar magnetism on the larger scales. He supposed that rays seen near the solar poles are open "lines of force discharging coronal matter from the body of the Sun". Bigelow also concluded that the "long equatorial wings" of the corona, detected at periods of minimum activity, are "due to the closing of the lines of force about the equator".

At times of reduced magnetic activity, prominent coronal streamers, detected during a solar eclipse, are indeed restricted to near the solar equator. The streamers have bulb shapes near the Sun, and extend into long, thin stalks further out (Fig. 2.2). At activity minimum, the streamers are molded by the available magnetism into extended shapes that point along the equatorial plane, often forming a ring or belt of hot gas that extends around the Sun. At the base of a streamer, in the low corona, electrified matter is then densely concentrated within closed magnetic loops or arches that straddle the equator. Farther out in the extended corona, the equatorial streamers narrow and stretch out tens of billions of meters in space.

Near a minimum in the activity cycle, the corona is relatively dim at the poles where faint plumes diverge out into interplanetary space, apparently outlining a global, dipolar magnetic field of about 10 Gauss in strength. This large-scale magnetism becomes pulled outward at the solar equator, confining hot material in the streamer belt. The streamers are sandwiched between regions of opposite magnetic direction or polarity, and are confined along an equatorial current sheet that is magnetically neutral.

Near the maximum in the activity cycle, the shape of the corona and the distribution of the Sun's extended magnetism can be much more complex. The corona then becomes crowded with streamers that can be found closer to the Sun's poles.

At times of maximum magnetic activity, the width and radial extension of a streamer are smaller and shorter; near minimum it is wide and well developed along the equator. The shape, location and tilt of extended sheets of opposing magnetic polarity are variable near sunspot maximum. At about this time, the Sun swaps its north and south magnetic poles, so a much more volatile corona exists at solar maximum.

Throughout the solar atmosphere, a dynamic tension is set up between the charged particles and the magnetic field (Focus 2.4). In the photosphere and below, the gas pressure dominates the magnetic pressure, allowing the magnetic field to be carried around by the moving gas. Because the churning gases are ionized, and hence electrically conductive, they sweep the magnetic field along. The situation is reversed in the low corona within active regions. Here strong magnetism wins and the hot particles are confined within coronal loops. Nevertheless, the loops are themselves tied into the underlying photosphere which is stirred up by mass motions. The gas pressure dominates the situation once again further out in the corona, where the magnetic field decreases in strength.

At large distances from the Sun, the charged particles in the expanding corona, or solar wind, take over and drag the magnetic fields with them. The theoretician, Eugene Parker, first speculated in 1958 that the solar wind carries both the Sun's particles and magnetic fields outward to the far reaches of the solar system.

Parker and his colleagues had already argued that interplanetary magnetism can account for an 11-year modulation in the amount of cosmic rays reaching the Earth. Low-energy cosmic ray protons are significantly more numerous when solar activity is minimal. When solar activity is at the peak of its 11-year cycle, the flux of low-energy cosmic rays detected at the top of the Earth's atmosphere is least. This unexpected anti-correlation is often called the Forbush effect after its discovery by Scott Forbush in the early 1950s.

Interplanetary magnetic fields act as a barrier to electrically-charged cosmic rays that are inbound from the depths of space, preventing them from reaching the Earth. During the maximum in the solar cycle, stronger solar magnetic fields are carried out into interplanetary space by the Sun's wind, deflecting more cosmic rays from Earth-bound paths. Less extensive interplanetary magnetism, during a minimum in the 11-year cycle of magnetic activity, lowers the barrier to the cosmic particles and allows more of them to arrive at Earth.

By 1958 Parker also concluded that interplanetary magnetism ought to have a spiral shape. This results from the combined effects of solar rotation and the radial expansion of the solar wind. While one end of the interplanetary magnetic field remains firmly rooted in the Sun, the other end is extended and stretched out by the solar wind. The Sun's rotation bends this radial pattern into an interplanetary spiral shape, coiling the magnetism up like a tightly-wound spring (Fig. 2.5). Because it is wrapped up into a spiral, the magnetic field's strength only falls off linearly with distance from the Sun, in contrast to the wind density that decreases much more rapidly with distance as it fills a larger volume. Of course, Parker's model was a theoretical speculation, with many skeptics. At the time, even the solar wind had yet to be directly measured, and evidence for Parker's spiral awaited the observations.

Studies of the interplanetary magnetic field were pioneered using America's Interplanetary Monitoring Platform 1, or IMP 1 for short, launched on 26 November 1963. Its magnetic field experiment, under the direction of Norman F. Ness from NASA's Goddard Space Flight Center, measured both the strength and direction of the magnetic field in interplanetary space near the Earth. The measurements confirmed that the interplanetary magnetic field has a spiral shape due to the combined effects of the solar wind and the rotation of the Sun. The magnetometers aboard IMP 1 also provided important results

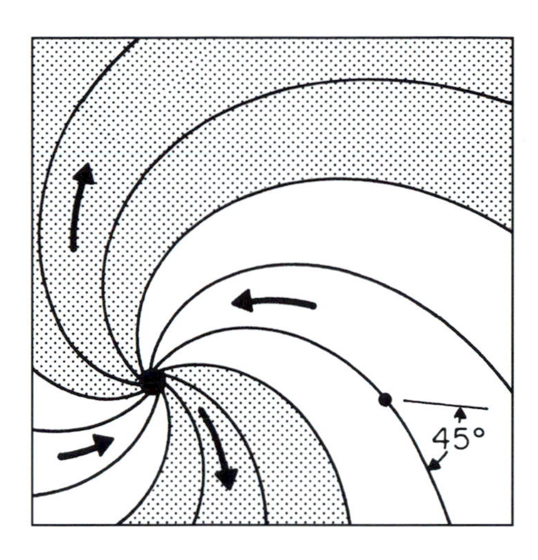

Fig. 2.5. Interplanetary Magnetic Spiral. If viewed from above, the magnetic field in the Sun's equatorial plane would be drawn out into a spiral pattern that is divided into magnetic sectors that point in opposite directions (*arrows*, *gray* and *white*). The rotating Sun sweeps these sectors of opposite magnetic polarity past the Earth, represented by the small black dot. At the Earth's location, the pulling effect of the solar wind is about equal to the twisting effect of the Sun's rotation, so the stretched-our field makes an angle of about 45 degrees with the radial direction from the Sun. (Adapted from L. Svalgaard and J. M. Wilcox, 1976)

on the confinement and distortion of the Earth's magnetic field by the solar wind.

Ness teamed up with John M. Wilcox to investigate the relation between solar and interplanetary magnetism, discovering large interplanetary regions of alternating magnetic polarity (Fig. 2.5). They showed that the magnetic field borne by the solar wind is divided into several longitudinal magnetic sectors directed toward or away from the Sun within the plane of the solar equator. For every field line leaving the Sun, another threads its way back through an adjacent sector of opposite magnetic polarity, forming enormous, greatly distended loops.

The large-scale regions of opposite magnetic polarity are separated by a magnetically neutral layer, or electric current sheet, that lies approximately in the plane of the solar equator. This dividing sheet is not precisely flat and is instead warped. So, when the Sun rotates the current sheet wobbles up and down, like the folds in the skirt of a whirling dervish, sweeping regions of opposite magnetic polarity past the Earth.

Magnetic measurements from the Pioneer 11 spacecraft, reported by Edward J. Smith and his colleagues in 1978, showed that the magnetic sectors disappeared when the spacecraft rose above the equatorial current sheet, where they were replaced by a magnetic field of one polarity. Pioneer traveled up to a heliographic latitude of 16 degrees after its encounter with Jupiter, whereas all previous spacecraft remained near the ecliptic plane at less than 7 degrees solar latitude. The new measurements supported a model in which the current sheet separates magnetic fields of opposite polarity in each hemisphere of the Sun, associated with its north and south magnetic poles.

Measurements over the subsequent decades have confirmed the equatorial magnetic sectors and repeatedly demonstrated that the interplanetary magnetic field is oriented into a simple spiral pattern, if averaged over a sufficiently long time. When a spacecraft is below the equatorial current sheet, it detects magnetic fields pointed in one direction; when located above the sheet the observed fields point in the opposite direction. The magnetic field on one side of the sheet points toward the Sun during one 11-year activity cycle, and away from the Sun throughout the next cycle, but the fields on the two sides of the current sheet always point in opposite directions.

Nevertheless, the magnetic structure of the solar wind must depend on position in the wind, as well as the level of solar activity, and until very recently no spacecraft has ventured far out of the ecliptic to see what happens way above the equatorial plane of the Sun. The Ulysses spacecraft has now traveled above the Sun's poles, observing the Parker spiral from above (Sect. 6.7, Fig. 6.20), and showing that Parker's spiral is generally preserved to the highest latitudes – but with significant departures that are attributed to magnetic waves in the wind (Sect. 5.4).

2.5 Invisible Radiation from the Sun

To complete our introductory survey of the discovery of space, we turn once more to the Earth's atmosphere, which is modified on its entire sunlit side by intense X-ray and ultraviolet radiation from the Sun. The discovery of this invisible high-energy component of solar radiation grew out of early studies of radio communication around the world. Although radio waves travel in straight lines, and cannot pass through the solid Earth, they get around the Earth's curvature by reflection from the ionosphere, or spherical shell of ions, an electrically charged layer in the upper atmosphere located about 100 thousand meters above your head.

The ionosphere's ability to mirror radio waves is disrupted during transient eruptions on the Sun, called solar flares, that occur more frequently during the maximum of the Sun's 11-year activity cycle. During moderately intense solar flares, radio communication can be silenced over the Earth's entire sunlit hemisphere, and it does not return to normal until the flaring activity stops. This was a major concern during World War II, when scientists would use the number of sunspots to forecast such interruptions.

How can the Sun produce changes in the Earth's ionosphere? The molecules in our atmosphere are ionized by invisible, short-wavelength radiation from the Sun, but not by the sunlight we see with our eyes. This energetic component of the solar spectrum includes ultraviolet, or UV, radiation, with wavelengths just shorter than blue light, and even shorter radiation called X-rays. When this radiation reaches the upper atmosphere, it rips electrons off the nitrogen and oxygen molecules there, producing free electrons and atomic and molecular ions.

The energy of radiation is inversely proportional to its wavelength, so invisible waves with short wavelengths have more energy than visible light with longer ones (Focus 2.5). This can be understood by looking at your hand in different ways. In visible light, you see the skin of your hand, but shorter ultraviolet rays enter the skin to give you a sunburn. The even shorter, more energetic X-rays penetrate through the skin and muscles of your hand to delineate its bones.

Because a solar flare raises the temperature of a localized region in the corona to tens of million of degrees, the bulk of a flare's radiation is emitted as X-rays. These catastrophic explosions can briefly outshine the entire Sun at X-ray wavelengths. Just eight minutes after such a solar flare, a strong blast of X-rays reaches our air, breaking some of its constituents into pieces and transforming the ionized layer of the upper atmosphere.

But that's not all. The Sun is always emitting X-rays. The million-degree corona radiates most of its energy at X-ray wavelengths (Focus 2.5). Although this emission waxes and wanes in step with the 11-year activity cycle, it is always there at some level. So the Sun not only alters the ionosphere, it creates it!

Solar ultraviolet radiation is partially blocked by the Earth's atmosphere, and the Sun's X-rays are completely absorbed in our air. These components of solar radiation must therefore be observed with telescopes and other instruments in space. Such investigations were pioneered by scientists at the United States (US) Naval Research Laboratory, or NRL for short. Richard Tousey and his colleagues at NRL used V-2 rockets, captured from the Germans at the end of World War II, to obtain the first photographs of the solar ultraviolet spectrum in 1946, and to first detect the Sun's X-rays in 1948.

By the late 1940s, scientists knew that the corona had a million-degree temperature, and should therefore produce X-rays, but it appeared so dilute that its radiant flux might not have a major effect on the Earth's atmosphere. In the early 1950s, Herbert Friedman's team at NRL used instruments aboard Aerobee rockets to show that there are enough solar X-rays, coming in from outside our atmosphere, to create the ionosphere. Sounding rocket measurements made by this group over the course of a full sunspot cycle subsequently indicated that the intensity of the Sun's X-rays rises and falls with the number of sunspots, as does the temperature in the ionosphere.

The first X-ray image of the Sun was obtained by the NRL group in 1960 during a brief 5-minute rocket flight. This primitive picture set the stage for the detailed X-ray images of the Sun taken from NASA sounding rockets in the early 1970s. During this time, solar X-ray instruments were also developed using NASA's series of small satellites known as the Orbiting Solar Observatories, or OSOs, launched from 1962 (OSO 1) to 1975 (OSO 8).

Focus 2.5

The Energy of Light

The Sun continuously radiates energy that spreads throughout space. This radiation is called "electromagnetic" because it propagates by the interplay of oscillating electrical and magnetic fields in space. Electromagnetic waves all travel through empty space at the same constant speed – the velocity of light. It is usually denoted by the lower case letter c, and has the value of $c = 2.9979 \times 10^8$ meters per second. No energy can be transported more swiftly than the speed of light.

Astronomers describe radiation in terms of its wavelength, frequency or energy. When light propagates from one place to another, it often seems to behave like waves or ripples on a pond. The light waves have a characteristic wavelength, often denoted by the symbol λ, the separation between wave crests. Sometimes radiation is described by its frequency, denoted by the symbol ν, instead of its wavelength. Radio stations are, for example, denoted by their call letters and the frequency of their broadcasts. The product of the wavelength and frequency is equal to the velocity of light, or

$$\lambda \nu = c\,.$$

When light is absorbed or emitted by atoms, it behaves like packages of energy, called photons. The photons are created whenever a material object emits electromagnetic radiation, and they are consumed when radiation is absorbed by matter. Moreover, each elemental atom can only absorb and radiate a very specific set of photon energies. The photon energy is given by:

$$\text{Photon Energy} = h\nu = \frac{hc}{\lambda}\,,$$

where Planck's constant $h = 6.6261 \times 10^{-34}$ Joule second. From this expression we see that photons of radiation at shorter wavelengths have greater energy. This is the reason that short, energetic X-rays can penetrate inside your body, while longer, less-energetic visible light just warms your face.

A hot gas at temperature, T, will emit radiation at all wavelengths, but the peak intensity of that radiation is emitted at a maximum wavelength, λ_{max}, that varies inversely with the temperature. This can be seen by equating the thermal energy of the gas to the photon energy of its radiation, or by:

$$\text{Gas Thermal Energy} = \frac{3}{2}kT = \frac{hc}{\lambda_{\text{max}}} = \text{Radiation Photon Energy}\,,$$

where Boltzmann's constant $k = 1.38066 \times 10^{-23}$ Joule per degree Kelvin. Collecting terms and inserting the values for the constants, this expression gives $\lambda_{\text{max}} = 2hc/(3kT) \approx 0.01/T$ meters. The exact relationship, known as the Wien displacement law, is given by

$$\lambda_{\text{max}} = \frac{0.0028978}{T}\ \text{meters}\,.$$

This tells us that a solar flare, whose temperature rises to 10 million degrees, and the Sun's corona, with a temperature of about 2 million degrees, will emit most of their radiation at soft X-ray wavelengths. The wavelength of maximum intensity for their thermal radiation is 10^{-9} to 10^{-10} meters, with photon energies of 1 to 10 keV, where $1\,\text{keV} = 1.602 \times 10^{-16}$ Joule. In contrast, the solar photosphere with a temperature of 5,780 degrees will radiate most of its energy at a visible wavelength of about 5×10^{-7} meters.

Our perspective of the X-ray Sun was forever changed with NASA's orbiting Skylab, launched on 14 May 1973, and manned by three-person crews until 8 February 1974. Skylab's X-ray telescope, and those to follow, could be used to look back at the Sun's million-degree corona, viewing it across the Sun's face for hours, days and months at a time. This is because hot material – such as that within the million-degree corona – emits most of its energy at these wavelengths. Also, the photosphere is too cool to emit intense radiation at these wavelengths, so it appears dark under the hot gas.

The Skylab X-ray pictures showed that the million-degree corona is far from uniform, and described its fundamental three-fold magnetic structure – coronal holes, coronal loops and X-ray bright points (Fig. 2.6). They indicated that sunspots are stitched together by magnetized coronal loops that mold, shape and constrain the high-temperature, electrified gas (Fig. 2.7). Since the hot coronal gases are almost completely ionized, they cannot readily cross the intense, closed magnetic field lines. Typically the electrons contained in coronal loops have temperatures of a few million degrees and a density of up to 100 million billion (10^{17}) electrons per cubic meter.

Fig. 2.6. X-ray and Eclipse Corona. A composite photo showing the white-light corona seen from the ground at the 30 June 1973 eclipse and the on-disk X-ray corona observed from Skylab at the same time. The X-ray image shows large, bright active regions, a dark coronal hole extending from the North Pole down through the middle of the disk, and X-ray bright points nearly uniformly distributed over the solar surface. The eclipse photograph, taken from Loyengalani, Kenya, shows bright helmet streamers concentrated near the solar equator and faint polar plumes. [Courtesy of the Solar Physics Group, American Science and Engineering Inc. (*X-ray image*) and the High Altitude Observatory, National Center for Atmospheric Research (*eclipse photograph*)]

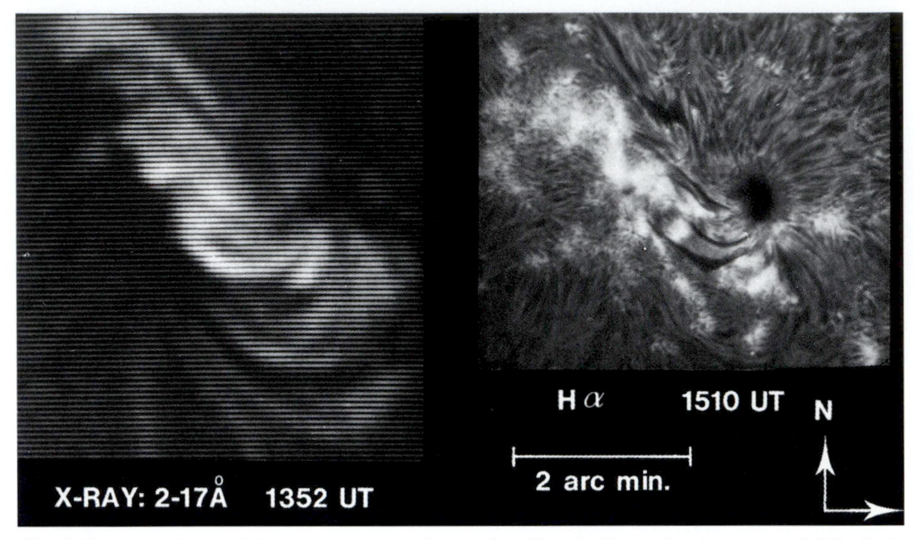

Fig. 2.7. X-ray Coronal Loops. A comparison of a Skylab X-ray image and a visible light (Hα) picture of a solar active region taken at the same time on 6 July 1973. The bright coronal loops shown in the X-ray image contain hot (million degree), dense (10 million billion, or 10^{16}, electrons per cubic meter) electrons trapped within magnetic loops that connect underlying regions of opposite magnetic polarity. The coronal loops are about 100 million meters in extent, or about nine times the Earth's diameter. (Courtesy of the Solar Physics Group, American Science and Engineering, Inc.)

Although Skylab demonstrated the ubiquitous nature of the X-ray emitting coronal loops, its X-ray telescope provided blurred images without fine detail. It lacked the resolution, sensitivity and rapid, uniform imaging required to fully understand them. As described in Chap. 4, this has now largely been accomplished by the X-ray telescope aboard the Yohkoh spacecraft.

Skylab's inquisitive X-ray eyes also recorded black, seemingly-empty voids, called coronal holes (Figs. 2.6 and 2.8). Such regions were first observed from the ground by Max Waldmeier, who in 1957 named them coronal holes. He noticed them when mapping out the distribution of intense, visible emission lines in the corona.

Coronal holes are nearly always present at the poles of the Sun and sometimes extend toward the equator. Such dark regions were noted in X-ray images obtained on NASA sounding rockets in 1970 and extensively described by Skylab; they are routinely detected by the Yohkoh and SOHO spacecraft (Sects. 5.2 and 5.3). The polar coronal holes disappear at activity maximum, reappearing soon after with opposite magnetic polarity. Between times of maximum and minimum of the 11-year cycle, coronal holes are not confined to the Polar Regions and often span the solar equator.

Coronal holes are not empty. They are just more rarefied and cooler than other places in the corona, so their X-ray emission is faint. The electrons in coronal holes have densities a factor of ten smaller than the typical value in coronal loops. This is because most of the magnetic field lines in coronal

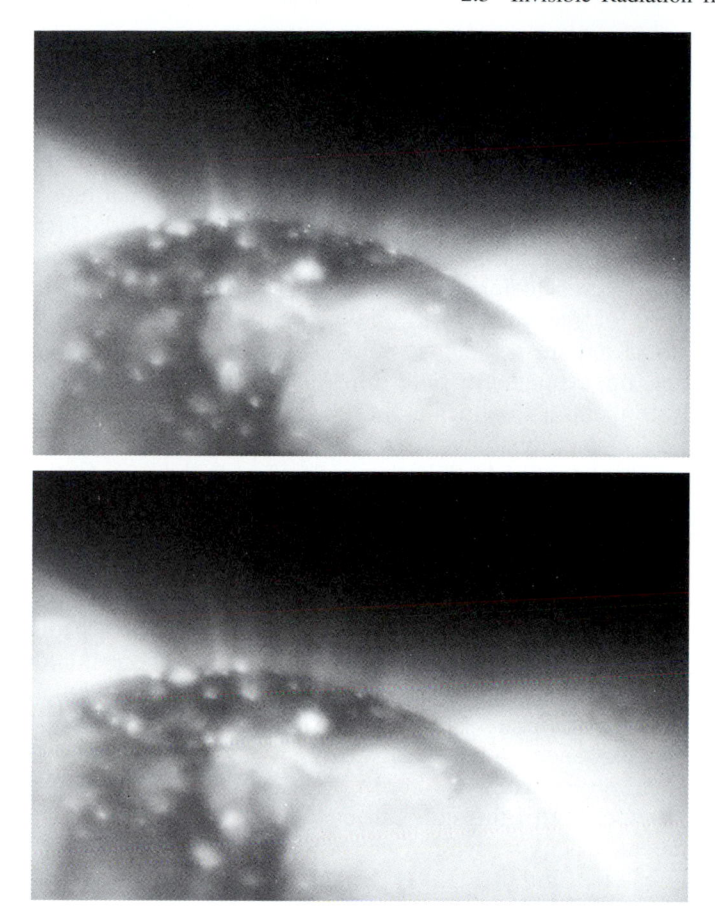

Fig. 2.8. Coronal Hole and Polar Plumes. These X-ray images, taken from the Skylab spacecraft in July 1973, show several small, bright polar plumes that project radially outward from the solar surface in a dark coronal hole at the Sun's north pole. The polar plumes are less than 10 million meters in width and have temperatures in excess of a million degrees. (Courtesy of the Solar Physics Group, American Science and Engineering, Inc.)

holes do not form locally closed loops. Coronal holes are instead characterized by open magnetic fields that do not return directly to another place on the Sun, allowing the charged particles to escape the Sun's magnetic grasp and flow outward into surrounding space. Although closed in a technical sense, the magnetic fields in coronal holes extend so far out into interplanetary space that they are effectively open.

Allen S. Krieger, Adrienne F. Timothy and Edmond C. Roelof concluded in 1973 that coronal holes are the sources of at least some of the recurrent high-speed streams in the solar wind. They did this by comparing an X-ray photograph of the Sun, taken during a five-minute rocket flight, with satellite measurements of the solar wind velocity. This result was substantiated in the following year by detailed comparison of coronal holes mapped by

Skylab and *in situ* measurements of the solar wind near the Earth. This conclusively demonstrated that the dark coronal holes can provide a fast lane for the high-speed solar wind. The recurrent high-speed streams were associated with equatorward extensions of the polar coronal holes, rotating at the 27-day equatorial rate.

Sounding rocket observations of extreme ultraviolet lines in coronal holes fully confirmed that they are the unknown font of the high-speed solar wind. As shown by Gary J. Rottman, Frank Q. Orrall and James A. Klimchuk in 1982–83, these spectral features have shorter wavelengths than their counterparts on the rest of the solar disk, indicating an outflow velocity of 7 thousand to 8 thousand meters per second. That meant that material was moving out along open magnetic field lines in coronal holes. It remained for SOHO instruments to delineate the exact site of escape within holes (Sect. 5.2, Fig. 5.7), and to show how the liberated gases are subsequently accelerated to create winds that move one hundred times faster (Sect. 5.3).

The fast component of the solar wind squirts out of the nozzle-like holes at about 800 thousand meters per second. Equatorial coronal holes can expel high-speed particles directly toward the Earth, but the fast solar wind also gushes out of the Polar Regions and can spread out to cross the Earth's path. Since spacecraft measurements were always made near the ecliptic, one could only speculate about what the flows looked like in three dimensions – that awaited the Ulysses spacecraft (Sect. 5.1).

There is also a slow component to the solar wind, moving at about half the speed of the fast one, and no one knew for sure where it came from. Some of the slow component dribbles out of the stalks of equatorial streamers, at least near a minimum in the solar activity cycle (Sect. 5.2).

Ubiquitous small features, dubbed X-ray bright points, glow like tiny, blinking lights in X-ray images of the Sun. Hundreds and even thousands of them appear each day. Like the coronal holes, they were first observed during rocket flights, but extensively studied during Skylab's coverage. They are concentrated regions of intense magnetism, some almost as large as the Earth. Many of them overlie magnetic regions that have both positive and negative polarity, and they fluctuate in brightness like small solar flares. Bright points come and go, usually lasting for hours, but some flare up and fade in just a few minutes.

X-ray bright points appear all over the Sun, in active regions, coronal holes and the places in between (Fig. 2.8). Skylab showed that ubiquitous bright points dot the ultraviolet Sun as well. The varying UV and X-ray features have a temperature of about a million degrees Kelvin, and may play a role in the heating of the corona (Sect. 4.4).

The X-rays provide information on the low corona near the visible solar disk. To look further out in the corona, astronomers use a special telescope, called the coronagraph, that has a small occulting disk, or miniature moon, to mask the Sun's face and block out the photosphere's light. A coronagraph detects photosphere sunlight scattered off coronal electrons, providing an edge-on, side view of the corona.

The first coronagraph was developed in 1930 by the French astronomer, Bernard Lyot, and soon installed by him at the Pic du Midi observatory in the Pyrenees. As Lyot realized, such observations are limited by the bright sky to high-altitude sites where the thin, dust-free air scatters less sunlight. The best coronagraph images with the finest detail are obtained from high-flying satellites where almost no air is left and where the daytime sky is truly and starkly black.

Powerful solar eruptions, now known as coronal mass ejections, were discovered using space-borne, white-light coronagraphs in the 1970s, first with the seventh Orbiting Solar Observatory (OSO 7, launched on 29 September 1971), then in greater profusion by Skylab. In spite of their relatively late discovery, thousands of mass ejections have been observed routinely during the past few decades with coronagraphs aboard several solar satellites (Sect. 6.5). These gigantic magnetic bubbles rush away from the Sun at speeds of millions of meters per second, expanding to become larger than the Sun (Fig. 2.9). They carry billions of tons of material out into space, and produce powerful gusts in the solar wind.

The physical size of the mass ejections dwarfs that of solar flares and even the active regions in which flares occur. Solar flares are nevertheless sometimes associated with coronal mass ejections, and similar processes probably cause both of them (Sect. 6.6). Like solar flares, the rate of occurrence of coronal mass ejections varies with the 11-year cycle of magnetic activity, ballooning out of the corona several times a day during activity maximum (Sect. 6.5).

A powerful mass ejection, with its associated shocks, energetic particles and magnetic fields, can seriously disrupt the Earth's environment in space, endangering astronauts and satellites and causing geomagnetic storms, intense auroras and electrical power blackouts (Sect. 7.2). Intense solar flares can disrupt radio communications, wipe out satellites and threaten astronauts in space (Sect. 7.2). All of these effects are of such vital importance that national centers now employ space weather forecasters to warn of threatening solar activity.

In summary, there is a lot more going on at the Sun than meets the eye (Fig. 2.10). Hot gases are caught within looping magnetic cages. Long-lasting magnetized holes, found at the Sun's poles, are locked open to continuously expel a high-speed wind. The corona's magnetic energy is abruptly released in other places, to power violent flares or mass ejections with threatening effects for the Earth. The entire solar atmosphere seethes and writhes in tune with the Sun's ever-changing magnetism.

Yet, even a decade ago, we did not understand how the Sun's magnetism originates and is regenerated in an 11-year cycle. We did not even have a firm grasp on the origins of sunspots. As we shall next see, the ancient mysteries of the sunspots and the Sun's cyclic magnetic variations have at least been partly solved by pinpointing the elusive dynamo that generates the solar magnetic fields. This has been accomplished by looking deep inside the Sun where internal motions amplify, compress and slowly transform the Sun's spotty, ever-changing magnetism.

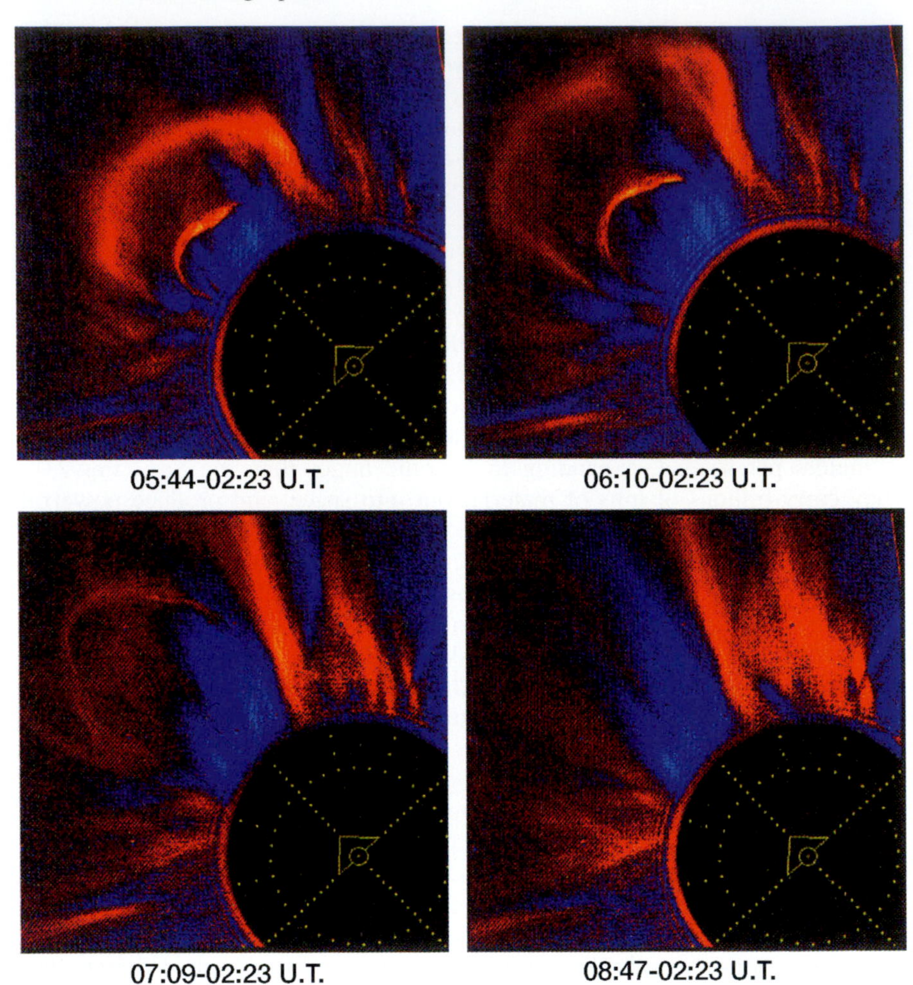

Fig. 2.9. Coronal Mass Ejection. A huge magnetic bubble rushes outward from the Sun, carrying billions of tons of coronal material and growing larger than the Sun in just a few hours. Such coronal mass ejections are detectable only from space-borne coronagraphs whose black occulting disks block out the bright light of the photosphere so that the dim light from the coronal mass ejection can be observed. This is a time sequence of differences between four coronagraph images and a single pre-event image at 02 hours 23 minutes Universal Time (UT), all taken on 14 April 1980 with the coronagraph aboard the Solar Maximum Mission (SMM) spacecraft. Brightening of the corona (positive differences) are shown in red and orange, negative differences in blue. The earliest difference images illustrate the classical three-part structure of some coronal mass ejections; the presence of a bright, red outer loop or frontal structure, followed by a dark, blue coronal cavity, within which is a bright, red loop-like structure identified with an erupting prominence. Both the mass ejection and the erupting prominence have moved beyond the field of view in the later difference images, leaving a wedge of depleted (blue) corona behind. (Courtesy of Arthur J. Hundhausen, High Altitude Observatory and NASA)

Fig. 2.10. Magnetic Fields Near and Far. In the low corona, strong magnetic fields are tied to the Sun at both ends, trapping hot, dense electrified gas within magnetized loops. Far from the Sun, the magnetic fields are too weak to constrain the outward pressure of the hot gas, and the loops break open to allow electrically-charged particles to escape, forming the solar wind and carrying magnetic fields away. (Courtesy of Newton Magazine, the Kyoikusha Company)

Table 2.8. Key events in the discovery of space*

Date	Event
1600	William Gilbert, physician to Queen Elizabeth I of England, publishes a small treatise demonstrating that "the terrestrial globe is itself a great magnet".
1610–13	Galileo Galilei first systematically studies sunspots through a telescope.
1716	Edmund Halley suggests that aurora rays delineate terrestrial magnetic field lines, and that the auroras are due to a magnetized fluid that circulates poleward along the Earth's dipole magnetic field lines.
1724	George Graham discovers large, irregular fluctuations in compass needles, later called magnetic storms; they were seen at about the same time by Anders Celsius.

1733 Jean Jacques d'Ortous de Mairan argues for a connection between the occurrence of auroras and sunspots.

1799–1804 Alexander von Humboldt makes regular, precise measurements of the strength, dip and inclination of the Earth's magnetic field during his voyage to South America. The term "magnetic storm" came into common usage as the result of Humboldt's scientific analyses and reports.

1802–15 In 1802 William Hyde Wollaston discovers that the solar spectrum is cut by several dark gaps now called absorption lines; by 1815 Joseph von Fraunhofer has catalogued the wavelengths of more than 300 of them.

1838 Carl Friedrich Gauss publishes a mathematical description of the Earth's dipolar magnetic field, using it with observations to show that the magnetism must originate deep down inside the Earth's core.

1844 Samuel Heinrich Schwabe demonstrates that the number of sunspots varies from a maximum to a minimum and back to a maximum again in a period of about 11 years.

1852 Edward Sabine demonstrates that global magnetic disturbances of the Earth, now called geomagnetic storms, vary in tandem with the 11-year sunspot cycle.

1858 Richard C. Carrington shows that sunspots move from high-latitudes to the solar equator during the 11-year sunspot cycle.

1859–60 In 1859 Richard C. Carrington and Richard Hodgson independently observe a solar flare in the white light of the photosphere and in 1860 publish the first account of such a flare. Seventeen hours after the flare a large magnetic storm begins on the Earth.

1859–61 By comparing the Sun's absorption lines with emission lines of elements vaporized in the terrestrial laboratory, Gustav Kirchhoff and Robert Bunsen identify in the solar atmosphere several elements known on Earth, including sodium, calcium and iron.

1868 During a solar eclipse on 18 August 1868 several observers noted a chromospheric line in the yellow. Most of them identified it as the sodium D line in emission, but Norman Lockyer pointed out that there was a discrepancy in the wavelength. Jules Janssen and Lockyer independently realized how to observe the chromosphere outside the eclipse, resulting in a more leisurely method of measuring the wavelengths. It was probably not until the following year that Lockyer convinced himself that the yellow line could not be matched with any known terrestrial element, and coined the name "helium" for the new element. Helium was not identified on the Earth until 1895 when William Ramsay detected the line in gases given off by a mineral called clevite.

1869 At the solar eclipse of 7 August 1869, Charles A. Young, and, independently, William Harkness, discover a single, bright, green emission line in the spectrum of the solar corona. This conspicuous feature remained unidentified with any known terrestrial element for more than half a century, but it was eventually associated with highly ionized iron atoms missing 13 electrons (Fe XIV), indicating that the corona has a million-degree temperature (see 1939–41).

1873 Elias Loomis demonstrates a correlation between intense auroras and the number of sunspots, both recurring with an 11-year periodicity and with similar times of maximum.

1889–90 Frank H. Bigelow argues that the structure of the corona detected during solar eclipses provides strong evidence for large-scale solar magnetic or electric fields. He correctly speculated that polar rays delineate open field lines along which material escapes from the Sun, and that equatorial elongations of the corona mark closed magnetic field lines.

1896–1913 Kristian Birkeland argues that polar auroras and geomagnetic storms are due to beams of electrons.

1908 George Ellery Hale measures intense magnetic fields in sunspots, thousands of times stronger than the Earth's magnetism.

1911 Arthur Schuster shows that a beam of electrons from the Sun cannot hold itself together against the mutual electrostatic repulsion of the electrons.

1912 Victor Franz Hess discovers high-energy, penetrating rays coming from outer space, using balloon measurements of ionization in the air.

1919 George Ellery Hale and colleagues show that sunspots occur in bipolar pairs with an orientation that varies over 22 years.

1919 Frederick Alexander Lindemann (later Lord Cherwell) suggests that an electrically neutral plasma ejection from the Sun is responsible for non-recurrent geomagnetic storms.

1926 Robert A. Millikan confirms that penetrating radiation comes from beyond the Earth, and gives it the name cosmic rays.

1927 Jacob Clay shows that cosmic rays are less intense near the Earth's equator than at higher latitudes, suggesting that they are charged particles deflected toward the Earth's magnetic poles.

1928–32 In 1928 Albrecht Unsöld uses absorption lines to show that hydrogen is at least a million times more abundant than any other element in the solar photosphere. In the following year William H. Mc Crea similarly demonstrates the overwhelming abundance of hydrogen in the chromosphere, and by 1932 Bengt Strömgren shows that hydrogen is the most abundant element in the entire star.

1931–40 Sydney Chapman and Vincent C. A. Ferraro propose that magnetic storms are caused when an electrically neutral plasma cloud ejected from the Sun envelops the Earth.

1932–33 Arthur H. Compton demonstrates that cosmic rays are charged particles deflected to high latitudes by the Earth's dipolar magnetic field.

1932 Carl D. Anderson discovers the positron, or anti-electron, in cloud-chamber observations of secondary particles produced by cosmic rays in our air.

1938 Thomas H. Johnson uses the east-west effect to show that cosmic rays are positively charged.

1939–41 Walter Grotrian and Bengt Edlén identify coronal emission lines with highly ionized elements, indicating that the Sun's outer atmosphere has a temperature of millions of degrees Kelvin. The conspicuous green emission line was identified with Fe XIV, an iron atom missing 13 electrons.

1946 Vitaly L. Ginzburg, David F. Martyn and Joseph L. Pawsey independently confirm the existence of a million-degree solar corona from observations of the Sun's radio radiation.

1946 Richard Tousey and his colleagues at the U. S. Naval Research Laboratory use a V-2 rocket to obtain the first extreme ultraviolet spectrum of the Sun on 10 October 1946.

1948–49 Soft X-rays from the Sun are first detected on 5 August 1948 with a V-2 rocket experiment performed by the U. S. Naval Research Laboratory, reported by T. R. Burnight in 1949. Subsequent sounding rocket observations by the NRL scientists revealed that the Sun is a significant emitter of X-rays and that the X-ray emission is related to solar activity.

1948–50 High-altitude balloon measurements by H. L. Bradt and Bernard Peters, and independently by Phyllis Freier and colleagues show that cosmic rays consist of hydrogen nuclei (protons), with lesser amounts of helium nuclei (alpha particles) and heavier nuclei.

1951–52 Herbert Friedman and his colleagues at the U. S. Naval Research Laboratory use instruments aboard sounding rockets to show that the Sun emits enough X-ray and ultraviolet radiation to create the ionosphere.

1951–57 Ludwig F. Biermann argues that a continuous flow of solar corpuscles is required to push comet ion tails into straight paths away from the Sun, correctly inferring solar wind speeds as high as 0.5 to 1.0 million meters per second.

1954–58 Scott E. Forbush demonstrates the inverse correlation between the intensity of cosmic rays arriving at Earth and the number of sunspots over two 11-year solar activity cycles.

1955 Leverett Davis Jr. argues that solar corpuscular emission will carve out a cavity in the interstellar medium, now known as the heliosphere, accounting for some observed properties of low-energy cosmic rays.

1955 Horace W. Babcock and Harold D. Babcock use magnetograms, taken over a two-year period from 1952 to 1954, to show that the Sun has a general dipolar magnetic field of about 10^{-4} Tesla, or 1 Gauss, in strength, usually limited to heliographic latitudes greater than ± 55 degrees. Bipolar magnetic regions are found at lower latitudes. Occasional extended unipolar areas, of only one outstanding magnetic polarity, are also found; they speculated that these unipolar regions might be related to 27-day recurrent terrestrial magnetic storms.

1956 Peter Meyer, Eugene N. Parker and John A. Simpson argue that enhanced interplanetary magnetism at the peak of the solar activity cycle deflects cosmic rays from their Earth-bound paths.

1957 Max Waldmeier observes intense coronal emission lines, calling attention to seemingly vacant places that he called coronal holes.

1957–59 Sydney Chapman shows that a hot, static corona should extend to the Earth's orbit and beyond.

1957 The first artificial satellite, Prosteyshiy Sputnik, was launched by the Soviet Union on 4 October 1957.

1958 Eugene N. Parker suggests that a perpetual supersonic flow of electric corpuscles, that he called the solar wind, naturally results from the expansion of a very hot corona. He also demonstrates that the solar magnetic field will be pulled into interplanetary space, modulating the amount of cosmic rays reaching Earth and attaining a spiral shape in the plane of the Sun's equator due to the combined effects of the radial solar wind flow and the Sun's rotation.

1958–59 The first American satellite, Explorer 1, was launched into orbit on 1 February 1958, followed by Explorer 3 on 26 March 1958. Instruments aboard these spacecraft, provided by James A. Van Allen and colleagues, first detected belts of charged particles that girdle the Earth's equator.

1960–61 Konstantin I. Gringauz reports that the Soviet spacecraft, Lunik 2, launched on 12 September 1959, has measured high-speed ions in interplanetary space outside the Earth's magnetic field with a flux of 2 million million (2×10^{12}) ions (protons) per square meter per second.

1960–61 The first, primitive X-ray picture of the Sun is obtained during a brief rocket flight, by Herbert Friedman and colleagues at the US Naval Research Laboratory.

1961 Cosmic-ray electrons are discovered, but in much lower flux than protons, in cloud chamber observations by James A. Earl and by the balloon measurements of Peter Meyer and Rochus Vogt.

1961–62 A plasma probe aboard the American spacecraft Explorer 10, launched on 25 March 1961, is used by Herbert A. Bridge and colleagues to provide rough measurements of the density, speed and direction of the solar wind, but never reaching the undisturbed interplanetary medium.

1962–67 Mariner 2 was launched on 7 August 1962. Using data obtained during Mariner's voyage to Venus, Marcia Neugebauer and Conway W. Snyder demonstrate that a low-speed solar wind plasma is continuously emitted by the Sun, and discover high-speed wind streams that recur with a 27-day period within the orbital plane of the planets.

1964–66	Magnetometers aboard NASA's Interplanetary Monitoring Platform 1, launched on 27 November 1963, are used by Norman F. Ness and John M. Wilcox to measure the strength and direction of the interplanetary magnetic field. They show that it is pulled into a spiral shape by the combined effects of the radial solar wind flow and the Sun's rotation. They also discover large scale magnetic sectors in interplanetary space that point toward or away from the Sun.
1973	Allen S. Krieger, Adrienne F. Timothy and Edmond C. Roelof compare an X-ray photograph, obtained during a rocket flight on 24 November 1970, with satellite measurements of the solar wind to show that coronal holes are the source of recurrent high-speed streams in the solar wind.
1973	Giuseppe S. Vaiana and his colleagues use solar X-ray observations taken from rockets during the preceding decade to identify the three-fold magnetic structure of the solar corona – coronal holes, coronal loops and X-ray bright points.
1973–77	X-ray photographs of the Sun taken with the Apollo Telescope Mount on the manned, orbiting Skylab satellite, launched on 14 May 1973, fully confirm coronal holes, the ubiquitous coronal loops and X-ray bright points. Detailed comparisons of Skylab X-ray photographs and measurements of the solar wind, made from Interplanetary Monitoring Platforms (IMPs) 6, 7 and 8, confirm that solar coronal holes are the source of the high-velocity solar wind streams as well as 27-day-recurrent geomagnetic disturbances. John Wilcox had previously suggested that the fast streams might originate in magnetically open, unipolar regions on the Sun, but the Skylab X-ray photographs definitely identified the place.
1974–86	The Helios 1 and 2 spacecraft, respectively launched in December 1974 and in January 1976, measured the solar wind parameters as close as 0.3 AU from the Sun for a whole 11-year solar cycle.**
1978	Edward J. Smith, Bruce T. Tsurutani and Ronald L. Rosenberg use observations from Pioneer 11 to show that the solar wind becomes unipolar, or obtains a single magnetic polarity, at high heliographic latitudes near 16 degrees.
1982–83	Gary J. Rottman, Frank Q. Orrall and James A. Klimchuk obtain rocket observations of extreme ultraviolet resonance lines formed in the low corona and transition region, showing that the lines are systematically shifted to shorter wavelengths in large polar coronal holes with well developed, low-latitude extensions. Outflow velocities of 7 to 8 thousand meters per second are inferred from these Doppler shifts.
1980–1999	See the key events at the end of Chaps. 5, 6 and 7.

* See the References for complete references to these seminal papers. Also see the tables at the end of subsequent chapters for greater detail on the history of specific unsolved problems, including contributions after 1980 and the main results of the SOHO, Ulysses and Yohkoh missions.

** An AU is the mean distance of the Earth from the Sun, or about 146 billion (1.46×10^{11}) meters.

3. Exploring Unseen Depths of the Sun

Overview

The Sun is energized by the fusion of abundant hydrogen nuclei within a hot, dense core where the temperature rises to 15.6 million degrees. We know the sequence of nuclear reactions that occur down there, and have used models to describe how the energy is generated and works its way out by radiation and convection. There was one nagging problem involving neutrinos produced in vast quantities by the nuclear reactions. Massive underground instruments detected only one-third to one-half the number of neutrinos that theory said they should. The theoretical models have now been tested by directly probing the solar interior, showing that the neutrino calculations are not in error, so we seem to have an incomplete knowledge of neutrinos.

The inner structure and dynamics of the Sun are being studied by a new science called helioseismology. It uses observations of ripples or oscillations in the visible solar gases to detect low-pitched sound waves and thereby look deep down inside the Sun, in much the same way that an ultrasonic scanner can peer inside a mother's womb and map out the shape of an unborn infant. Helioseismologists have taken the temperature of the Sun's energy-generating core, showing that it agrees with model predictions. This strongly disfavors any astrophysical solution for the solar neutrino problem.

Helioseismology instruments aboard the SOlar and Heliospheric Observatory, or SOHO, have shown how the Sun rotates inside, and suggested the internal location of the dynamo that sustains its magnetism. They have also described hot, gaseous rivers that circulate beneath the photosphere, detected internal tremors, or sunquakes, generated by flaring explosions in the corona, and discovered how to look through the Sun to observe active regions on the hidden back side of the Sun.

3.1 What Makes the Sun Shine?

To understand what makes the Sun shine, we must first describe the star's physical characteristics (Table 3.1). The entire Sun is nothing but a big, hot luminous ball of gas, concentrated at the center and more tenuous further out. It is the most massive and largest object in our solar system – 333,000 times Earth's mass and 109 times its diameter. We know the Sun's mass with high precision from the motions of the planets, and the Sun's size can be inferred from its distance and angular extent (Focus 3.1).

Table 3.1. Vital statistics of the Sun*

Mass	1.989×10^{30} kilograms (332,946 Earth masses)
Radius	6.955×10^{8} meters (109 Earth radii)
Volume	1.412×10^{27} m^3 (1.3 million Earths)
Density (center)	151,300 kg m^{-3}
(mean)	1,409 kg m^{-3}
Pressure (center)	2.334×10^{11} bars
(photosphere)	0.0001 bar
Temperature (center)	15.6 million degrees Kelvin
(photosphere)	5,780 degrees Kelvin
(corona)	2 million to 3 million degrees Kelvin
Luminosity	3.854×10^{26} Joule s^{-1}
Solar constant	1,366 Joule s^{-1} m^{-2} = 1,366 W m^{-2}
Mean distance	1.4959787×10^{11} m
Age	4.55 billion years
Principal chemical constituents (by number of atoms)	Hydrogen 92.1 percent Helium 7.8 percent All others 0.1 percent

* See Focus 3.1 for calculations of many of these quantities. Mass density is given in kilograms per cubic meter, or kg m^{-3}; the density of water is 1,000 kg m^{-3}. The unit of pressure is bars, where 1.013 bar is the pressure of the Earth's atmosphere at sea level. The unit of luminosity is Joule per second, power is often expressed in watts, where 1.0 Watt = 1.0 Joule per second.

From the Sun's size and luminous output, we can infer the temperature of its visible disk, 5780 degrees Kelvin (Focus 3.2). The gas we see there is extremely rarefied, about ten thousand times less dense than the air we breathe. The pressure of the tenuous gas is less than that beneath the foot of a spider.

The material deep down inside the Sun must become hotter and more densely concentrated to keep our star extended, support its weight and keep it from collapsing. Calculations show that the temperature reaches 15.6 million (1.5×10^{7}) degrees Kelvin at the center of the Sun (Focus 3.2). The outward pressure generated by this hot gas is 233 billion (2.33×10^{11}) times Earth's air pressure at sea level. The center is also extremely compacted with a density that is 13 times that of solid lead, but it is still a gas.

Under the extreme conditions inside the Sun, atoms lose their identity. The temperature is so high, and the particles are moving so fast, that atoms are ripped apart by innumerable collisions. They tear whole atoms into their constituent pieces, negatively charged free electrons and positively charged fragments called ions. These sub-atomic particles are so small that they can move about freely as in a gas, even at high density.

The lightest known element, hydrogen, is the most abundant ingredient of the Sun (Sect. 2.1), about 75 percent by mass and 92 percent by number of atoms. Each hydrogen atom consists of one proton and one electron, and they are both liberated from their atomic bonds in the hot solar gas. Helium accounts for almost all the rest of the solar material; the heavier elements only amount to 0.1 percent by number. Helium and heavier elements are ionized inside the Sun, with some of their electrons set free. So the Sun is a seething, incandescent plasma, a collection of negative electrons, protons and heavier ions, with no net electric charge.

The Sun provides, directly or indirectly, almost all of the energy on Earth, and has been doing so for several billion years. No ordinary fire can make the Sun shine so brightly for such long periods of time. The ultimate source of the Sun's energy is nuclear fusion in its hot, dense core.

In a sequence of nuclear reactions at the center of the Sun, four protons are successively fused together into one heavier helium nucleus. The helium is slightly less massive (by a mere 0.7 percent) than the four protons that combine to make it. This mass difference, Δm, is converted into energy, ΔE, according to Einstein's famous equation $\Delta E = \Delta mc^2$; since the velocity of light, c, is a large number, about 300 million meters per second, a small mass difference, Δm, can account for the Sun's awesome energy.

The details of just how helium nuclei are synthesized from protons had to await the discovery of fundamental particles, such as the neutron and positron in 1932. (The positron, denoted as e^+, is the positive electron, or the antimatter version of the electron, denoted by e^-, with the same mass and a reversed charge.) It wasn't until 1938 that Hans A. Bethe and Charles L. Critchfield demonstrated how a sequence of nuclear reactions, called the proton-proton chain, makes the Sun shine (Focus 3.3). Bethe was awarded the Nobel Prize in Physics in 1967 for this and other discoveries concerning energy production in stars.

The center of the Sun is so densely packed that a proton can only move about one centimeter before encountering another proton. Moreover, at a central temperature of 15.6 million degrees Kelvin, the protons are darting about so fast that each one of them collides with other protons about 100 million times every second. Yet, the protons nearly always bounce off each other without triggering a nuclear reaction during a collision.

Since protons have the same electrical charge, they repel each other, and this repulsion must be overcome for protons to fuse together. Only a tiny fraction of the protons are moving fast enough to break through this barrier. Even then, they have to be helped along by a bizarre tunneling phenomenon. An uncertainty in determining the exact position of the two protons helps them tunnel through the electrical obstruction between them.

Focus 3.1

The Sun's Distance, Mass and Radius

All the planets are under the gravitational control of the Sun, and orbit it with a speed equal to the Sun's escape velocity at their distance. The escape velocity formula derived in Focus 2.1 is:

$$V_{\mathrm{P}} = \frac{2\pi D_{\mathrm{P}}}{P_{\mathrm{P}}} = \left(\frac{2GM_{\odot}}{D_{\mathrm{P}}}\right)^{1/2}$$

for the velocity, V_P, of a planet with distance, D_P, from the Sun and orbital period P_P around it. The gravitational constant $G = 6.6726 \times 10^{-11}\,\mathrm{N\,m^2\,kg^{-2}}$, and $M_\odot = 1.989 \times 10^{30}$ kilograms is the mass of the Sun. It is derived later in this Focus 3.1.

The equation can be rearranged to give Kepler's third law that relates the square of the orbital periods to the cube of their distances:

$$P_P^2 = 5.9165 \times 10^{11} \frac{D_P^3}{M_\odot} .$$

The planets move in elliptical orbits with the Sun at one focus, and not in an exact circle. When the distance, D_P, is the semi-major axis of the ellipse, the relevant constant is $4\pi^2/G = 5.9165 \times 10^{11}$. This expression tells us that a planet further away from the Sun will move around it at a slower pace. For example, Jupiter is 5.2 times as far away from the Sun as the Earth is, and it takes Jupiter 11.86 Earth years to travel once around the Sun.

Once the distance of any planet is known, we can infer the distances of all the rest. The Earth's mean distance from the Sun, the astronomical unit or AU, can thus be determined from the orbital distance, D_V, of Venus using:

$$\mathrm{AU} = D_E = \left(\frac{P_E}{P_V}\right)^{2/3} D_V = 1.496 \times 10^{11} \text{ meters}$$

where the orbital period of the Earth is $P_E = 1$ year $= 3.1557 \times 10^7$ seconds, and Venus orbits the Sun at a distance of $D_V = 0.723$ AU and a period of $P_V = 0.615$ years. An accurate distance to Venus is inferred by measuring the round-trip time to send a radar (radio range and detection) pulse off the planet and receive the echo here at Earth. The distance is equal to one half the travel time times the velocity of light.

Once we know the Earth's orbital period and mean distance from the Sun, we can determine the mass of the Sun using Kepler's third law:

$$M_\odot = 5.9165 \times 10^{11} \frac{(\mathrm{AU})^3}{P_E^2} = 1.989 \times 10^{30} \text{ kilograms} .$$

The linear radius of the Sun, $R_\odot$, can be determined from its angular diameter, θ, using

$$\theta = \frac{2R_\odot}{\mathrm{AU}} = 0.0093 \text{ radians} ,$$

to obtain:

$$R_\odot = \frac{\theta \mathrm{AU}}{2} = 6.955 \times 10^8 \text{ meters} .$$

A full circle subtends 2π radians and 360 degrees, and since there are 60 minutes of arc in a degree, we can express the angular diameter of the Sun in minutes of arc, $\theta = 60 \times 360 \times 0.0093/(2\pi) = 31.97$ minutes of arc, where $\pi = 3.14159$. This is about the same angle as that subtended by the thumb of your outstretched arm.

The forces of nature therefore conspire against nuclear fusion. It is not hot enough at the center of the Sun for the vast majority of protons to fuse together, for most of them do not move fast enough. Electrical repulsion between protons keeps them from joining together, and tunneling does not happen very often. As a result, nuclear reactions proceed at a slow, stately pace inside the Sun. That is a very good thing! Frequent fusion could make the Sun blow up like a giant hydrogen bomb.

The Sun is consuming itself at a prodigious rate. Every second its central nuclear reactions burn about 700 million tons of hydrogen into helium ashes. In doing so, 5 million tons (0.7 percent) of this matter disappears as pure energy, and every second the Sun becomes that much less massive (Focus 3.3). Yet, the loss of material is insignificant compared to the Sun's total mass; it has lost only 1 percent of its original mass in 4.5 billion years.

The central temperature of the solar crucible has adjusted itself to be just right, permitting nuclear fusion to proceed at a steady, leisurely rate. If the temperature inside the Sun was a lot higher, rapid nuclear fusion would cause the Sun to blow up like an immense hydrogen bomb. If it was a lot colder, the reactions would occur less often and make the Sun shine feebly, like a flashlight with a worn out battery.

All the nuclear energy is released in the dense, high-temperature central core. Outside of the core, where the overlying weight and compression are less, the gas is cooler and thinner, and nuclear fusion cannot occur. It requires a high-speed collision which can only be obtained with a very high temperature in excess of 5 million degrees Kelvin. The energy-generating core therefore only extends to about one quarter the distance from the center of the Sun to the visible photosphere, or 2.4 times the size of the giant planet Jupiter, accounting for only 1.6 percent of the Sun's volume. But about half the Sun's mass is packed into its dense core.

Astronomers have built mathematical models to deduce the Sun's internal structure. These models use the laws of physics to describe a self gravitating sphere of highly ionized hydrogen, helium and small quantities of heavier elements. At every point inside the Sun, the force of gravity must be balanced precisely by the gas pressure, which itself increases with the temperature. The energy released by nuclear fusion in the Sun's core heats the gas, keeps it ionized, and creates the pressure. This energy must also make its way out to the surface of the Sun to keep it shining at the presently observed rate.

Detailed models of the interiors of the Sun and other stars were developed in the 1950s and 1960s, and then continued to be refined for decades using extensive computer programs. Such models must include and explain the evolution of the Sun, from the time it began to shine by nuclear reactions about 4.55 billion years ago. The reason for tracking the evolving Sun is that its structure changes as more and more hydrogen is converted into helium. The evolutionary calculations are reworked until the simulated Sun ends up with its known mass, radius, energy output, and effective temperature (Table 3.1, Focus 3.1 and Focus 3.2). According to the models, the Sun has now used up about half the available hydrogen in its central parts since nuclear reactions began, and should continue to shine by hydrogen fusion for another 7 billion years (Sect. 7.5, Fig. 7.24).

Focus 3.2

The Luminosity, Effective Temperature and Central Temperature of the Sun

Satellites have been used to accurately measure the Sun's total irradiance just outside the Earth's atmosphere, establishing the value of the solar constant

$$f_{\odot} = 1{,}366 \text{ Joule per second per square meter},$$

where one Joule per second is equivalent to one watt. The solar constant is defined as the total amount of radiant solar energy per unit time per unit area reaching the top of the Earth's atmosphere at the Earth's mean distance from the Sun, or at $1 \text{ AU} = 1.496 \times 10^{11}$ meters. We can use it to determine the Sun's absolute luminosity, $L_{\odot}$, from:

$$L_{\odot} = 4\pi f_{\odot} (\text{AU})^2 = 3.8 \times 10^{26} \text{ Joule s}^{-1}.$$

The effective temperature, $T_{e\odot}$, of the visible solar disk, called the photosphere, can be determined using the Stefan-Boltzmann law:

$$L_{\odot} = 4\pi\sigma R_{\odot}^2 T_{e\odot}^4,$$

where the Stefan-Boltzmann constant $\sigma = 5.670 \times 10^{-8} \text{ Joule m}^{-2}\text{ K}^{-1}\text{ s}^{-1}$, and $R_{\odot} = 6.955 \times 10^8$ meters, so

$$T_{e\odot} = [L_{\odot}/(4\pi\sigma R_{\odot}^2)]^{1/4} = 5{,}780 \text{ degrees Kelvin}.$$

Incidentally, the Stefan-Boltzmann law applies to other stars, indicating that at a given temperature bigger, giant stars have greater luminosity.

The temperature, T_C, at the center of the Sun can be estimated by assuming that a proton must be hot enough and fast enough to counteract the gravitational compression it experiences from all the rest of the star. That is:

$$\text{Thermal Energy} = \frac{3}{2}kT_C = \frac{Gm_p M_{\odot}}{R_{\odot}} = \text{Gravitational Potential Energy},$$

where Boltzmann's constant $k = 1.38066 \times 10^{-23} \text{ J K}^{-1}$ and the mass of the proton is $m_p = 1.6726 \times 10^{-27}$ kg. Solving for the central temperature we obtain:

$$T_C = \frac{2Gm_p M_{\odot}}{3kR_{\odot}} = 1.56 \times 10^7 \text{ degrees Kelvin},$$

so the temperature at the center of the Sun is 15.6 million degrees.

As confident as we are about understanding energy production in the Sun, there is one thorn in the side of our standard model. It concerns the neutrinos that are produced in prodigious quantities in the Sun's central furnace. Two hundred trillion, trillion, trillion (2×10^{38}) of them are created every second (Focus 3.3). These tiny, ghostlike particles have no charge and little mass, moving at nearly the velocity of light and traveling almost unimpeded through the Sun, Earth and nearly any amount of ordinary matter.

A handful of these neutrinos can be captured using massive detectors that are buried deep underground so that only neutrinos can reach them. By finding solar neutrinos in roughly the predicted numbers, four pioneering detectors have now demonstrated that the Sun is indeed energized by hydrogen fusion. However, these experiments detect only one-third to one half of the number of neutrinos that theory predicts; a discrepancy known as the solar neutrino problem (Focus 3.4, Table 3.2).

The ghostly solar neutrinos may be transforming themselves on the way out of the Sun to a form we cannot detect. There are three types, or flavors,

Table 3.2. Solar neutrino experiments*

Target	Experiment	Threshold Energy (MeV)	Measured Neutrino Flux (SNU)*	Predicted Neutrino Flux (SNU)	Ratio: Measured/ Predicted
Chlorine 37	HOMESTAKE	0.814	$2.56 \pm 0.16 \pm 0.14$	$9.5^{+1.2}_{-1.4}$	0.27 ± 0.022
Water	KAMIOKANDE*	7.5	$2.80 \pm 0.19 \pm 0.35$	$6.62^{+0.93}_{-1.12}$	0.42 ± 0.060
Gallium 71	GALLEX	0.2	$69.7 \pm 6.67^{+3.9}_{-4.5}$	136.8^{+8}_{-7}	0.51 ± 0.058
Gallium 71	SAGE	0.2	72^{+12+5}_{-10-7}	136.8^{+8}_{-7}	0.53 ± 0.095

* Adapted from Lang (1999). Here the uncertainties are one standard deviation, or 1σ, and the first and second measurement uncertainties correspond, respectively, to the statistical and systematic uncertainties. The neutrino reaction rate within even massive, underground detectors is so slow that a special unit has been invented to specify the experiment-specific flux. This solar neutrino unit, or SNU, is equal to one neutrino interaction per second for every 10^{36} atoms, or $1\,\mathrm{SNU} = 10^{-36}$ neutrino absorptions per target atom per second. The units for the measured and predicted values for the Kamiokande experiment are $10^{6}\,\mathrm{cm}^{-2}\,\mathrm{s}^{-1}$, while the SNU or solar neutrino unit, is used for the other three experiments.

Focus 3.3

The Proton–Proton Chain

The hydrogen-burning reactions that fuel the Sun are collectively called the proton–proton chain. It begins when two of the fastest moving protons, designated by the symbol p, collide head on. They move into each other and fuse together to make a deuteron, D^2, the nucleus of a heavy form of hydrogen known as deuterium. Since a deuteron consists of one proton and one neutron, one of the protons entering into the reaction must be transformed into a neutron, emitting a positron, e^+, to carry away the proton's charge, together with a low-energy electron neutrino, ν_e, to balance the energy in the reaction. This first step in the proton–proton chain is written:

$$p + p \rightarrow D^2 + e^+ + \nu_e \, .$$

The deuteron next collides with another proton to form a nucleus of light helium, He^3, together with energetic gamma ray radiation, denoted by the symbol γ. In the final part of the proton–proton chain, two such light helium nuclei meet and fuse together to form a nucleus of normal heavy helium, He^4, returning two protons to the solar gas. These reactions are written:

$$D^2 + p \rightarrow He^3 + \gamma \, ,$$

$$He^3 + He^3 \rightarrow He^4 + 2p \, .$$

Additional gamma rays are in the meantime produced when the positrons, e^+, combine with electrons, e^-, in the annihilation reaction:

$$e^+ + e^- \rightarrow 2\gamma \, .$$

The net result is that four protons have been fused together to make one helium nucleus, or that:

$$4p \rightarrow He^4 + 6\gamma + 2\nu_e \, ,$$

with the creation of an amount of energy, ΔE, given by:

$$\Delta E = \Delta mc^2 = (4m_p - m_{He})c^2 = 0.007(4m_p)c^2 = 0.42 \times 10^{-11} \text{ Joule}$$

where $m_p = 1.6726 \times 10^{-27}$ kilograms and $m_{He} = 6.6465 \times 10^{-27}$ kilograms respectively denote the mass of the proton and the helium nucleus, and $c = 2.9979 \times 10^8$ meters per second is the velocity of light.

The number, N, of helium nuclei that are formed every second is $N = L_\odot / \Delta E \approx 10^{38}$ where the solar luminosity is $L_\odot = 3.854 \times 10^{26}$ Joule per second. That is, a hundred trillion, trillion, trillion helium nuclei are formed each second to make the Sun's energy at its present rate. In the process, the amount of mass, ΔM, consumed by this hydrogen burning every second is $\Delta M = 10^{38} \Delta m = 4.7 \times 10^9$ kilograms ≈ 5 million tons, where one ton is equivalent to 1,000 kilograms.

Each second 2×10^{38} neutrinos are released from the Sun, moving out in all directions in space at the velocity of light. The Earth will intercept a small fraction of these, but it is still a large number, about $4 \times 10^{29} = 2 \times 10^{38} (R_E / AU)^2$, where the Earth's radius is $R_E = 6.371 \times 10^6$ meters and its mean distance from the Sun is $AU = 1.496 \times 10^{11}$ meters.

of neutrinos, named the electron, muon and tau neutrinos. Existing instruments only taste one flavor, called the electron neutrino. This is the type produced during nuclear fusion inside the Sun. If some of the Sun's electron neutrinos change flavor *en route* to Earth, it would explain the solar neutrino problem. Measurements with the Super-Kamiokande instrument indicate that muon neutrinos, produced when energetic cosmic-rays hit the Earth's atmosphere, exhibit a dependence on arrival direction that can be explained by the metamorphosis of muon neutrinos into tau neutrinos.

This atmospheric neutrino data has nothing to do with electron neutrinos, the only kind made by nuclear reactions in the Sun, and therefore has no direct bearing on the solar neutrino problem. It nevertheless suggests that some kinds of neutrinos have at least a tiny amount of mass and are capable of transformation into another type. If this is happening to solar electron neutrinos, then the underground experiments will detect fewer neutrinos than expected and the problem may be resolved.

Experiments currently under way should settle the question of whether solar neutrinos switch identities on there way out of the Sun, providing stringent limits to, or a measurement of, the neutrino mass. The Sudbury Neutrino Observatory, located 2.0 thousand meters underground in a working nickel mine near Sudbury, Ontario, is poised to search for such a neutrino change. The detector is filled with 1,000 tons of heavy water, a form of H_2O that contains deuterium, or heavy hydrogen, ^2H, with a neutron and a proton in its nucleus. Unlike the other solar neutrino detectors, which are sensitive only to electron neutrinos, Sudbury's heavy water will be sensitive to all three types of neutrinos.

So, the solar neutrino problem may be eventually resolved without affecting our models of the solar interior. These models also describe how radiation gets from the energy-producing core to the sunlight that illuminates and warms our days. In addition to the core, there are two distinct regions – a radiative zone and a convection zone (Fig. 3.1). The radiative zone surrounds the core out to 71.3 percent of the Sun's radius. As the name implies, energy moves through this region by radiation. It is a relatively tranquil, serene and placid region, somewhat like the deeper parts of Earth's oceans. Above the radiative zone lies the convection zone, within which energy is transported in a churning, wheeling motion called convection; it is a lively, turbulent and vigorous place.

Even though light is the fastest thing around, radiation does not move quickly through the Sun's innards. The core radiation, liberated in the form of gamma rays, is continuously absorbed and re-emitted at lower and lower temperatures as it travels through the radiative zone. The radiation diffuses slowly outward in a haphazard, zigzag pattern, like a drunk staggering down a sidewalk, so that by the time it nears the surface it is primarily visible light. The calculations of this continued ricocheting and energy loss in the radiative zone require a detailed knowledge of atomic physics, which describes how radiation interacts with matter. Recent computations indicate that it takes about 170 thousand years, on average, for the radiation to work its way out from the Sun's core to the overlying convection zone.

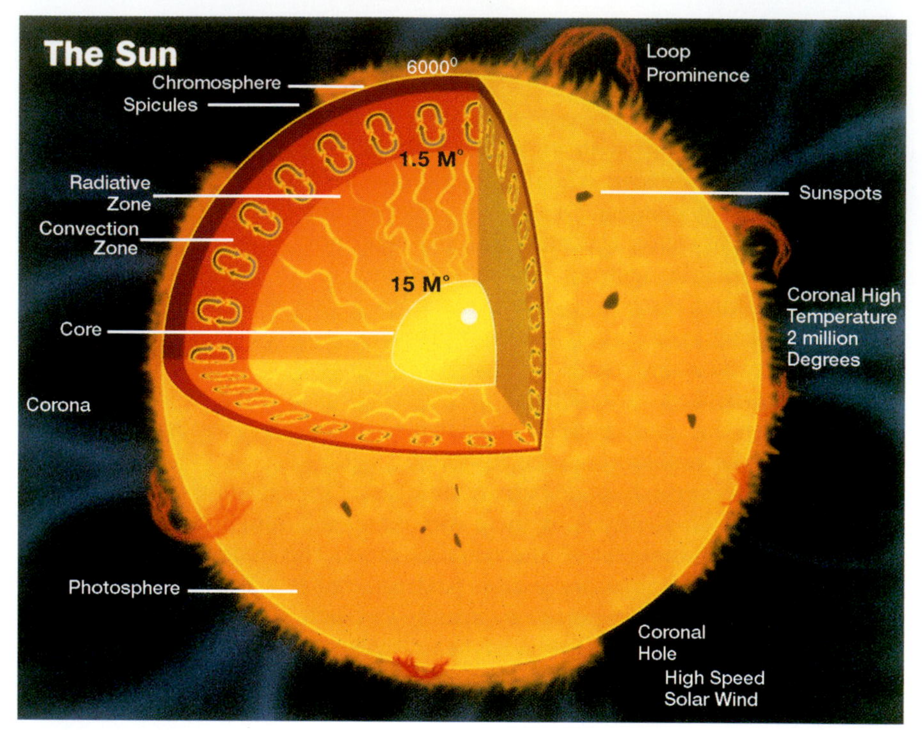

Fig. 3.1. Anatomy of the Sun. The Sun is an incandescent ball of ionized gas powered by the fusion of hydrogen in its core. As shown in this interior cross-section, energy produced by nuclear fusion is transported outward, first by countless absorptions and emissions within the radiative zone, and then by convection. Internal motions generate and sustain magnetic fields that are seen at the visible solar disk, or photosphere, as sunspots. The magnetism molds hot million-degree gas within coronal loops and cooler material in prominences

Throughout the radiative zone, the gas is hot enough to keep all the nuclei completely stripped of electrons. At the bottom of the convection zone, the temperature has become cool enough, at about 2 million degrees Kelvin, to allow some heavy nuclei to capture electrons; because of their lighter weight and greater speed, the abundant hydrogen and helium remain fully ionized.

The rare heavy particles absorb light and block the flow of heat more efficiently than bare nuclei or electrons do. The increased light-absorbing ability, or opacity, of the solar gas obstructs the outflowing radiation, like dirt on a window, and causes the gas to become hotter than it would otherwise be. The pent-up-energy is transported by convection, which carries energy through the outer 28.7 percent of the Sun.

Convective motion occurs when a gas or fluid is heated from below. In response to heating, gases in the bottom layer expand and thereby become less dense than the gas in the overlying layers. The heated material, due to its low density, rises, just as a balloon does. On Earth you can see hot air rising when watching the smoke above a fire or hawks riding on upward currents of heated air.

In the Sun's convection zone, hot gases rise to the visible solar disk, in roughly 10 days, and cool by radiation. The cooled gas then sinks, because it is denser than the hotter gas, only to be re-heated and rise again. Such wheeling convective motions occur in a kettle of boiling water or a simmering pot of oatmeal, with hot rising bubbles and cooler sinking material. Similar convective flow also produces towering thunder clouds, the great ocean currents and the shimmering of air currents over a paved road on a sunny day.

High-resolution images of the Sun taken in white light, or in all the visible colors of the Sun combined, show a granular pattern that marks the top of the convection zone (Fig. 3.2). The bright center of each granule, or convection cell, is the highest point of a rising column of hot gas. The dark edges of each grain are the cooled gas beginning its descent to be heated once again. These granules have diameters of about 1.5 million meters, and more than a million of them cover the photosphere. They constantly evolve and change on time scales of minutes, like the seething surface of an immense boiling caldron.

Supergranules mark the tops of large-scale wheeling motions that are detected as horizontal motions in the photosphere. They were discovered by A. B. Hart in 1954–56 and extensively studied by Robert B. Leighton and his collaborators in the early 1960s. The mean size of a supergranule cell is 35 million meters, so there are roughly 1,500 supergranules on the visible solar disk. The velocity cells exhibit horizontal velocities as large as 500 meters per second, and persist for several hours or more. In the process of studying these motions, Leighton discovered the Sun's five-minute oscillations (Sect. 3.2).

Focus 3.4

Detecting Solar Neutrinos

Neutrinos move at or very near the velocity of light, have no electrical charge, and contain almost no mass. They are so insubstantial that they pass right through almost anything in their path, true ghost riders of the Universe. It is nevertheless possible to snag small amounts of the elusive neutrinos using massive detectors placed beneath mountains or deep inside mines; the intervening rocks shield them from deceptive signals caused by cosmic rays. The neutrino detectors consist of large amounts of material, literally tons of it, because of the very small interaction cross section between neutrinos and matter.

The first solar neutrino detection experiment, constructed by Raymond Davis, in 1967, is a 615-ton tank containing 100 thousand gallons of C_2Cl_4 (perchloroethylene, a cleaning fluid), located 1.5 thousand meters underground in the Homestake Gold Mine near Lead, South Dakota. The neutrinos interact with the chlorine, ^{37}Cl, in the cleaning fluid to produce argon atoms, ^{37}Ar, by the reaction:

$$\nu_e + {}^{37}\mathrm{Cl} \rightarrow \mathrm{e}^- + {}^{37}\mathrm{Ar} \,,$$

where ν_e denotes an electron neutrino and e^- is an electron. By counting the number of argon atoms, which appear on the average of one atom every 2.17 days, one can determine the number of neutrino interactions and compare it with the expected amount. Due to its high-energy threshold, the Homestake experiment only detects relatively rare, energetic neutrinos produced during infrequent nuclear reactions involving boron 8.

A second experiment, that began operating in 1987, is located in a mine at Kamioka, Japan. It consists of a 3,000-ton tank of pure water (680-ton fiducial volume). This neutrino–electron scattering experiment is based upon the reaction

$$\nu_e + e^- \rightarrow \nu_e' + e^{-\prime} .$$

Here the unprimed and primed sides respectively denote the incident and scattered neutrinos and electrons. A passing electron neutrino occasionally knocks a high-speed electron from a water molecule. The electron moves through the water faster than light travels in water, generating an electromagnetic shock wave and a light cone of Cherenkov radiation. Thousands of light detectors lining the water tank measure the axis of the light cone, which tells the direction of the incoming neutrino. The observed signals indicate that the neutrinos are indeed coming from the Sun. These are also relatively rare high-energy neutrinos.

More recently, two gallium, ^{71}Ga, experiments have been developed, using the reaction

$$\nu_e + {}^{71}\mathrm{Ga} \rightarrow e^- + {}^{71}\mathrm{Ge} ,$$

that produces radioactive germanium, ^{71}Ge. The gallium detects low-energy neutrinos produced by the main proton–proton reaction in the Sun's core. In 1990, the Soviet–American Gallium Experiment, or SAGE, began operation in a long tunnel, some 2 thousand meters below the summit of Mount Andyrchi in the northern Caucasus. A second multi-national experiment, dubbed GALLEX for gallium experiment, started operating in 1991, located in the Gran Sasso Underground Laboratory some 1.4 thousand meters below a peak in the Appenine Mountains of Italy.

Each of these experiments has a different threshold, and is therefore sensitive to neutrinos of different energies (Table 3.2). The details depend on subtle differences in the terminations of the hydrogen burning reactions, but the main thing is that all four of the experiments detect fewer neutrinos than expected.

There are two possible explanations to the solar neutrino problem. Either we don't really know exactly how the Sun creates its energy, or we have an incomplete knowledge of neutrinos. Astronomers have used sound waves to probe the internal constitution of the Sun with results that verify their models (Sect. 3.3). They therefore suspect that there is something wrong with our understanding of the neutrino.

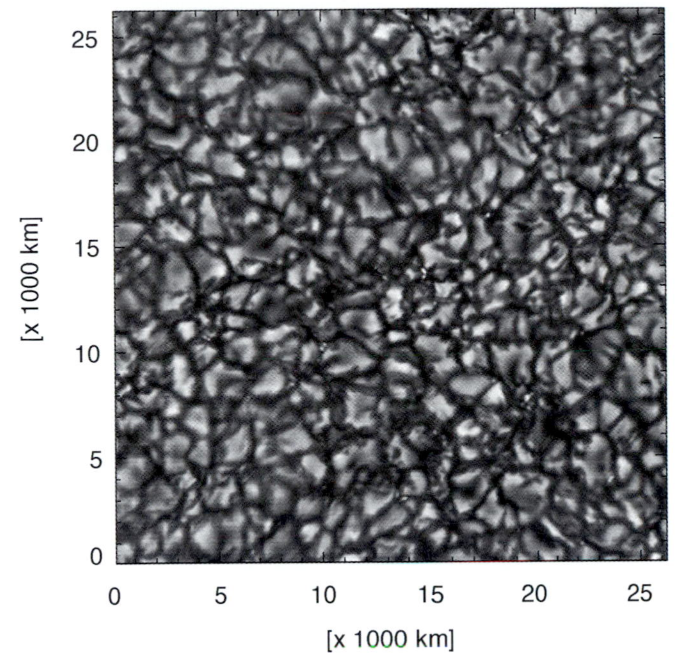

Fig. 3.2. Double, Double, Toil and Trouble. At high magnification, the photosphere appears completely covered with an irregular, strongly-textured pattern of cells called granules. The hot granules, each about 1.5 million meters across, rise at speeds of 500 meters per second, like supersonic bubbles. The rising granules radiate their energy and cool material then sinks down along the dark, intergranular lanes. This photograph was taken with an exceptional angular resolution of 0.2 seconds of arc, or 150 thousand meters at the Sun, using the National Solar Observatory's Vacuum Tower Telescope at the Sacramento Peak Observatory. [Courtesy of Thomas R. Rimmele, Association of Universities for Research in Astronomy (AURA), National Optical Astronomy Observatories (NOAO), and the National Science Foundation (NSF)]

The convection zone is capped by a thin veneer of solar radiation, called the photosphere (from *photos*, the Greek word for light), at a temperature of 5,780 degrees Kelvin (Focus 3.2). About 500 thousand meters thick, the photosphere is a zone where the gaseous material changes from being completely opaque to radiation to being transparent. It is the layer from which the light we actually see is emitted and where most of the Sun's energy finally escapes into space.

The existing computer models are simple ones that provide a good description of the Sun's global properties. However, they omit some of the details by assuming complete spherical symmetry and ignoring internal rotation, magnetic fields and possible mixing of core material to surrounding gas. Now we can improve these models by using sound waves to open a window into the Sun's interior.

The new probes of the internal workings of our star have been compared with the computer models that show, among other things, how the pressure, temperature, mass density, sound speed, and hydrogen content vary as a function of radius inside the Sun. They have further refined our understanding of the properties of matter under conditions that we cannot possibly reproduce in laboratories on Earth. In addition, the new observations have shown how the Sun's gases rotate and flow inside, and pinpoint where its magnetism is probably generated.

3.2 Taking the Sun's Pulse

We can explore the unseen depths of the Sun by watching the widespread throbbing motions of its visible disk, the photosphere (Fig. 3.3). These oscillations, which can be tens of kilometers high and travel a few hundred meters per second, arise from sounds that course throughout the solar interior. The sounds are continuously excited by vigorous turbulent motion in the convection zone, somewhat like the deafening roar of a jet aircraft or the hissing noises made by a pot of boiling water.

The sound waves are trapped inside the Sun; they cannot propagate through the near vacuum of space. Nevertheless, when these sounds strike the photosphere and rebound back down, they disturb the gases there, causing them to rise and fall in oscillating reverberations. They are superficially similar to ripples formed when a stone hits a pond or when a floating log has been pushed below a pond's surface and let go. However, unlike a water ripple, the

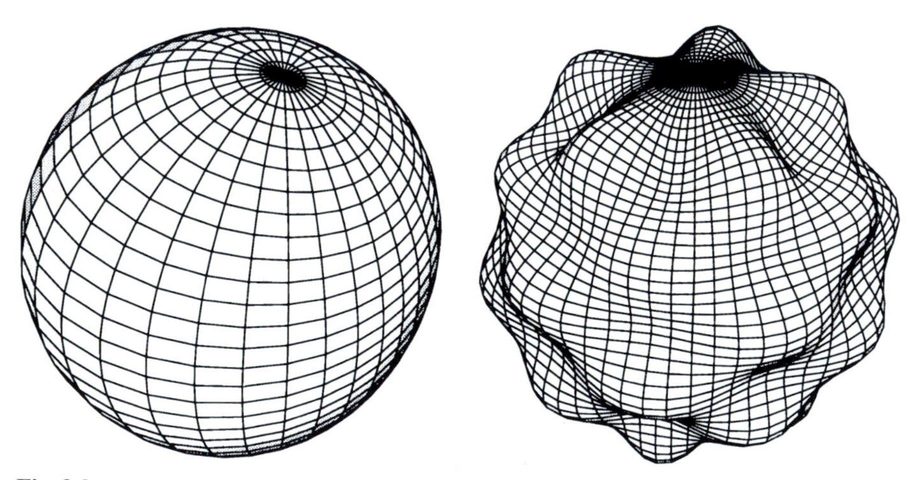

Fig. 3.3. Internal Contortions. Sound waves resonating within the solar interior cause the photosphere to move in and out, rhythmically distorting the shape of the Sun. The Sun exhibits over a million shapes produced by its internal oscillations. Two of these shapes are illustrated here with an exaggerated amplitude. (Courtesy of Arthur N. Cox and Randall J. Bos, Los Alamos National Laboratory)

photosphere's heaving motions never stop, and the entire Sun can move in and out.

Local areas of the photosphere swell and contract, slowly and rhythmically at five-minute intervals, like the tides in a bay or a beating heart. Each five-minute period is the time it takes for the localized motion to change from moving outward to moving inward and back outward again. That interval is similar to the separation between the most intense contractions during child birth, at least during the birth of my children.

Even if the solar song could reach the Earth, the periods are too long, and the notes are too low-pitched, for the human ear to hear. At a frequency of 0.003 vibrations a second, they are way below the lowest note audible to humans. Sound waves that a healthy youth hears vibrate mainly from 25 to 20,000 vibrations a second, the latter is about equivalent to a bad telephone connection. Nevertheless, we can "listen" to the Sun's vibrating notes indirectly. They are detected as local rhythmic motions that cause the entire photosphere to throb with constant motion, like waves on the surface of a stormy sea. These undulations can be used to record the Sun's hidden melody, and decipher the inner workings of our life-sustaining star.

When discovered by Robert B. Leighton in 1960, the five-minute vertical oscillations of the photosphere seemed to be a chaotic, short-lived and purely local effect, with each small region moving independently of nearby ones. However, subsequent observations and theoretical considerations, in the early 1970s, demonstrated that all the local motions are driven by sound waves that echo and resonate through the solar interior. Eventually observers showed that the entire Sun is vibrating with ponderous rhythms that extend to its very core.

Astronomers look under the Sun's skin, and down to its heart, by examining sounds with different paths within the Sun. They use the term helioseismology to describe such investigations of the solar interior. It is a hybrid name combining the Greek word *helios* for Sun or light, the Greek word *seismos* meaning quake or tremor, and *logos* for reasoning or discourse. So literally translated helioseismology is the logical study of solar tremors. Geophysicists similarly unravel the internal structure of the Earth by recording earthquakes, or seismic waves, that travel to different depths; this type of investigation is called seismology.

Sound waves, produced by hot gas churning in the convection zone, circle within resonant cavities inside the Sun (Fig. 3.4). Starting near the photosphere, a sound wave moves into the Sun toward the center. Since the speed of sound increases with temperature (Focus 3.5), which in the Sun increases with depth, the wavefront's deeper, inner edge travels faster than its shallower outer edge, so the inner edge pulls ahead. Gradually an advancing wavefront is refracted, or bent, until the wave is once again headed toward the Sun's surface. At the same time, the enormous drop in density near the photosphere reflects waves traveling outward back in.

Thus, the sounds move within solar cavities. They resemble the noise of a flying jet aircraft that gets trapped near the ground on a hot day, or the shouts of a child that can travel far across a cloud-covered lake on a summer night.

Each of these resonating sounds has a well-defined trajectory (Fig. 3.4). Those that travel far into the Sun move nearly perpendicular to the photosphere when they hit it, and they touch the photosphere only a few times during each internal circuit around the Sun. These deep sound waves cause the entire Sun to ring like a bell. Sound waves with shorter trajectories strike the photosphere at a glancing angle, and travel through shallower and cooler layers. They bounce off the visible photosphere more frequently, and describe motions and structure in the outer shell of the Sun. Sounds with both deep and shallow turning points go around and around within cavities inside the Sun, like hamsters caught in an exercise wheel.

Focus 3.5

The Speed of Sound

Sound waves are produced by perturbations in an otherwise undisturbed gas. They can be described as a propagating change in the gas mass density, ρ, which is itself related to the pressure, P, and temperature, T, by the ideal gas law:

$$P = nkT = \frac{\rho kT}{m_u \mu},$$

where n is the number of particles per unit volume, Boltzmann's constant $k = 1.38066 \times 10^{-23}$ Joule per degree Kelvin, and the atomic mass unit $m_u = 1.66054 \times 10^{-27}$ kilograms. The mean molecular weight, μ, is given by $\mu = 2/(1 + 3X + 0.5Y)$ where X, Y, and Z represent the concentration by mass of hydrogen, helium and heavier elements and $X + Y + Z = 1$. For a fully ionized hydrogen gas, $\mu = 1/2$, and for a fully ionized helium gas $\mu = 4/3$.

The wavelength, λ, and frequency, ν, of the sound waves are related by

$$\lambda \nu = s,$$

and the velocity of sound, s, is given by

$$s = \left(\frac{\partial P}{\partial \rho}\right)^{1/2} = \left(\frac{\gamma kT}{m_u \mu}\right)^{1/2},$$

where the symbol ∂ denotes a differentiation under conditions of constant entropy. For an adiabatic process, the pressure, P, and mass density, ρ, are related by $P = K\rho^{\gamma}$, where K is a constant. For adiabatic perturbations of a monatomic gas the index $\gamma = 5/3$ and for isothermal perturbations $\gamma = 1$. For both adiabatic and isothermal perturbations the sound speed is on the order of the mean thermal speed of the ions of the gas, or numerically:

$$s \approx 10^4 (T/10^4\,^{\circ}\mathrm{K})^{1/2} \text{ meters per second}.$$

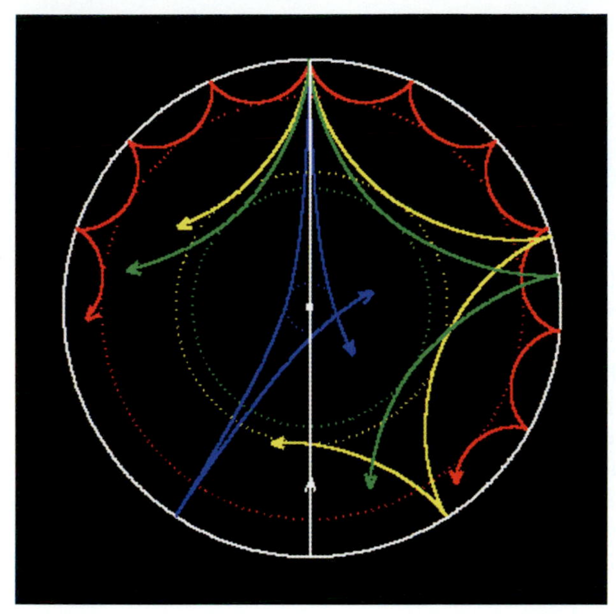

Fig. 3.4. Resonating Sound Paths. Sound waves inside the Sun, like seismic waves in the Earth, do not travel in straight lines. The sound rays are bent, like light within the lens of an eye, and circle the solar interior in spherical shells or resonant cavities. Each solar cavity is bounded at the top by a large density drop near the photosphere and bounded at the bottom by an increase in sound speed with depth that refracts a downward propagating wave back toward the surface. The bottom turning points occur along the dotted circles shown here. How deep a wave penetrates and how far around the Sun it goes before it hits the surface depends on the harmonic degree, l. The white curve is for $l = 0$, the blue one for $l = 2$, green for $l = 20$, yellow for $l = 25$ and red for $l = 75$. (Courtesy of Jørgen Christensen-Dalsgaard and Philip H. Scherrer)

Although the photosphere's undulations cannot be seen with the unaided eye, instruments aboard the SOlar and Heliospheric Observatory (SOHO) detect them with unsurpassed clarity. Two of these telescopes, the Michelson Doppler Imager (MDI) and the Global Oscillations at Low Frequency (GOLF), record small periodic changes in the wavelength of light, measuring the oscillation speeds using the Doppler effect (Focus 3.6). This effect occurs whenever a source of light or sound moves with respect to the observer, and results in a change in the wavelength and frequency of the waves.

A similar thing happens to sunlight. When part of the Sun approaches us, the wavelength of light emitted from that region becomes shorter, the wave fronts or crests appear closer together, and the light therefore becomes bluer; when the light source moves away the wavelength becomes longer and the light redder. The magnitude of the wavelength change, in either direction, establishes the velocity of motion along the line of sight, which is called the radial velocity (Focus 3.6).

A single, very narrow spectral feature, called a line, is observed, which is sharply defined in wavelength. MDI watches the motions of nickel atoms, in their red spectral line (Ni I at 676.8 nanometers), as they bob like buoys on the undulating Sun. The nickel line is one of the dark absorption lines formed by the relatively cool gas of the photosphere. Such spectral lines appear when the intensity of sunlight is displayed as a function of wavelength like a rainbow.

Focus 3.6

The Doppler Effect

Just as a source of sound can vary in pitch or wavelength, depending on its motion, the wavelength of electromagnetic radiation shifts when the emitting source moves with respect to the observer. This Doppler shift is named after Christiaan Doppler who discovered it in 1842. If the motion is toward the observer, the shift is to shorter wavelengths, and when the motion is away the wavelength becomes longer. You notice the effect when listening to the changing pitch of a passing ambulance siren. The tone of the siren is higher while the ambulance approaches us and lower when it moves away from us.

If the radiation is emitted at a specific wavelength, λ_{emitted}, by a source at rest, the wavelength, $\lambda_{\text{observed}}$, observed from a moving source is given by the relation:

$$z = \frac{\lambda_{\text{observed}} - \lambda_{\text{emitted}}}{\lambda_{\text{emitted}}} = \frac{V_r}{c},$$

where V_r is the radial velocity of the source along the line of sight away from the observer. The parameter z is called the redshift since the Doppler shift is toward the longer, redder wavelengths in the visible part of the electromagnetic spectrum. When the motion is toward the observer, V_r is negative and there is a blueshift to shorter, bluer wavelengths.

The Doppler effect applies to all kinds of electromagnetic radiation, including X-rays, visible light and radio waves, and it also occurs in sound waves. Because everything in the Universe moves, the Doppler effect is a very important tool for astronomers, determining the radial velocity of all kinds of cosmic objects.

Our equation is strictly valid for radial velocities that are much smaller than the velocity of light, or for V_r less than c, where $c = 2.9979 \times 10^8$ meters per second. The largest redshifts measured for the most distant galaxies reach $z = 5$ or more, but this does not mean that the radial velocity exceeds c; nothing can move faster than the velocity of light. For objects that move at speeds comparable to that of light, a somewhat more complicated expression describes the Doppler effect:

$$1 + z = \left[\frac{c + V_r}{c - V_r}\right]^{1/2},$$

or equivalently

$$\frac{V_r}{c} = \frac{(z+1)^2 - 1}{(z+1)^2 + 1}.$$

The photosphere motions can be inferred by taking solar images at different wavelengths on both the long-wavelength and short-wavelength sides of the absorption line, and fitting a model to them in order to determine the Doppler shift (Focus 3.6). Sequences of such measurements, taken at regular, successive intervals of time, can be used to determine the periodicity of the motions.

A single measurement determines the velocity with an accuracy of about 5 meters per second. When a lot of MDI's measurements are averaged over adjacent places or long times, the velocity can be determined with a remarkable precision of better than one millimeter per second – when the Sun's photosphere is heaving and dropping about a million times faster.

Moreover, the MDI telescope detects these motions at a million points across the Sun every minute! Because the instrument is positioned well above Earth's obscuring atmosphere, MDI continuously resolves fine detail that cannot always be seen from the ground. The blurring effects of our air, cloudy weather and the day-night cycle prohibit long, continual, spatially-resolved velocity measurements, such as those made with MDI for oscillations of relatively small scales.

Still, spacecraft have by no means rendered ground-based helioseismology obsolete. The Sun is also observed around the clock by a worldwide network of observatories, known by the acronym GONG for the Global Oscillations Network Group. They form an unbroken chain that follows the Sun as the Earth rotates.

SOHO's GOLF and its Variability of solar IRradiance and Gravity Oscillations (VIRGO) instrument, observe the entire disk of the Sun, without resolving fine surface details, and thereby garner long, undisturbed views of global, whole-Sun oscillations that probe the deep solar interior. GOLF detects light from the entire disk of the Sun, while VIRGO resolves the disk coarsely into 12 elements. Their telescopes gather and integrate light over the visible solar disk and tend to average out the peaks and troughs of the smaller localized undulations that occur randomly at many different places and times within the field of view.

GOLF uses the Doppler effect of absorption lines (the sodium Na D1 and Na D2 lines at 589.6 and 589.0 nanometers) to measure the net velocity fluctuations and pulsating motions of the entire Sun. It showed that the background noise due to random turbulent motions on the Sun is somewhat less than was thought based on measurements from ground-based helioseismology, and that the confusing noise is mainly due to the overlapping edges or wings of adjacent spectral lines.

Our atmosphere makes some kinds of solar oscillations difficult to record, especially very low frequencies and oscillations of exceptionally low or high degree. For these sounds, ground-based observations are something like trying to listen to a Mozart piano concerto when your son is blasting rap music on his boom box. Out in quiet, peaceful and tranquil space, unperturbed by terrestrial interference, SOHO has a long, clear undistorted view. It obtains recordings of both large-scale global oscillations and small-scale ones with unprecedented quality.

A third device tracks another change caused by the sound waves. As these vibrations heat and cool the gases in light-emitting regions of the Sun, the entire orb regularly flickers. When the sound waves alternately compress and relax the photosphere, they alter the brightness of the whole Sun, over periods of about five minutes, by a tiny fraction of a few millionths, or a few times 10^{-6}, of the Sun's average brightness. The minuscule but regular changes in the Sun's total output correspond to a change in temperature of just 0.005 degrees Kelvin. SOHO's VIRGO instrument records these intensity changes, precisely determining variations in the Sun's total irradiance of Earth (Sect. 7.4). They can provide a sensitive record of global, long-period oscillations, and refine our knowledge of the physical and dynamical properties of the central regions of the Sun.

The precise, long-duration measurements from GOLF and VIRGO might also provide the first unequivocal detection of gravity waves. These are slow sloshings of material that are forced or restored by gravity, like waves in the deep sea, rather than by pressure as in the case of sound waves. They are expected to be sensitive to density variations in the Sun's core and will have longer periods, of an hour or so, than the oscillations measured so far. We might use them to penetrate the core of the Sun where its energy is produced.

3.3 Looking Within the Sun

The combined sound of all the notes reverberating inside the Sun has been compared to a resonating gong in a sandstorm, being repeatedly struck with tiny particles and randomly ringing with an incredible din. The Sun produces order out of this chaos by reinforcing certain notes that resonate within it, like the plucked strings of a guitar. They are called standing waves.

Individual vibrations have velocity amplitudes of no more than 0.1 meters per second, but when millions of them are superimposed, they produce oscillations that move with thousands of times this speed. The low-amplitude components reinforce each other, producing the well-known, five-minute oscillations that grow and decay as numerous vibrations go in and out of phase to combine and disperse and then combine again, somewhat like groups of birds or schools of fish that gather together, move apart and congregate again.

Scientists have examined the power in the various photosphere oscillations, or how often each and every note is played, confirming that the power is concentrated into such resonant standing waves. Instead of meaningless, random fluctuations, orderly motions are detected with specific combinations of size and period, or wavelength and frequency (Fig. 3.5). Destructive interference filters out all but the resonant waves that combine and reinforce each other. Yet, there are still millions of such notes resonating in the Sun, so prolonged observations with high spatial resolution and detailed computer analysis are required to sort them all out.

The spatial patterns of oscillations detected at the photosphere can be separated into their component standing waves, each characterized by nodes at

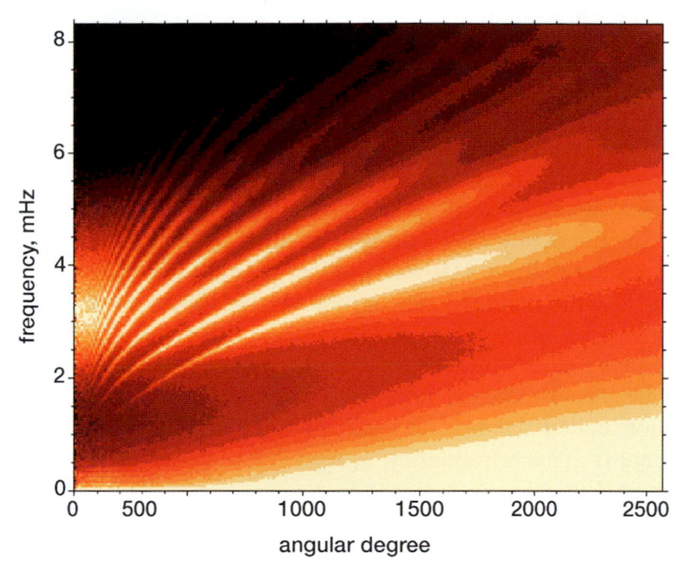

Fig. 3.5. An *l*-nu Diagram. Sound waves resonate deep within the Sun, producing photosphere oscillations with a frequency near 3 milliHertz, or 0.003 cycles per second, which corresponds to a wave period of five minutes. Here the frequency, nu, is plotted as a function of the spherical harmonic degree, *l*, for just eight hours of high-resolution data taken with the SOHO MDI instrument. The degree, *l*, is the inverse of the spatial wavelength, or surface size; an *l* of 400 corresponds to waves on the order of 10 million meters in size. The oscillation power is contained within specific combinations of frequency and degree, or period and wavelength, demonstrating that the observed oscillations are due to standing waves confined within resonant cavities. The depth of the resonating cavity depends on the oscillation frequency, nu, and the degree, *l*. What is happening near the core of the Sun is described by oscillations in the lower left corner of the diagram. Moving up in frequency or degree tells more about what is happening near the photosphere. (Courtesy of the SOHO MDI/SOI consortium. SOHO is a project of international cooperation between ESA and NASA)

which the motion vanishes. For example, if you shake one end of a rope that is fixed at the opposite end, a standing wave can be created. The nodal points are the places on the rope that are standing still while the rest of the rope moves up and down. On a two-dimensional surface, such as a drum skin or the solar photosphere, the nodal points are arranged in nodal lines.

Since the oscillations are due to sound waves that reverberate in three dimensions within the Sun, three whole numbers are needed to specify each standing wave. They are called spherical harmonics and are designated by three integers l, m and n. The spherical harmonic degree, l, indicates the total number of node lines on the solar surface. The azimuthal number, m, describes the number of node lines that cross the equator. The radial order n is the number of nodes in the radial direction inside the Sun, on a line from the center of the Sun to its surface. The number m is always less than or equal to the degree l. It turns out that oscillations with $m = l$ are concentrated close to the equator, while oscillations with low m reach high latitudes.

The solar oscillations are the combined effect of about 10 million separate resonating notes – each of which has a unique path of propagation and samples a well-defined section inside the Sun. So to trace the star's physical landscape all the way through – from its churning convection zone into its radiative zone and core – we must determine the exact pitch of all the notes and measure the precise frequency of every one of them. The frequency is the number of wave crests passing a stationary observer each second, and it therefore tells us how fast the sound wave moves up and down in a given cavity.

Each frequency is like a note on a musical scale, and helioseismologists are interested in the frequency of each note. They are portrayed in a two-dimensional power plot, called the l–ν or l–nu diagram, that displays how much sound energy there is at each frequency, ν, for every one of the different spatial scales or l values (Fig. 3.5).

Most of the power is concentrated near a frequency of 0.003 vibrations a second or 3 mHz. That corresponds to a period of 5 minutes, the time interval between wave peaks. Sound waves have been observed with periods ranging from about 1.5 minutes to about 20 minutes and spatial extents of between a few million meters to a thousand million meters. For $l = 0$, the whole solar globe resonates, across its radius of 696 million meters.

At constant ν, there is a relationship between the degree, l, and the depth of propagation in the sense that low-degree modes travel to the solar core, while high-degree modes are confined closer to the surface. For $l = 0$ the waves actually go to the solar center; such waves correspond to purely radial oscillations in which the Sun expands and contracts but keeps its spherical shape. The $l = 0$ oscillations are "breathing modes" where the whole Sun moves in and out at the same time. All the other higher-order values of l correspond to non-radial oscillations that deform the surface into non-spherical shapes. The higher the degree the smaller the spatial scale of the distortion.

The deepest parts of the Sun are probed by the oscillations with the largest sizes and the lowest spherical harmonic degree l's, thus requiring the least spatial resolution. Full-Sun, integrated-light observations provide this information. The SOHO instruments GOLF and VIRGO respectively observe $l = 0$ to 3 and $l = 0$ to 7, that are able to penetrate the deep solar interior.

In contrast, a large number of shallower depths are accessible to MDI. It observes sound waves with very high values of l up to $l = 4500$ that can dive just a short distance below the surface and return quickly.

The main goal of helioseismology is to infer the internal properties of the Sun from the oscillation frequencies of many different sound waves. The precise frequency depends on the propagation speed of the sound wave and the thickness of its resonant cavity. If we know the spherical harmonics of a particular sound wave, we can specify the path that the wave followed in the Sun. The observed frequencies for many different waves, that penetrate to different depths, can then be inverted to determine, for instance, how the speed of sound varies with depth. To be precise, the observed frequencies are integral measures of the speed along the path of the sound wave; helioseismologists have to use complex mathematical techniques and powerful computers to invert this measured data and get the sound speed.

The dominant factor affecting each sound is its speed, which in turn depends on the temperature and composition of the solar regions through which it passes (Focus 3.5). The sound waves move faster through higher-temperature gas, and their speed increases in gases with lower than average molecular weight. So, by investigating many different waves we can build up a very detailed three-dimensional picture of the physical conditions inside the Sun, including the temperature and chemical composition, from just below the photosphere down to the very core of the Sun. For example, early measurements of the sound speed were higher than specified by solar models. This led to improvements in our understanding of the internal opacity that blocks the flow of radiation and governs the Sun's temperature structure.

More recently, scientists have used inversion methods to determine the difference between the observed sound speed and that of a numerical model. These differences are obtained from discrepancies between the detected and calculated sound wave frequencies. Relatively small differences between the computer calculations and the observed sound speed are used to fine-tune the model and establish the Sun's radial variation in temperature, density and composition. A small but definite change in sound speed, for example, marks the lower boundary of the convection zone, pinpointing it at a radius of 71.3 percent of the radius of the visible solar disk, or at a depth of 28.7 percent of the photosphere's radius.

The sound speed measurements have important implications for the solar neutrino problem. The difficulty is that massive, underground neutrino detectors observe fewer neutrinos than expected (Focus 3.4). The discrepancy might be resolved if the Sun's central temperature is lower than current models predict, for the number of neutrinos produced by nuclear reactions is a sensitive, increasing function of the temperature. However, the measured and predicted sound velocities do not now differ from each other by more than 0.2 percent from 0.95 solar radii down to 0.05 solar radii from the center. Since the square of the sound speed is proportional to the temperature, this close agreement confirms the predicted core temperature to 15.6 million degrees Kelvin with a high degree of accuracy, and strongly disfavors any astrophysical explanation of the solar neutrino problem.

Perhaps neutrinos have an identity crisis on their way to us from the center of the Sun. Neutrinos exist in three varieties, associated with the electron, the muon, and the tau particle. The first of these, the type generated in the Sun's core, is (for now) the only kind of solar neutrinos that our detectors respond to. Some of these electron neutrinos might be undergoing metamorphosis to another type during their 8.3-minute journey from the center of the Sun, thereby escaping detection (Sect. 3.1).

The small sound-speed discrepancies between measurements and theory are significant (Fig. 3.6). Just below the convection zone, there is an increase in the observed sound speed compared to the model, suggesting that turbulent material is mixing in and out within this base layer. Since the

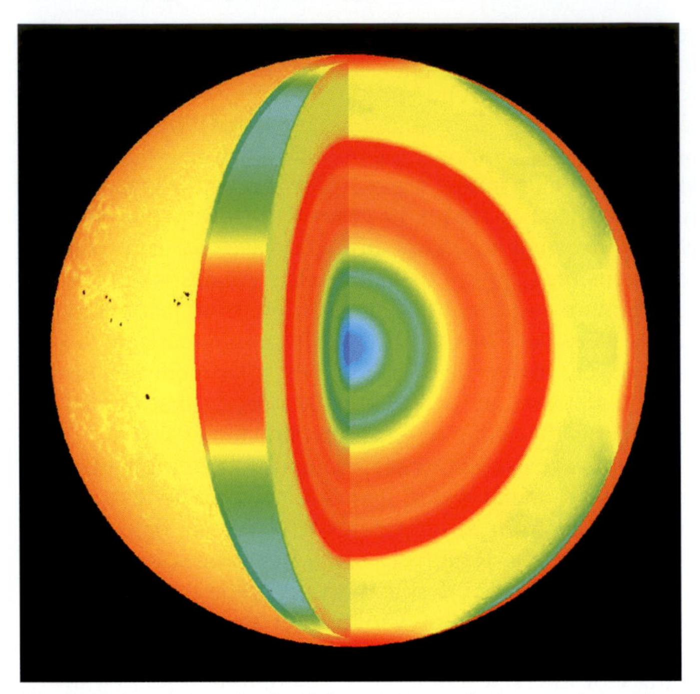

Fig. 3.6. Sound Speeds. Radial variations of the sound speed are shown in this right-hand cutaway image of the solar interior. Red and blue correspond to faster and slower sound speeds, respectively, relative to a standard solar model (*yellow*). When the sound travels faster than predicted by theory, the temperature is higher than expected; slow sound waves imply temperatures that are colder than expected. The conspicuous red layer, about a third of the way down, shows unexpected high temperatures at the boundary between the turbulent outer region (convection zone) and the more stable region inside it (radiative zone). The very core of the Sun may be 0.1 percent cooler than expected, perhaps implying slower thermonuclear reactions. Latitudinal variations in temperature are seen near the photosphere (*center*). These speed of sound measurements were made by two instruments aboard the SOHO spacecraft. The MDI/SOI measures vertical motions, due to sound waves reverberating through the Sun, at a million points on the visible photosphere. VIRGO detects the solar oscillations by rhythmic variations in the Sun's brightness. (Courtesy of Alexander G. Kosovichev, the SOHO MDI/SOI consortium, and the SOHO VIRGO consortium. SOHO is a project of international cooperation between ESA and NASA)

speed of sound depends on temperature as well as composition, the temperature might also increase in this place. There is a sharp decrease of the observed speed relative to theoretical expectations at the boundary of the energy-generating core, hinting that either cooler material or turbulent churning motions might occur there. Scientists speculate that they could be due to unstable nuclear burning processes. If substantiated by further studies, this could be very important for studies of stellar evolution; they usually assume that nuclear reactions proceed without any mixing of fresh material into the core.

SOHO helioseismologists are also using measurements of sound wave frequencies to infer rotational and other motions inside the Sun. The moving material produces slight changes in the frequency of a sound wave that passes through it, as expected from the well-known Doppler effect (Focus 3.6).

3.4 How the Sun Rotates Inside

For at least a century, astronomers have known from watching sunspots that the photosphere rotates faster at the equator than at higher latitudes, and that the speed decreases smoothly toward each pole. Yet, until recently, we had no knowledge of how fast the Sun spins inside. This can now be done using photosphere oscillations driven by internal sound waves.

The Sun's sidereal rotation period, from east to west against the stars, is usually specified as a function of heliographic latitude, the angular distance north or south of the solar equator in a spherical coordinate system. For instance, the sidereal rotation period of the photosphere is about 25 days at the equator where the latitude is zero degrees; the rotation period reaches 34 days at about ±75 degrees latitude. These rotation periods can be converted into velocities – just divide the circumference at the latitude by its period (Table 3.3). Regions near the poles rotate with very slow speeds, in part because the rate of rotation is smaller, but also because the surface near the poles is closer to the Sun's axis and the distance around the Sun is shorter.

The varying rotation of the photosphere is known as differential rotation, because it differs with latitude. In contrast, every point on the surface of the Earth rotates at the same speed, so a day is 24 hours long everywhere on the Earth. Such a uniform spin is called solid-body rotation. Only a gaseous or liquid body can undergo differential rotation; it would tear a solid body into pieces.

Internal rotation imparts a clear signature to sound waves. When a sound wave travels in the direction of the rotating gas, the wave will speed up, like a jet airplane or bird traveling with the wind. A sound wave propa-

Table 3.3. Differential rotation at the photosphere*

Solar Latitude (degrees)	Rotation Period (days)	Rotation Speed (miles per hour)	Rotation Speed (kilometers per hour)
0 (Equator)	25.67	4,410	7,097
15	25.88	4,230	6,807
30	26.64	3,680	5,922
45	28.26	2,830	4,544
60	30.76	1,840	2,961
75	33.40	880	1,416

* Data from the MDI instrument aboard the SOHO spacecraft.

gating against the rotation will slow down. Thus, waves moving around the Sun in the direction of rotation appear slightly speeded up in frequency, while those running in the opposite direction from the rotation appear slightly slowed down in frequency. These opposite effects split the single frequency of a sound wave into a pair of close, but distinct, observed frequencies that beat together like two notes that are not quite at the same pitch.

The size of the expected frequency splitting can be estimated from the visible rotation. The solar oscillations have a period of about five minutes and the photosphere's equatorial rotation period is about 25 days. So the rotational splitting is roughly five minutes divided by 25 days, or about one part in ten thousand. The oscillation frequencies have to be measured ten or a hundred times more accurately to determine subtle variations in the Sun's rotation, or as precisely as one part in a million.

Inversion techniques using the measured frequency splitting have enabled researchers to determine the radial and latitude dependence of the internal rotation (Fig. 3.7). Such observations, from both the Earth and space, have confirmed that the general pattern of differential rotation persists through the convection zone. Furthermore, the rotation speed becomes uniform from pole

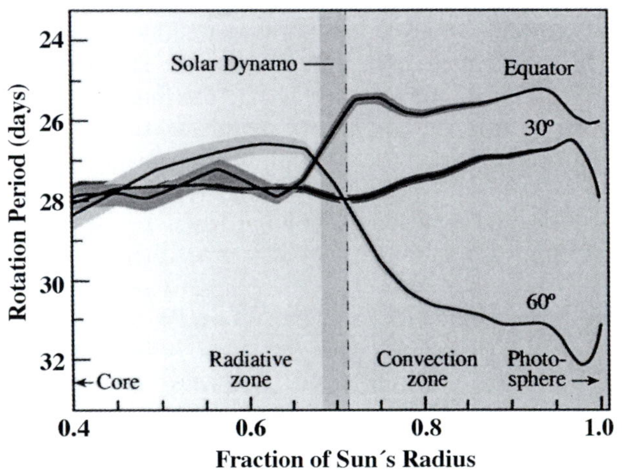

Fig. 3.7. Internal Rotation. The rotation rate inside the Sun at latitudes of zero (solar equator), 30 and 60 degrees has been inferred using data from the Michelson Doppler Imager (MDI) aboard the SOHO spacecraft. Just below the convection zone, the rotational speed changes markedly. Shearing motions along this interface may be the dynamo source of the Sun's magnetism. Unlike the solid Earth, the outer parts of the gaseous Sun rotate far faster at the equator than at the poles; this differential rotation persists to the bottom of the convection zone at 28.7 percent of the way down. Below that, uniform rotation appears to be the norm. Detailed scrutiny indicates that near the equator the spin is actually fastest just below the photosphere, at about 0.95 of the solar radius. We don't know much about the rotation of the central parts of the Sun, within its energy-generating core, since most of the observed sound waves do not reach that far. (Courtesy of Alexander G. Kosovichev and the SOHO SOI/MDI consortium. SOHO is a project of international cooperation between ESA and NASA)

to pole about a third of the way down. Thus, the rotation velocity changes sharply at the base of the convection zone. There the outer parts of the radiative interior, which rotates at one speed, meet the overlying convection zone, which spins faster in its equatorial middle. The difference in speed between these adjacent regions causes a sliding or shearing force on the intervening gas.

A solar dynamo amplifies and regenerates the Sun's magnetic field. The dynamo is located somewhere in the solar interior where the hot circulating gases, which are good conductors of electricity, generate electrical currents and convert the energy of motion into magnetic energy. The electric currents create magnetic fields; these fields in turn sustain the generation of electricity, just as in a power-plant dynamo. The same thing also happens in the fluid interiors of many planets.

Magnetic fields, spawned by the Sun's internal dynamo, are carried along by the electrified material inside the Sun, which deforms, stretches and twists the embedded magnetism. As a result, the magnetic fields thread their way out of the Sun to form bipolar sunspots in the photosphere and to mold the million-degree solar atmosphere into an ever-changing shape. If we could understand how this dynamo works in the Sun, we might explain the winding up and relaxation of solar activity over its 11-year cycle.

The exact position of the solar dynamo has remained something of a mystery for several decades, leading some to doubt the dynamo theory. In the 1970s it was believed that the dynamo was located somewhere in the convection zone, but in the early 1980s scientists realized that the observed magnetic features are inconsistent with any dynamo located within the convection zone, where differential rotation is maintained. Theorists temporarily avoided this problem by placing the dynamo down near to the Sun's center, where no one could observe it and thereby contradict or confirm its hypothetical properties.

Now helioseismologists have discovered an enticing clue to the dynamo problem, suggesting its probable location in a sharp, thin, base layer just below the convection zone. At this boundary layer, there is a strong, depth-related rotational shear, due to a large gradient in rotation velocity. The shallow transition layer features a large (positive) radial gradient in rotation near the equator, and large negative gradients near the poles, within which the dynamo must operate. Moreover, the MDI team found that sound waves speed up more than expected in this shear layer, indicating that turbulence and mixing associated with a dynamo are most likely present.

The Sun's magnetism is probably generated at the interface between the deep interior, that rotates with one speed, and the overlying gas that spins faster in the equatorial middle. Below this boundary layer, the Sun rotates like a solid object, with too little variation in spin to drive a solar dynamo. Closer to the photosphere, the rotation rates at different latitudes diverge over broad areas that are not focused enough to play much of a role in the dynamo.

Ongoing helioseismology may demonstrate how internal motions amplify, compress and slowly transform the Sun's magnetism, leading to an improved understanding of solar activity. Scientists have, for example, recently used helioseismology observations from both the ground (GONG) and space (MDI) to detect latitude-dependent temporal variations in the rotation of the Sun near the base of its convection zone close to the presumed site of the solar dynamo. The distinct 1.3-year-period oscillation in equatorial regions, and a more complicated variation with a dominant 1.0-year period at higher latitudes, may provide major clues to how the dynamo operates.

A more esoteric implication of the rotation results involves tests of Einstein's theory of gravity. According to Einstein's General Theory of Relativity, space is distorted or curved in the neighborhood of matter, and this distortion is the cause of gravity. The result is a gravitational effect that departs slightly from Newton's expression, and the planetary orbits are not exactly elliptical. Instead of returning to its starting point to form a closed ellipse in one orbital period, a planet moves slightly ahead in a winding path that can be described as a rotating ellipse. Einstein first used his theory to describe such a previously unexplained twisting of Mercury's orbit.

Mercury's anomalous orbital shift, of only 43 arcseconds per century, is in almost exact agreement with Einstein's prediction, but this accord depends on the assumption that the Sun is a nearly perfect sphere. If the interior of the Sun is rotating very fast, it will push the equator out further than the poles, so its shape ought to be somewhat oblate rather than perfectly spherical. The gravitational influence of the outward bulge, called a quadrupole moment, will provide an added twist to Mercury's orbital motion, shifting its orbit around the Sun by an additional amount and lessening the agreement with Einstein's theory of gravity.

Fortunately, helioseismology data indicate that most of the inside of the Sun does not rotate significantly faster than the outside, at least down to the energy-generating core. This is not enough to produce a substantial asymmetry in its shape, even if the core of the Sun is spinning rapidly. So, we may safely conclude that measurements of Mercury's orbit confirm the predictions of General Relativity. In fact, the small quadrupole moment inferred from the oscillation data, about one ten millionth rather than exactly zero as Einstein assumed, is consistent with a very small improvement in Mercury's orbit measured in recent times. So, the Sun does have an extremely small, middle-aged bulge after all.

3.5 Internal Flows

By recording sounds that have passed through the Sun at different angles, one can use triangulation to obtain detailed, high-resolution knowledge. Such tomography, as the method is known, can be used to trace out the non-rotational motions in the Sun's outer shells. It is somewhat like using Computed Axial Tomography (CAT) scans to derive clear views of the insides of living bodies from the numerous readings of X-rays criss-crossing the body from different directions.

The MDI instrument on SOHO can continuously obtain clear, sharp images with fine detail. For this reason, it has proved particularly useful in time-distance helioseismology, a new tomographic technique that directly measures travel times and distances for sound waves that probe regions just below the photosphere.

The method is quite straightforward: the telescope records small periodic changes in the wavelength of light emitted from a million points across the Sun every minute. By keeping track of them, it is possible to determine how long it takes for sound waves to skim through the Sun's outer layers. This travel time tells of both the temperature and gas flows along the internal path connecting two points on the visible solar disk. If the local temperature is high, sound waves move more quickly – as they do if they travel with the flow of gas. This data is then inverted in a computer to chart the three-dimensional internal structure and dynamics of the Sun, including the sound speed, flow speed, and direction of motion.

The MDI has provided travel times for sounds crossing thousands of paths, linking myriad points on the photosphere. Researchers have fed one year of nearly continuous observations into a supercomputer, using it to work out temperatures and flows along these intersecting paths. After weeks of number crunching, the SOHO scientists have identified vast new currents coursing through the Sun and clarified the form of previously discovered ones (Fig. 3.8). These are not fast movements such as rotations, but rather slower ones with rotational motions removed from the data. Their speeds only reach tens of meters per second, or thousands of meters per hour.

White-hot rivers of plasma are flowing beneath the visible solar disk, circulating about the Sun's equatorial regions and running between the equator and poles. These incendiary tempests describe a solar meteorology that resembles weather patterns on Earth, but on a much larger scale with hotter temperatures and no rain in sight. Broad bands of higher-velocity gas ring the solar perimeter in its equatorial regions, reminding us of the Earth's trade winds that blow westward almost every day of the year. Currents of hot gas also flow toward both of the Sun's poles from the equator, like the air currents that circulate primarily in the north-south direction in Earth's tropical zone.

Great zonal bands of gas that lie parallel to the Sun's equator sweep along the photosphere at different speeds relative to each other (Fig. 3.8). The faster flows remind us of the trade winds in the Earth's tropics. The Sun's "trade winds" are detected to a considerable depth below the photosphere. The banding is apparently symmetric about the solar equator, with at least two zones of faster rotation and two zones of slower rotation in each hemisphere of the Sun. A single zonal flow band is broad and deep, more than 65 million meters wide and penetrating as far as 20 million meters below the photosphere.

The velocity of the faster zonal flows is about 5 meters per second higher than gases to either side. This is substantially smaller than the mean velocity of rotation, which is about 2,000 meters per second, so the fast zones glide along in the spinning gas, like a wide, lazy river of fire. Helioseismologists speculate that sunspots might originate at the boundaries of zones moving at different speeds, where the shearing force and turbulence might twist the magnetic fields and intensify magnetic activity.

Considerable stress and turbulence is expected at the edges of the faster bands where they rub against the slower moving gas, like two nearby speed

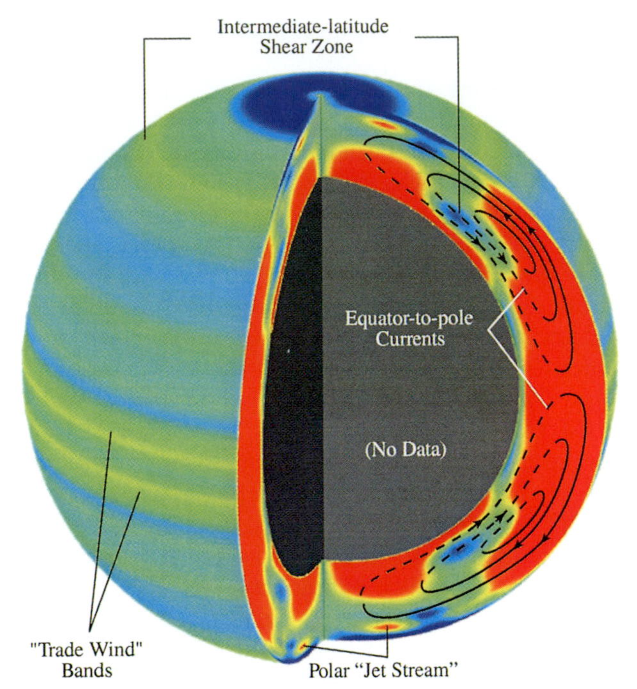

Fig. 3.8. Solar Meteorology. A surprising variety of large-scale flows occur in the Sun's interior. Red corresponds to faster-than-average flows, yellow to slower than average, and blue too slower yet. On the left side, deeply rooted zones (*yellow bands*), analogous to the Earth's trade winds, travel slightly faster than their surroundings (*blue regions*). Sunspots may tend to form at the edges of these zones. The right-hand cutaway reveals speed variations in the outer 30 percent of the Sun's interior; below that the variations are uncertain. Controversial polar "jet streams" (*dark blue ovals*) could move approximately 10 percent faster than their surroundings (*light blue*). There is a slow movement poleward from the equator shown by the streamlines; the return flow below it is inferred. This image is the result of computations using one year of continuous observation, from May 1996 to May 1997, with the Michelson Doppler Imager (MDI) instrument aboard SOHO. (Courtesy of Philip H. Scherrer and the SOHO SOI/MDI consortium. SOHO is a project of international cooperation between ESA and NASA)

boats moving at different velocities and churning the water between them. The Sun's zonal flows might resemble the colorful bands in Jupiter's roiling atmosphere, where cyclonic eddies churn along the edges of zones and belts that move with different speeds in opposite directions relative to the average flow. However, the zonal motions are very evident at Jupiter, while they only become apparent on the Sun after subtracting out the much larger effects of average rotational motions.

The existence of bands of fast and slow rotation was first noticed in 1980, as the result of photosphere observations from the Mount Wilson Observatory. However, only with the new techniques of helioseismology could we realize that the zonal flow bands are not a superficial phenomenon, but instead extend deeply into the Sun. The full extent of this component of the Sun's stormy weather could never have been seen looking at the visible layer of the solar atmosphere.

Gas in the Sun's photosphere has also been observed to slowly flow away from the equator toward both poles, carrying the magnetic remnants of former sunspots and decayed active-region magnetic fields to high latitudes. This could be responsible for the formation of some of the Sun's polar magnetic fields. Helioseismologists have demonstrated that this so-called meridional surface flow penetrates deeply. The entire outer layer of the Sun, to a depth of at least 25 million meters, is slowly but steadily flowing from the equator to the poles with a speed of about 20 meters per second. At this rate, an object would be transported from the equator to a pole in a little more than one year. Of course, the Sun rotates at more than 100 times this rate, completing one revolution at the equator in about 25 days.

Thus, as the material in the outer layers of the Sun rotates rapidly, it also flows slowly from equator to poles in about a year. At the same time, while currents near the surface are heading poleward, the zonal bands and sunspots are moving in the opposite direction toward the equator, completing the trip from mid-latitudes to low ones in about 11 years. The flows from the equator to both poles could descend inside at high latitudes, returning to the equator at greater depths. An equatorward return flow with a speed of about 1 meter per second at the bottom of the convection zone could account for the 11-year progression of sunspot activity from high to low latitudes, known as the butterfly diagram (Fig. 2.3). Researchers suspect that such a return flow exists, but they have not yet observed detailed motions down there. Perhaps the Sun's deep magnetic dynamo carries or pushes the broad rivers and dark sunspots against the poleward flow, like a slow barge moving upstream.

The newly discovered internal motions may also help explain the ancient mystery of why sunspots exist. The MDI team compared horizontal motions just beneath the visible solar disk, at a depth of only 1.4 million meters, to a magnetic image of the overlying photosphere (Fig. 3.9). They found that strong magnetic concentrations tend to lie in regions where the subsurface gas flow converges. Thus, the churning gas probably pushes and forces the magnetic fields together and concentrates them into sunspots near the photosphere, thereby overcoming the outward magnetic pressure that ought to make such localized concentrations expand and disperse.

Future high-resolution helioseismology should help show exactly what sunspots are and how solar activity originates. Sunspots absorb as much as half the power of sound waves that propagate through them, and the Sun's oscillation frequencies vary slightly with the 11-year cycle of sunspot activity. Detailed observations over several years, during various parts of the activity cycle, should tell us more about how the churning motions of hot solar gas interact with concentrations of the Sun's magnetism and give rise to its explosive activity. These incredible explosions threaten astronauts and satellites, and can affect Earth with power and communications disruptions (Sect. 7.2). They can also move down into the Sun, producing seismic waves that propagate deep within it.

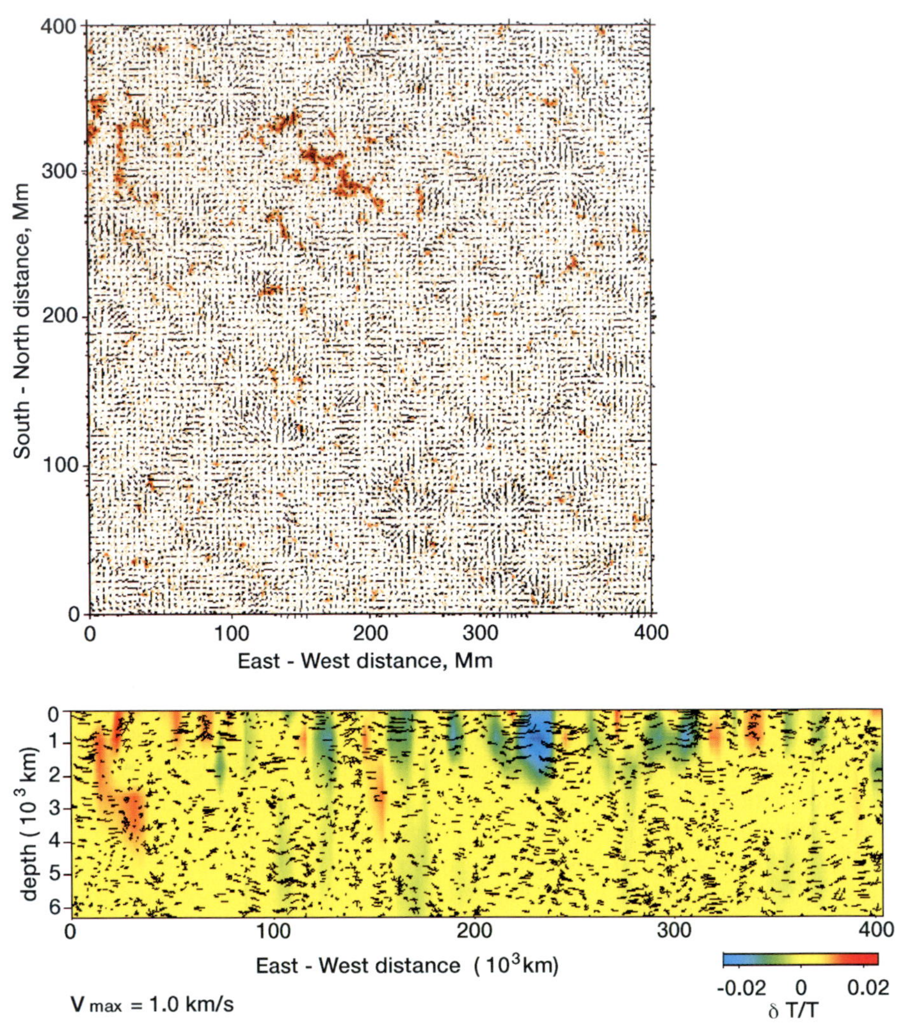

Fig. 3.9. Material Flows. The flow and temperature of material below the photosphere, or visible solar disk. The vertical cut (*bottom*) shows flow and temperature changes in the outer 1 percent, or the top 7 million meters, of the Sun. Color shading indicates changes from cool temperatures (*blue*) to hot ones (*red*). Long cool tongues of material spiral down to greater depths, like water from an unplugged bath. The horizontal cut (*top*) is at a depth of 1.4 million meters; it is compared with photosphere magnetic fields (*dark concentrations*). Dark magnetic spots are confined by converging flows below the photosphere. In both cases, the arrows indicate the direction and relative speeds of the material, which reach a maximum velocity of one thousand meters per second. (Courtesy of Thomas L. Duvall, Jr., Alexander G. Kosovichev and Philip H. Scherrer, and the SOHO SOI/MDI consortium. SOHO is a project of international cooperation between ESA and NASA)

3.6 Sunquakes and Active Regions

The relatively calm solar atmosphere can be torn asunder by brief, catastrophic outbursts of awesome power and violence, releasing stored magnetic energy equivalent to billions of nuclear bombs on Earth. These awesome explosions on the Sun, called solar flares, can occur just above the tops of coronal magnetic loops connected to the underlying photosphere. High-energy electrons, accelerated near the loop tops, are rapidly channeled down the loops arching magnetism into the lower, denser reaches of the solar atmosphere. Here the electron beams are slowed by collisions, emitting hard X-rays and gamma rays.

The solar flare, which explodes above the photosphere, can also generate localized seismic disturbances in the photosphere and major seismic waves in the Sun's interior. Scientists have speculated that powerful shocks are produced when a flare-associated beam of electrons slams down into the dense solar atmosphere. Now SOHO's Michelson Doppler Imager, or MDI, has detected a powerful sunquake caused by an exploding solar flare higher up, and observed circular flare-generated seismic waves moving across the photosphere like ripples spreading out in all directions from a pebble thrown into a pond.

The explosion's particle beams and shocks came down at the Sun with such force that they caused waves 3 thousand meters high on the Sun's surface. These waves moved out to a distance of at least 120 million meters from the flare epicenter at an average speed of approximately 50 thousand meters per second (Fig. 3.10). Unlike water ripples, that travel outward at constant velocity, the solar waves accelerated from about 30 thousand meters per second to approximately 100 thousand meters per second as the waves moved from 20 to 120 million meters from the epicenter.

Seismic waves produced by sunquakes can shake the Sun to its very center, just as earthquakes can cause our entire planet to vibrate and ring like a bell. However, unlike their terrestrial counterparts, sunquakes involve unheard-of amounts of energy. The observed sunquake, which was produced by a perfectly ordinary solar flare, was equivalent to an earthquake of magnitude 11.3 on the Richter scale. That is 40,000 times more energy than the devastating San Francisco earthquake in 1906, at a magnitude of 8 on the same scale. Scientists remain somewhat perplexed about how such a relatively modest flare could have the downward thrust and power to generate such powerful seismic waves.

Helioseismologists have discovered how to look through the Sun and observe regions of solar activity on the hidden back side of the Sun. The technique, dubbed helioseismic holography, examines a wide ring of sound waves that emanate from a small region on the side of the Sun facing away from the Earth (the far side) and reach the near side that faces the Earth. Very strong magnetic fields on the far side speed up the sound waves, making them reflect within the Sun more quickly. When a large, strong active region is present on the back side of the Sun, the round-trip travel time from the near to far sides and back again is about 6 seconds shorter than the average 6 hours.

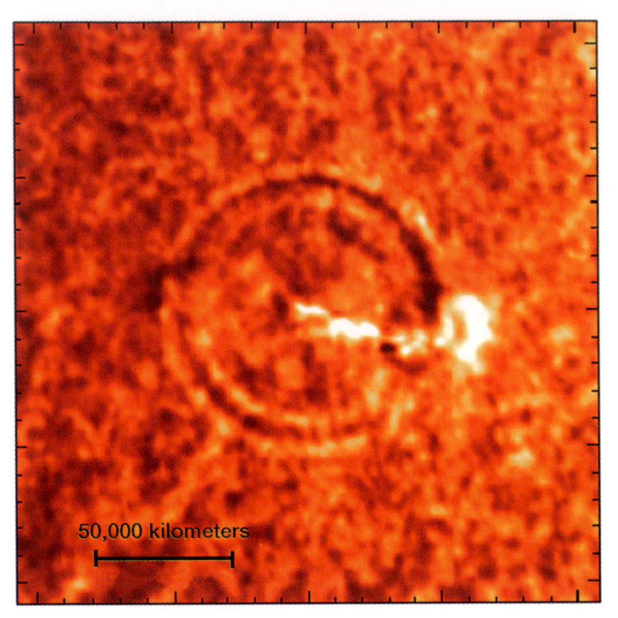

Fig. 3.10. Sunquakes. A powerful explosion, or flare, in the solar atmosphere sent shock waves into the underlying gases, causing massive sunquakes that resemble seismic earthquakes on our planet. The explosion, on 9 July 1996, produced concentric rings of a sunquake that spread away from the flare, somewhat like ripples from a rock dropped into a pool. Over the course of an hour, the solar waves traveled for a distance equal to 10 Earth diameters. This solar flare left the Sun quaking with about 40,000 times the energy released in the great earthquake that devastated San Francisco in 1906. (Courtesy of Alexander G. Kosovichev and Valentina V. Zharkova, and the SOHO SOI/MDI consortium. SOHO is project of international cooperation between ESA and NASA)

This acoustic lens can make the Sun transparent, enabling scientists to monitor the structure and evolution of large regions of magnetic activity as they cross the far side of the Sun. If this technique can reveal big active regions that are growing in magnetic strength before they rotate around to the side of the Sun that faces the Earth, it could give more than a week of advance warning of threatening active regions. Energetic phenomena, such as solar flares and coronal mass ejections, can be explosively released from these regions, emitting electromagnetic and particle radiation that interferes with telecommunications and power transmission on the Earth and can pose significant hazards to astronauts, satellites and space stations near Earth.

Table 3.4. Key events in the development of helioseismology*

Date	Event
1960–62	Discovery of the five minute vertical velocity oscillations in the solar photosphere by Robert B. Leighton.
1968	Edward N. Frazier demonstrates that the oscillation power exists over a broad range of frequencies (or periods) and horizontal wave numbers (or sizes).
1970–72	The resonant-cavity, spherical shell model for internal sound waves is further developed and used to make predictions for the solar five minute oscillations – by Roger K. Ulrich in 1970, John Leibacher and Robert F. Stein in 1971, and Charles L. Wolff in 1972.
1975	Franz-Ludwig Deubner confirms the acoustic cavity hypothesis, showing that the oscillating power of the observed five-minute oscillations at low degrees is concentrated into narrow ridges in a spatial-temporal display.
1976	Douglas O. Gough points out the diagnostic capabilities of helioseismology for the internal constitution of the Sun.
1975–78	Hiroyasu Ando and Yoji Osaki develop theoretical models of sound waves trapped in the Sun. H. M. Antia, S. M. Chitre, M. H. Gokhale and D. M. Kale advance the idea that the five-minute oscillations are excited thermodynamically by slow regular growth. Peter Goldreich and Douglas A. Keeley suggest that the oscillations are excited mechanically by the turbulent motions in the outer convection zone.
1979	Franz-Ludwig Deubner, Roger K. Ulrich and Edward J. Rhodes, Jr. publish the first study of internal differential rotation using solar oscillations.
1979	First observational studies of the global, low degree modes ($l \approx 1, 2, 3$) of the five-minute oscillations that penetrate deeply in the Sun by Andre Claverie, G. R. Isaak, C. P. McLeod, H. B. van der Raay and T. Roca-Cortés.
1980–83	High quality, low degree ($l \approx 0, 1, 2, 3$) spectra of solar oscillations obtained by Gérard Grec, Eric Fossat and Martin A. Pomerantz from continuous observations (120 hours) at the South Pole.
1982	Thomas L. Duvall, Jr. demonstrates a dispersion law for solar oscillations.
1983	Solar oscillations observed in the Sun's total irradiance of the Earth by Martin F. Woodard and Hugh S. Hudson using the Active Cavity Radiometer Irradiance Monitor (ACRIM) aboard the Solar Maximum Mission (SMM) satellite.
1983	Thomas L. Duvall, Jr. and John W. Harvey obtain the first observations connecting solar oscillations of low and high degree ($l = 1$ to 139).
1984	Thomas L. Duvall, Jr., Wojciech A. Dziembowski, Philip R. Goode, Douglas O. Gough, John W. Harvey and John W. Leibacher use solar oscillations to determine the Sun's internal rotation near its equatorial plane. They show that the rotation rate through the solar convection zone is very close to that observed at the photosphere. They infer a slow rotation for much of the solar interior, with very little solar oblateness (low quadrupole moment), increasing the accuracy of the confirmation of Einstein's General Theory of Relativity using Mercury's orbital motion.
1985	First determination of the speed of sound in the solar interior from inversions of the frequencies of the Sun's five minute oscillations by Jørgen Christensen-Dalsgaard Thomas L. Duvall, Jr., Douglas O. Gough, John W. Harvey and Edward J. Rhodes, Jr..
1986–89	Solar oscillation data is used by Thomas L. Duvall, Jr., John W. Harvey and Martin A. Pomerantz to show that the entire convection zone mimics the observed surface differential rotation, with slower rotation at higher latitudes. This was confirmed in greater detail by Timothy M. Brown, Jørgen Christensen-Dalsgaard, Wojciech A. Dziembowski, Philip R. Goode, Douglas O. Gough and Cherilynn A. Morrow.

1987–88 Discovery of the absorption of sound waves in and around sunspots by Douglas C. Braun, Thomas L. Duvall, Jr. and Barry J. LaBonte.

1989–90 Discovery that the frequencies of sound waves change during the 11-year solar activity cycle for both low degree oscillations and intermediate or high degree oscillations. The relevant observations and data analysis are reported by P. L. Pallé, C. Régulo and T. Roca-Cortés in 1989; Y. P. Elsworth, R. Howe, G. R. Isaak, C. P. McLeod and R. New in 1990; Kenneth G. Libbrecht in 1989; and Kenneth G. Libbrecht and Martin F. Woodard in 1990.

1990 Peter Goldreich and Pawan Kumar use theoretical arguments and oscillation data obtained by Kenneth Libbrecht to demonstrate that sound waves are excited by turbulent convection in the upper part of the solar convection zone.

1991 Accurate determination of the depth of the solar convection zone at 0.287 ± 0.003 solar radii by Jørgen Christensen-Dalsgaard, Douglas O. Gough and Michael J. Thompson from the observed frequencies of solar oscillations.

1993 Time-distance helioseismology applied by Thomas L. Duvall, Jr., Stuart Mark Jefferies, John W. Harvey and Martin A. Pomerantz.

1995–97 Observations of solar oscillations from the ground and space indicate that the deep solar interior, from the base of the convection zone down to about 0.2 solar radii, rotates like a rigid body with a uniform latitude-independent rate that is somewhat slower than the surface equatorial rate. The relevant observations and data analysis are reported by Y. P. Elsworth and colleagues in 1995, Steven Tomczyk, Jesper Schou and Michael J. Thompson in 1995; Michael J. Thompson and colleagues in 1996; and Alexander G. Kosovichev and colleagues in 1997.

1996–97 Helioseismological measurements of low and intermediate frequencies from the ground and space are inverted to obtain sound speeds that are in agreement with predictions of numerical (standard) solar models to within 0.2% throughout almost the entire Sun. This agreement indicates that the solar neutrino problem cannot be resolved by any nonstandard solar model or related astrophysical solutions. The relevant observations and data analysis are reported by Sarbani Basu and colleagues in 1996; Alexander G. Kosovichev and colleagues in 1997; John N. Bahcall and colleagues in 1997; and D. B. Guenther and Pierre Demarque in 1997.

1997 Alexander G. Kosovichev and Jesper Schou use the Michelson Doppler Imager on SOHO to demonstrate that zonal shear flows extend deep below the solar surface.

1997 Peter M. Giles, Thomas L. Duvall, Jr., Philip H. Scherrer and Richard S. Bogart use the Michelson Doppler Imager on SOHO to detect a subsurface flow of material from the Sun's equator to its poles.

1998 Alexander G. Kosovichev and Valentina V. Zharkova use the Michelson Doppler Imager aboard SOHO to detect seismic waves generated when a flare impacts the lower solar atmosphere.

2000 Charles Lindsey and Douglas C. Braun use their technique of helioseismic holography with data from the Michelson Doppler Imager on SOHO to look through the Sun and make images of an active region on the back side of the Sun. This could give more than a week of advance warning of strong active regions that will rotate around to the side of the Sun that faces the Earth.

2000 Rachel Howe and her colleagues use data from the Global Oscillation Network Group (GONG) and the Michelson Doppler Imager (MDI) on SOHO to detect temporal variations in the rotation of the Sun near the base of the convection zone and close to the presumed site of the solar dynamo. The observed oscillations include a prominent variation of 1.3 years in the equatorial regions and a probable 1-year one at higher latitudes.

---

* See the References for complete references to these seminal papers.

4. Solving the Sun's Heating Problem

Overview

Instead of growing colder at higher regions of its atmosphere, the Sun becomes unexpectedly hotter, soaring to a temperature of millions of degrees Kelvin in the low corona just above the photosphere at 5,780 degrees Kelvin. Heat should not emanate from a cold object to a hotter one anymore than water should flow up hill. The first plausible explanations for this heating paradox were in terms of either sound waves or magnetic waves, generated by motions in the turbulent convective zone. Spacecraft measurements in 1978, however, showed that although the lower chromosphere is probably heated by the dissipation of sound waves, there is not enough energy left over to heat the overlying corona by any substantial amount. Magnetic waves were also detected long ago, moving in the solar wind, but they seemed to travel through the corona without depositing significant amounts of energy into it.

Coronal heating is usually greatest where the magnetic fields are strongest. The Skylab, Yohkoh and SOHO spacecraft have demonstrated that the hottest, densest material in the low corona, with the most intense X-ray and extreme ultraviolet emission, is concentrated within thin, long and strongly magnetized loops located in solar active regions. Studies of loop physical parameters suggest that they are energized by stressed magnetism.

Most of the material we can see in the transition region, between the chromosphere and corona, is falling down into the Sun instead of moving out into the overlying corona, further complicating the heating problem.

Continued dynamic activity and forced magnetic connections are nevertheless ubiquitous features throughout the low solar corona. Magnetic concentrations merge together and cancel all the time and all over the Sun, providing a plausible explanation for heating the low corona outside active regions. Evidence for coronal heating by ongoing magnetic reconnection is obtained from rapid jets, X-ray bright points, ultraviolet blinkers, numerous low-level flares, the continual replenishment of the photosphere's small-scale magnetism every 40 hours, and the uniform heating of the diffuse corona.

4.1 Mysterious Heat

Scientists have been searching for the elusive heating mechanism of the corona ever since its million-degree temperature was measured more than half a century ago (Sect. 2.1). It is physically impossible to transfer thermal energy by conduction from the cooler underlying photosphere to the much hotter

In the case of zero resistivity or "infinite conductivity" we have a perfectly conducting medium and the magnetic field satisfies the relation:

$$\frac{\partial \boldsymbol{B}}{\partial t} + \nabla \times (\boldsymbol{v} \times \boldsymbol{B}) = 0 \, .$$

This equation expresses a condition in which the magnetic field is tied to, or frozen within, the plasma and moves with it.

For most of the Universe, the second term in our first equation is very much smaller than the first term, so the second equation is a good approximation. An important exception is in singularities called current sheets, where the magnetic gradient and electric current are extremely large. In such current sheets, the magnetic field lines can break and reconnect by slipping through the plasma and, in the process, magnetic energy is converted to heat, kinetic energy and fast-particle energy. This process, called magnetic reconnection, is important in heating the solar corona (this section) and in energizing solar flares (Sect. 6.6).

The other main equation of magnetohydrodynamics is the equation of motion:

$$\rho \frac{\mathrm{d}\boldsymbol{v}}{\mathrm{d}t} = -\nabla P + \boldsymbol{j} \times \boldsymbol{B} \, ,$$

where $\boldsymbol{j} \times \boldsymbol{B}$ is the force that the magnetic field exerts on a plasma of density, ρ, the pressure is denoted by P and the velocity by $\boldsymbol{v}$, and the electric current is given by Ampère's law:

$$\boldsymbol{j} = \frac{1}{\mu_0} \nabla \times \boldsymbol{B} \, .$$

The two main equations can be combined to describe perturbations in density that act as waves, with a velocity, v, given by

$$v = (s^2 + v_{\mathrm{A}}^2)^{1/2} \, ,$$

where s is the velocity of sound (Focus 3.5), and the Alfvén velocity, v_{A}, is given by

$$v_{\mathrm{A}} = B/(\mu_0 \rho)^{1/2} \, .$$

The waves represent alternating compression and rarefaction of the gas and field. They are called fast magnetoacoustic waves since they are faster than both the sound and Alfvén waves. When the velocity of sound, s, is much smaller than the Alfvén velocity, v_{A}, or when $s \ll v_{\mathrm{A}}$, we have a compressional Alfvén wave. These waves may propagate in a direction perpendicular to the magnetic field with gas particles oscillating in the direction of propagation.

A more general relation describes both fast, slow and Alfvén magnetohydrodynamic waves. When the direction of wave propagation is parallel to the magnetic field, one can have the slow wave and the Alfvén wave moving at the Alfvén velocity, where the particles oscillate in transverse motion to both the magnetic field and the direction of propagation.

Such waves are now called Alfvén waves after Hannes Alfvén who first described them mathematically and argued in 1947 that they might heat the corona. Alfvén pioneered a new field of study with the ponderous name of magnetohydrodynamics (Focus 4.1), and was awarded the Nobel Prize in Physics in 1970 for his discoveries in it.

Alfvén waves have been detected propagating away from the Sun in the outer solar corona. They were monitored by plasma and magnetic field instruments aboard Mariner 5 on its way to Venus in 1967. Alfvén waves have been directly observed within the ecliptic by several spacecraft since then. The Ulysses spacecraft has also measured Alfvén waves above the Sun's polar regions, propagating in the open-field regions of coronal holes at distances greater than the Earth's distance from the Sun (Sect. 5.4).

SOHO may have provided evidence for magnetic waves much closer to the Sun. Spectral features formed in the transition region, between the chromosphere and corona, are wider than would be expected from that due to a hot gas alone. Although the combined Doppler effects of all the randomly moving ions produce a broadening, that increases with temperature, there is an excess of roughly 10 thousand meters per second that cannot be explained by such thermal motions. The anomalous, non-thermal motions have also been found in ultraviolet lines detected by Skylab and OSO 8, as well as in X-ray lines observed from the Solar Maximum Mission satellite. They might be signatures of magnetic waves moving up from under the transition region to heat the corona, but we cannot be sure.

Once you get energy into an Alfvén wave it is difficult to get it out. So there may be a problem in depositing enough magnetic wave energy into the coronal gas to heat it up. Like radiation, the Alfvén waves seem to propagate right through the inner corona to remote distances from the Sun, without being noticeably absorbed or dissipated. This ability to move through the solar atmosphere without losing energy is just what allows Alfvén waves to penetrate the chromosphere and transition region to reach the corona in the first place. Yet, the dissipation of magnetic waves generated below the corona is still a viable candidate for coronal heating in some instances.

The most obvious place to search for the elusive heating mechanism is in solar active regions, the places in, around and above sunspots. They contain the greatest density of hot coronal material, shine brightly in X-rays and extreme ultraviolet radiation, and contain intense magnetic fields.

4.2 Pumping up the Corona in Active Regions

Some invisible agent is spiriting energy out into the corona, pumping it up and keeping it distended. Magnetism most likely plays a pivotal role. The corona glows in soft X-rays emitted by gases with temperatures in excess of a million degrees, and this radiation is brightest where the magnetic fields are strongest (Fig. 4.1). That is, coronal heating is generally greatest where the magnetism is the most intense.

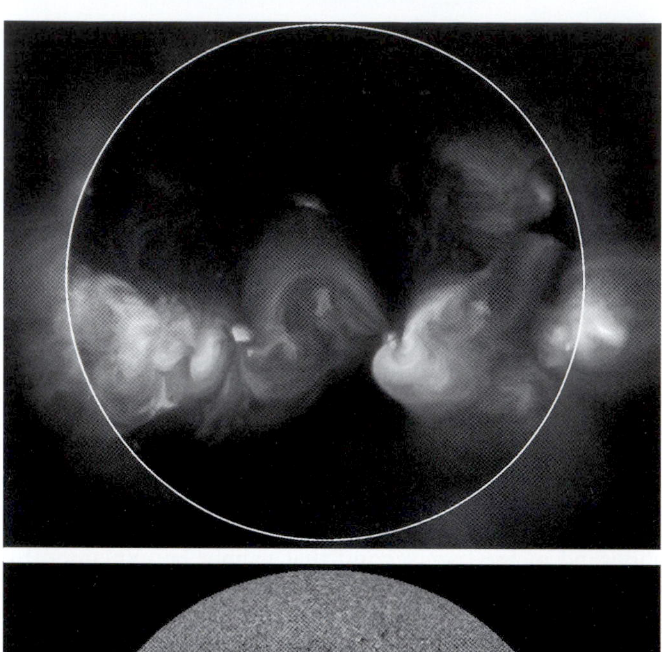

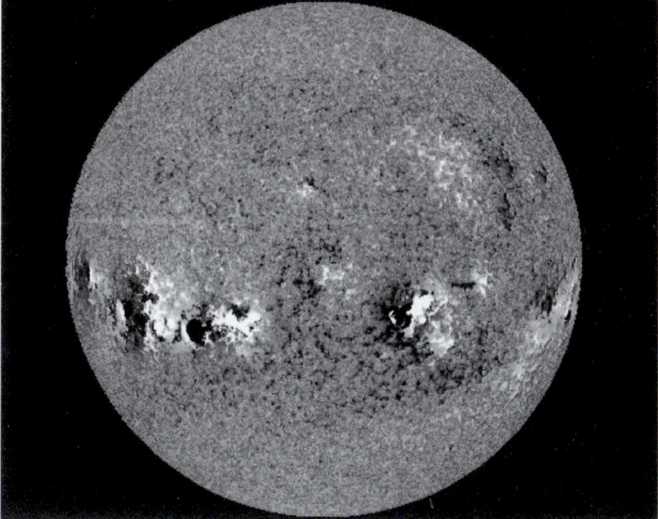

Fig. 4.1. Bright X-rays, Intense Magnetism. A coronal X-ray image (*top*) from the Yohkoh Soft X-ray Telescope on 26 December 1991, scaled to show the brightest parts, is compared with a magnetogram (*bottom*) from the National Solar Observatory, Kitt Peak, taken on the same day. In the magnetogram, black and white denote different magnetic directions or polarity, and the intensity is a measure of the magnetic field strength. Strong regions of opposite magnetic polarity are joined by magnetic loops that constrain the hot, dense gas that shines brightly in X-rays. This comparison shows that the structure and brightness of the X-ray corona are dictated by their magnetic roots in the photosphere. (Courtesy of David A. Falconer and Ronald L. Moore)

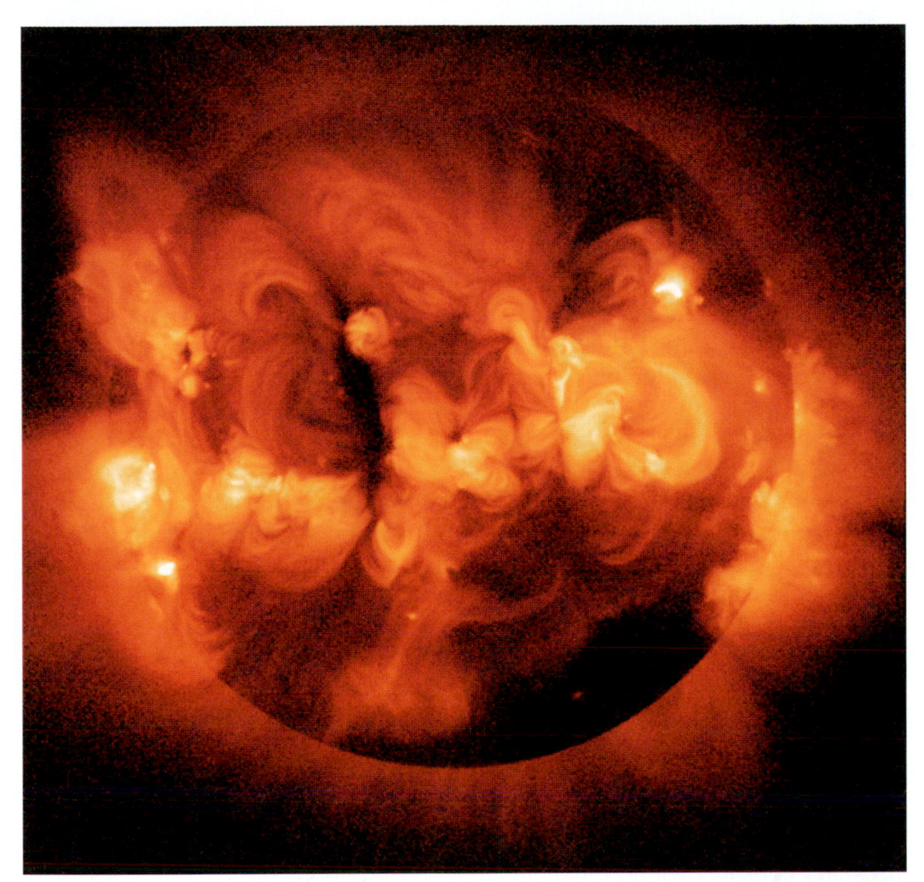

Fig. 4.2. Ubiquitous X-ray Loops. X-ray images of the Sun offer a new view of the nearby star that differs markedly from the view we obtain with our eyes that are tuned to visible light. This image of the Sun's corona was recorded by the Soft X-ray Telescope (SXT) aboard the Japanese Yohkoh satellite on 1 February 1992, near the maximum of the 11-year cycle of solar magnetic activity. The bright glow of X-rays is produced by ionized gases at a temperature of about a million degrees Kelvin. The intricate structures which can be seen in the image are defined by the magnetic fields which thread the corona and hold the hot gases in place. These ubiquitous magnetic loops seem to stitch together the Sun's hot atmosphere. The brightest features are called active regions and correspond to the sites of the most intense magnetic field strength. Subsequent SXT images, taken about five years later near activity minimum, show a remarkable dimming of the corona when the active regions associated with sunspots have almost disappeared, and the Sun's magnetic field has changed from a complex structure to a simpler configuration – see Sect. 7.3 and Fig. 7.12. (Courtesy of Gregory L. Slater, Gary A. Linford, and Lawrence Shing, NASA, ISAS, the Lockheed-Martin Solar and Astrophysics Laboratory, the National Astronomical Observatory of Japan, and the University of Tokyo)

Magnetic fields mold and shape the low corona into highly structured, inhomogeneous configurations. Rocket flights in the early 1970s, followed by the highly successful Skylab mission in 1973–74, sharpened our understanding of this morphology, showing that incandescent X-ray loops provide the woven fabric of the low corona (Sect. 2.5). The ubiquitous coronal loops outline the closed magnetic structures of the corona, seemingly holding it in place.

The satellites launched to study the Sun in the 1980s were primarily intended to study the solar flares and coronal mass ejections, respectively discussed in Sects. 6.3 and 6.5. They were followed by the Yohkoh satellite, launched on 30 August 1991, which also studied solar flares but additionally rediscovered the multitudinous coronal loops and examined them in greater detail (Fig. 4.2).

X-ray images taken by the Yohkoh spacecraft show dramatic changes in the corona from the maximum phase of the 11-year sunspot cycle, when active regions dominate the radiation, to the minimum in the activity cycle when the active magnetic regions associated with sunspots have almost disappeared (Sect. 7.3, Fig. 7.12).

When present, active regions dominate the X-ray corona. Their bright loops rise to heights of more than 100 million (10^8) meters above the photosphere, in long, thin tubular arches rooted in sunspots or other regions of opposite magnetic polarity. The X-ray radiation is strongest near the loop tops, where the magnetic field switches direction; the weakest X-ray emission is located at the loop "footpoints" just above the sunspots. The temperature reaches 3 to 8 million degrees near the apex of a loop, where the electron density is up to 100 million billion (10^{17}) electrons per cubic meter.

Why is the hot, dense coronal gas concentrated to its highest temperatures and densities within the elongated, arching loops of active regions? The electrified gas is constrained and trapped by the powerful magnetic fields. In other words, the magnetic pressure of the intense magnetism dominates the gas pressure, so the ionized gases can only move in their magnetic cage (Focus 2.4). The bright coronal loops therefore act like magnetic conduits, or pipes, through which the hot, dense gas flows but cannot escape, and the intense X-ray emission delineates this non-uniform magnetic structure.

How are magnetic loops in active regions related to the heating of material in them? Magnetic energy might provide the searing heat. Already in 1960, Thomas Gold and Fred Hoyle argued that explosive solar flares are powered by stressed magnetic loops that interact, dissipating their energy in the corona (Sect. 6.6). Then, in 1964, Gold suggested that the twisted magnetic field configurations might heat the corona. During the ensuing decades, Eugene N. Parker developed related heating models involving the winding and wrapping of coronal loops caused by the random, continuous motion of their footpoints in the photosphere. In the 1990s, David A. Falconer, Ronald L. Moore and their colleagues used comparisons of active-region coronal loops, observed with Yohkoh's soft X-ray telescope, and photosphere magnetic fields to demonstrate that enhanced loop heating is indeed aided by strong magnetic shear near the loop feet. These scientists used a vector magnetograph that measures both the longitudinal, or line of sight, and transverse, or horizontal, components of the magnetic field in the photosphere, permitting the determination of its size and direction.

Contorted and sheared coronal loops are commonly found in high-resolution Yohkoh soft X-ray images. Their twisted patterns may be imprinted by motions down below, and the complex, oddly-shaped magnetic geometry may be a signature of current flow up above. The free electrons in the coronal gas can move along the magnetic fields lines in thin sheets where oppositely directed magnetism is pressed together, creating very large electric currents that can generate their own magnetism and alter the overall magnetic shape. These electrical currents might heat the resistive gas in much the same way that currents heat the filaments of a light bulb.

Many coronal loops are steady or unchanging with respect to the relevant physical time scales, such as the radiative and cooling times of roughly ten minutes, and this permits one to measure their physical parameters and test various models for heating them (Focus 4.2). Relationships between parameters such as temperature, pressure and length were originally derived from Skylab X-ray photographs, and confirmed using Yohkoh soft X-ray data (Fig. 4.3). Coronal heating models based upon stressed coronal magnetic fields in active regions and a uniform deposition of heat in the diffuse corona are consistent with these results, but heating by sound waves is not.

Although the X-ray emitting active regions are the brightest things around, they provide only a localized explanation for coronal heating. The hot corona is present throughout the 11-year cycle of solar activity, both when the num-

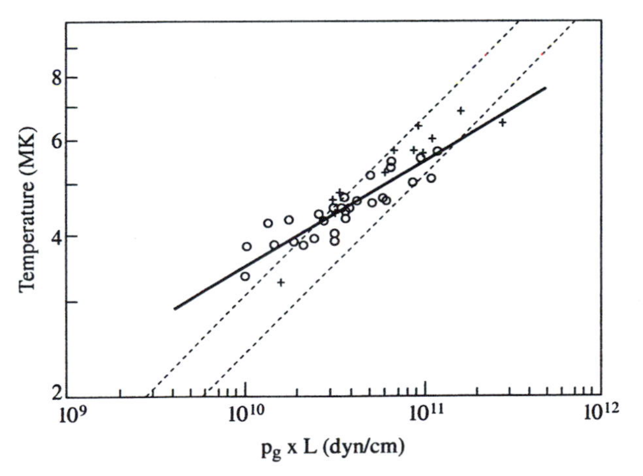

Fig. 4.3. Temperature Scaling Law. Correlation between the maximum temperature, $T_{\max}$, and the product of the peak gas pressure, p, and the loop length, L, for coronal loops observed with the Soft X-ray Telescope aboard Yohkoh. Such a scaling law relationship can be used to probe coronal heating mechanisms. The crosses and circles are measured data, the solid line denotes the best fit to the observations, with $T_{\max} = 3.8 \times 10^4 (pL)^{0.20}$, and the two dashed lines are theoretical scaling laws. Here the temperature is given in millions of degrees Kelvin, or MK. (Adapted from R. Kano and S. Tsuneta, 1996)

bers of active regions and sunspots are greatest and when those numbers are least. The temperature of the corona remains at a million degrees or more, even during an activity lull when there are no active regions and the X-ray radiation has diminished to nearly undetectable levels. Something else must

Focus 4.2

Testing Models for Heating Coronal Loops

Soft X-ray and extreme ultraviolet images of the Sun show that bright, arching loops are the dominant structural element of the low solar corona. By examining the statistical or individual properties of these coronal loops, it is possible to place important constraints on theories of how these structures are heated. Soft X-ray images can be used, for example, to compare the maximum temperatures, T_{max}, pressures, p, and full lengths, L, of many different coronal loops or to determine the temperature distribution along specific loops.

More than two decades ago, Robert Rosner, Wallace H. Tucker and Guiseppe Vaiana developed a simple model for steady coronal loops and derived a relation between their physical parameters inferred from Skylab X-ray photographs. By assuming a heat source located at the top of the loops, where the temperature is greatest, they derived a scaling law:

$$T_{max} = 1.4 \times 10^4 (pL)^{0.33} \text{ degrees Kelvin },$$

where the constant pressure is in dyn cm^{-2} and the length, L, of the loop is in centimeters. Similar relationships were derived at about the same time, also from Skylab data, by I. J. D. Craig, A. N. Mc Clymont and J. H. Underwood, and also by Eric R. Priest. Coronal heating models invoking magnetic fields are consistent with the scaling relation, while heating by sound waves is not.

A scaling law has been subsequently derived using Yohkoh Soft X-ray Telescope, or SXT, observations of coronal loops (Fig. 4.3). The maximum temperature of the loops similarly increases with the product of gas pressure and size, with a slightly different power given by Ryouhei Kano and Saku Tsuneta in 1995:

$$T_{max} = 3.8 \times 10^4 (pL)^{0.20} \text{ degrees Kelvin }.$$

The Yohkoh SXT data were also examined by James A. Klimchuk and Lisa J. Porter in 1995, showing that the heating rate scales inversely with the square of the loop length. A heating model of energy dissipation by stressed coronal magnetic fields is the most consistent with their observational result.

In 1998, Eric R. Priest and his colleagues used SXT observations of a single large loop in the extensive, diffuse corona to determine the temperature profile along the loop. Models that provide a uniform deposition of heat produce the best fit with their observations. This comparison favors turbulent breaking and reconnection of magnetic field lines in many small current sheets as the heating mechanism for the diffuse corona.

therefore be pumping energy into the corona and contributing to the heat. It can be found by ignoring the obvious and focusing on the subtler details, located far away from regions of concentrated activity and deep down near the base of the corona where its energy must come from. Such observations show that the corona outside active regions is not an isolated, inactive entity. It is highly dynamic with varying brightness and continual magnetic changes.

4.3 Tuning in and Dropping out

What you see in the solar atmosphere depends on how you look at it! Scientists focus on its diverse regions, with varying physical conditions, by isolating radiation at different wavelengths and forming images there. Each

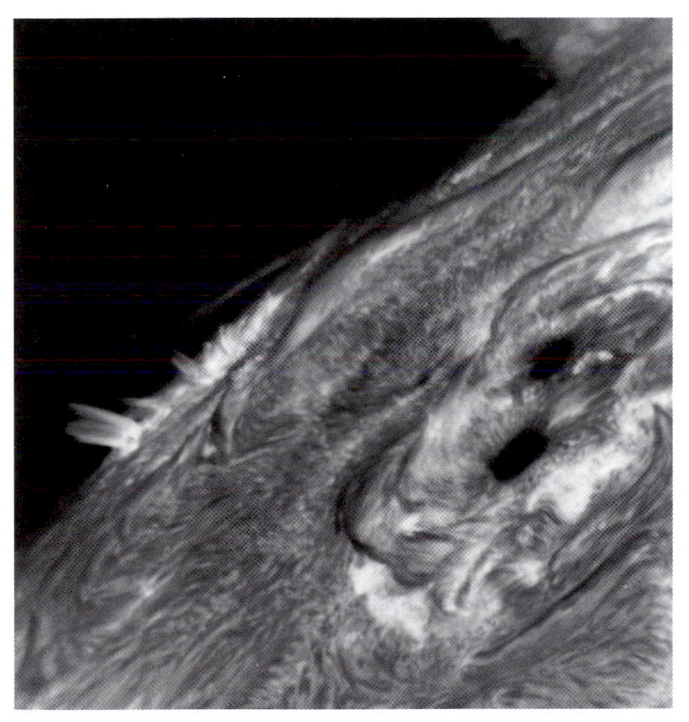

Fig. 4.4. The Red-Faced Sun. At optical wavelengths, solar activity is best viewed by tuning to the red line of atomic hydrogen – the hydrogen alpha line at 656.3 nanometers. Light at this wavelength originates in the chromospheric layers of the Sun, which lie just above the part we see with the eye. An active region, shown in the right half of this image, contains two round, dark sunspots, each about the size of the Earth, and bright plage that marks highly-magnetized regions. Long, dark filaments are held in place by arched magnetic fields. This image was taken on 26 April 1978. (Courtesy of Victor Gaizauskas, Ottawa River Solar Observatory, Herzberg Institute of Astrophysics, National Research Council of Canada)

responding to an average downflow velocity of about 5 thousand meters per second.

The almost universal Doppler redshifts in the transition region have been known for a long time. They were observed by George A. Doschek, Uri Feldman and J. David Bohlin in ultraviolet spectra taken with Skylab in the early 1970s, and have been confirmed by instruments aboard OSO 8, SMM and SOHO. These results imply that descending material produces more emission than ascending material at temperatures of about 100 thousand degrees.

The solar atmosphere is apparently falling down all over the transition region, like chicken-little's story about the sky. Hot gas is nevertheless always seen at higher levels in the corona, either held within magnetic loops in active regions or expanding outward in other places to fill the space between the planets. Since this hot gas must be replaced from below, it somehow breaks on through the transition region to the other side of the Sun. Scientists just could not see that happening, at least in the earliest investigations.

Perhaps we can only detect the denser, brighter material that moves down through the transition region and into the Sun. The downflowing material might be the bright residue of explosive phenomena occurring above the transition region, and it could be brighter because it is denser than anything else there. Lesser amounts of more tenuous material, that remains dark and unseen, could travel in the opposite direction. Or, the upflowing material might be at

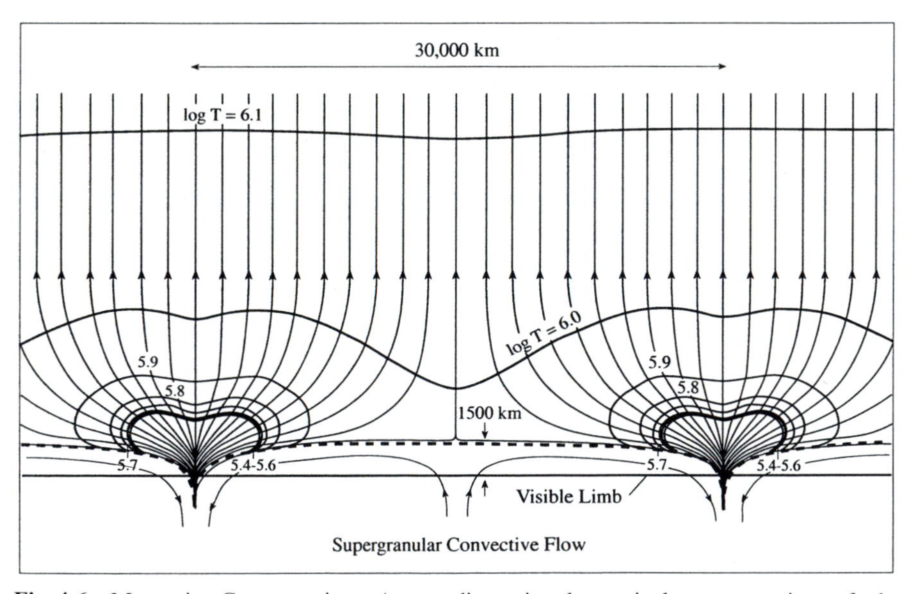

Fig. 4.6. Magnetic Concentration. A two-dimensional, vertical cross section of the magnetic-network model of the solar transition region. The motion of supergranular convective cells (*bottom*) concentrates magnetic fields at their boundaries. The magnetic fields (*arrowed lines*) are pushed together and amplified up to 0.1 Tesla at the cell edges, and expand with height in the corona. Temperature contours between log $T = 6.1$ (corona) and log $T = 5.4$ (upper transition region) are marked. (Adapted from A. H. Gabriel, 1976)

hotter coronal temperatures rather than cooler transition region temperatures, and hence invisible to the temperature-sensitive, transition-region lines. Another possible explanation is the low spatial or temporal resolution of early observations that might not have resolved the small, variable heating structures.

The transition region is not homogeneous and uniform, even in relatively calm places outside active regions. Wheeling, turbulent motions in the underlying convection zone produce large-scale horizontal motions in the upper photosphere, marking the overturning tops of supergranules, each about 30 million meters across (Sect. 3.1).

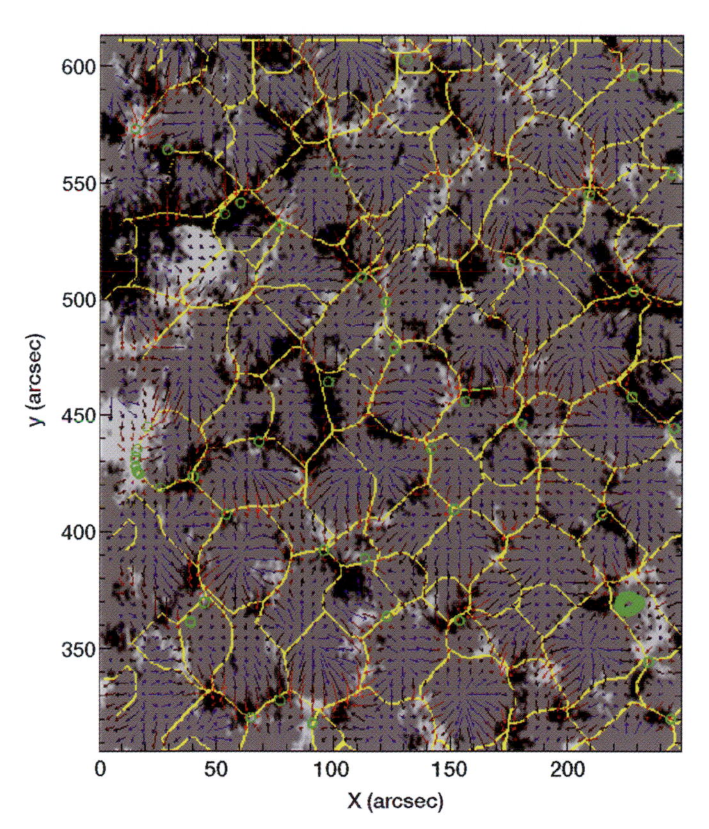

Fig. 4.7. Transitory Magnetic Elements. A high-resolution magnetic map of the quiet part of the solar disk, obtained using the Michelson Doppler Imager (MDI) aboard SOHO, is overlaid with lines of convergence of the horizontal flow, describing the magnetic network, and with green circles showing the convergence points. Red arrows describe the inferred downflow and blue ones describe the inferred upflow. The photospheric magnetic field is shown light gray for positive fields directed out of the Sun and black for negative fields pointing in. Magnetic flux disappears in collisions between opposite polarities so fast that all magnetic fields should disappear in a few days. New flux emerging in small ephemeral regions replaces the disappearing flux. (Courtesy of the SOHO MDI consortium. SOHO is a project of international cooperation between ESA and NASA)

The supergranules outline a network of convection cells with magnetic fields concentrated at their boundaries. As first modeled by Alan H. Gabriel in 1976, the magnetism in the transition region is amplified to high values at the narrow edges of this magnetic network (Fig. 4.6). SOHO's magnetograms reveal the magnetic network across the solar disk with great clarity, undistorted by the Earth's atmosphere. They show that the magnetic fields in the Sun's quiet regions are positioned within the downflowing boundaries of the supergranular convection, and that much of that magnetic flux originates in small bipolar regions (Fig. 4.7).

The SOHO data, reported by Carolus Schrijver and his colleagues in 1998, indicate that these tiny magnetic elements rise up out of the solar interior, merge together, and are replenished every 40 hours. So the entire small-scale magnetic slate in the photosphere and chromosphere is repeatedly wiped clean, losing its identity and reforming in less than two days. As we shall next see, the ephemeral magnetic regions may be related to coronal heating of quiet places outside active regions.

4.4 Magnetic Connections in the Quiet Corona

The apparently serene Sun, an unchanging disk of brilliant light to the casual eye, has no permanent features. The hot coronal gases are in a constant state of agitation, turmoil and motion, continuously brightening, fading and altering their form. This endless activity was noticed in Skylab's X-ray photographs, and viewed with greater clarity using Yohkoh's Soft X-ray Telescope, or SXT for short. Yohkoh conclusively demonstrated that the Sun's inner corona is never still. It has no fixed form and nothing is steady or unvarying there. The coronal material and magnetic fields change their intensity and shape on all possible spatial and temporal scales, with effects felt throughout the solar atmosphere, even in the so-called "quiet corona" outside active regions.

Thus, the steady corona frozen into one eclipse or X-ray photograph is an illusion. Using such a single picture to represent reality is a little like someone viewing the world through a hole in a wall, with no recognition of its varying complexity in space and time.

The key to much of this volatility is the Sun's ever-changing magnetism. The wide dynamic range, high spatial resolution and rapid, uniform sequence of images from the SXT have shown that the inner corona is nothing but magnetized loops, and that these loops are only temporary features. They continuously shift around, become deformed, break up and form new connections, responding to internal motions and continuously dumping energy into the corona. This continual emergence and rearrangement of the long, thin coronal loops has been substantiated by images from the Transition Region and Coronal Explorer, or TRACE (Figs. 4.8 and 4.9).

Coronal loops can suddenly appear out of nowhere, filling up with superheated material. Hot, X-ray emitting gas can be propelled to remote locations within well-collimated jets, even to or from the polar coronal holes; these mag-

Fig. 4.8. Coils of Flame. Thin magnetic loops constrain the hot, dense coronal plasma, seen at the edge of the Sun in this pioneering image with the Transition Region and Coronal Explorer (TRACE) spacecraft, taken at extreme ultraviolet wavelengths. The larger loops are about 200 million meters (15 Earths) across. False colors are used to indicate temperature: blue corresponds to roughly 100,000 degrees Kelvin, green to 900,000 degrees, and red to 2,700,000 degrees. The loops' thinness and differing temperatures suggest that coronal plasma is heated along small, very localized groups of field lines. (Courtesy of Alan Title, the Stanford-Lockheed Institute for Space Research, the TRACE consortium and NASA)

netic conduits are sometimes as long as the Sun is wide. The X-ray jets have an average length of 150 million meters and an average apparent velocity of 200 thousand meters per second; they are associated with small explosive flares.

Localized changes can trigger disturbances that cascade across the Sun like avalanches – as if the corona was poised on the brink of instability. The slightest magnetic disturbance can temporarily throw the entire corona out of control. It then globally adjusts to the change, relaxing to a less-agitated, lower-energy state.

The high resolution and sensitivity of the SXT demonstrated that coronal magnetic energy can be released, suddenly and catastrophically, when opposing magnetic fields merge and cancel each other. The magnetic fields move into each other but never end. They reform or reconnect in new magnetic orientations and in so doing dump energy into the corona. Such magnetic reconnections may occur when newly emerging magnetic fields rise through the photosphere to encounter pre-existing ones in the corona, or when the twisted, writhing coronal loops are forced together. In either case, the moving coils are charged with pent-up energy and ready to erupt.

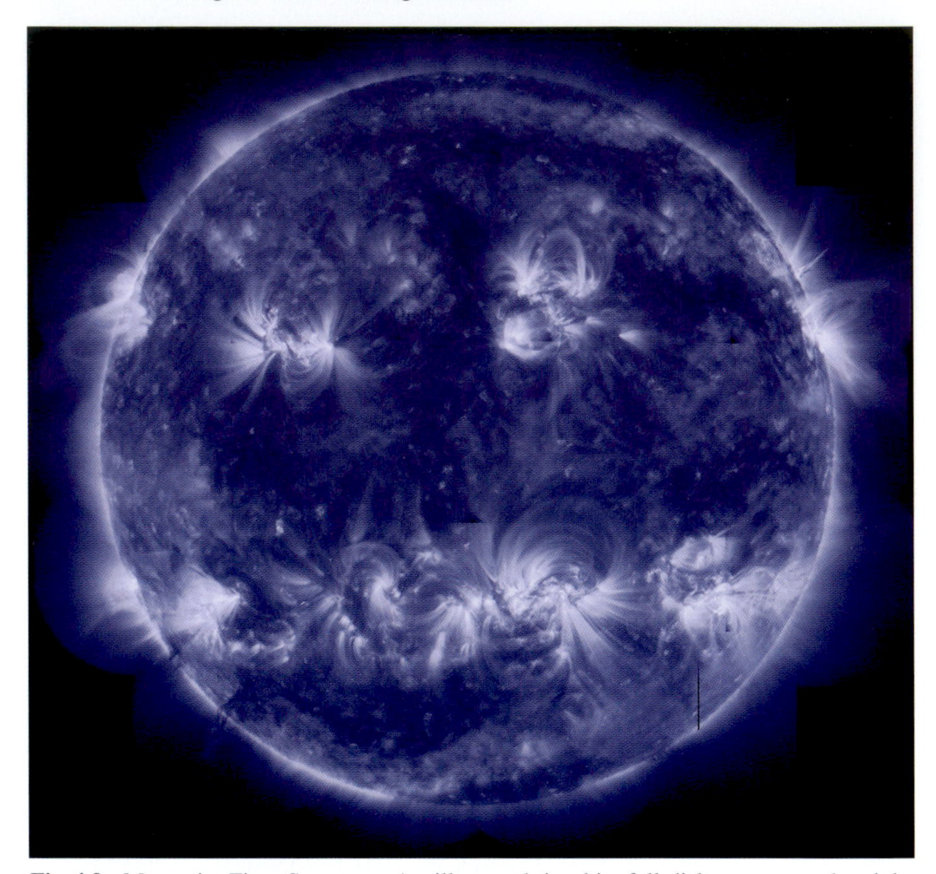

Fig. 4.9. Magnetic Fine Structure. As illustrated in this full-disk, extreme ultraviolet image, there is a great deal of fine structure in the low solar corona and transition region, associated with bright, thin magnetized loops. This image was taken in the light of eight times ionized iron (Fe IX at 17.1 nanometers), emitted by gas in the temperature range 600,000 to 1,200,000 degrees Kelvin. Bright loops with a broad range of lengths all have a fine, thread like substructure with widths as small as the telescope resolution of 725 thousand meters. This image was taken with the Transition Region and Coronal Explorer (TRACE) spacecraft on 10 October 1998. The body of TRACE observations support the idea of a carpet of constantly emerging magnetic fields on a range of scales emerging from the photosphere, causing a continuous rearrangement of magnetic fields in the corona and impressing on the outer atmosphere the fine structure in the magnetic fields. (Courtesy of the TRACE consortium, Stanford-Lockheed Institute for Space Research, and NASA)

Restructuring of magnetic fields is commonly observed throughout the low corona. As an example, Kazunari Shibata and his colleagues have demonstrated, in the 1990s, that many collimated X-ray jets display the morphology and physical characteristics of magnetic reconnection. They are sometimes propelled by magnetic connections involving new magnetic flux, coming up through the photosphere, and pre-existing magnetic loops (Fig. 4.10).

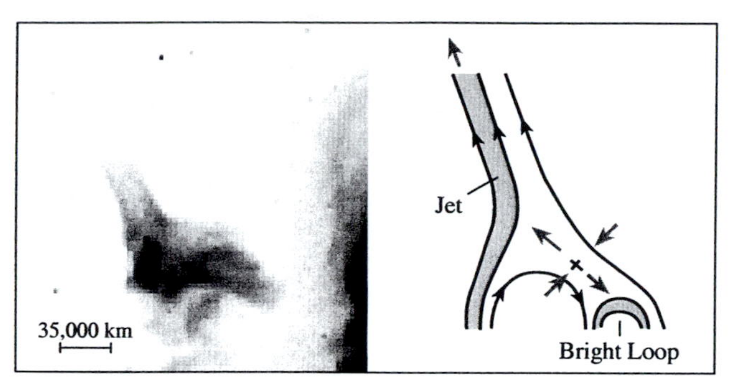

Fig. 4.10. X-ray Jet. This figure shows how an X-ray jet (*left*) is produced by an emerging magnetic loop that connects with pre-existing magnetism (*right*). This magnetic configuration has been called the anemone jet, since the active region at the footpoint of the jet looks like a sea anemone. The collimated X-ray jets can shoot across large distances in the solar corona, sometimes nearly all the way across the visible solar disk. The X-ray image was taken by the Soft X-ray Telescope on Yohkoh. (Courtesy of Kazunari Shibata, adapted from T. Yokoyama and K. Shibata, 1995)

In one common configuration for magnetic reconnection, suggested by Harry E. Petschek in 1964, antiparallel fields are pressed together by converging flows and carried inward from both sides (Fig. 4.11). Petchek's mechanism is now known to be a particular member of a much wider family of fast reconnection regimes discovered by Eric Priest and Terry Forbes in 1986.

During magnetic reconnection, the oppositely-directed field lines are effectively cut at the place where they touch and rejoin into a lower-energy configuration. The new, reconnected field lines are sharply bent, and so experience a strong magnetic tension force that snaps them apart. This accelerates and hurls material in opposite directions, like squeezing a tube of toothpaste open at both ends.

Such bi-directional, collimated jets were observed by Davina Innes and colleagues in 1997 (Fig. 4.11). Their SOHO ultraviolet spectra of explosive events in the chromosphere showed emission that was Doppler shifted to both longer wavelengths and shorter ones. This indicated material emanating in opposite directions from a common point, presumably driven by the magnetic reconnection process. The structure of these jets evolved in the manner predicted by theoretical models of magnetic reconnection, suggesting that it is a fundamental process for accelerating material on the Sun.

The ubiquitous X-ray bright points, first observed extensively by Leon Golub and his colleagues during the Skylab mission (Sect. 2.5), also apparently result from magnetic reconnection in the low corona. Unlike sunspots and active regions, X-ray bright points are uniformly distributed over the Sun, even appearing at the poles and in coronal holes. However, like coronal loops in active regions, the X-ray bright points occur above regions of opposite magnetic polarity in the photosphere. About one of the loop-like bright points appears per hour averaged over the whole Sun. They are about 10 million meters long, and have a mean lifetime of 8 hours.

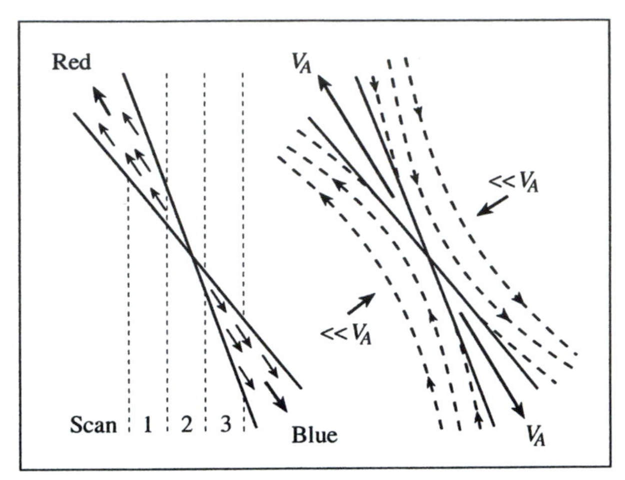

Fig. 4.11. Connecting Magnetism. Ultraviolet spectral line observations with the SOHO SUMER instrument reveal Doppler shifts that change as the spectrometer scans across a jet structure (*left*). This is explained by a magnetic reconnection model (*right*), involving magnetic field lines (*dashed lines with arrows for magnetic direction*) and plasma flow (*solid arrows*). Material flowing inwards from each side, at speeds much less than the Alfvén velocity, V_A, carry anti-parallel magnetic fields together. At the center X, magnetic fields that point in opposite directions meet and join together. This catapults jets that move in both directions away from the point of magnetic contact at about the Alfvén velocity. (Adapted from D. E. Innes, et al., 1997)

As shown by Eric R. Priest, Clare E. Parnell and their colleagues in 1994, the loop-like X-ray bright points can be energized by the magnetic interaction of emerging magnetic loops with existing coronal magnetic fields. The bright points are more often powered by the convergence of the photosphere's magnetic fragments, which drives the coronal encounter of previously unconnected magnetism. In the latter case, magnetism of opposite polarity comes together as the result of converging motion, releasing energy into the corona and accounting for the internal structure of particular bright points. The magnetic signposts then vanish from sight in the photosphere.

The X-ray bright points resemble small flares that occur more frequently than the larger ones. In 1988 Eugene N. Parker had, in fact, interpreted the X-ray emission from the entire corona in terms of a large number of small flares, due to bipolar magnetic fields driven together by underlying convective motion. The less-frequent, powerful flares had been previously explained in terms of magnetic interaction and magnetic energy release in the corona (Sect. 6.6).

Since intense solar flares occur relatively infrequently, providing an average total power that is at least 100 times below that radiated by the million-degree corona, the weaker flares have to occur much more frequently than the high-energy ones if they are going to significantly heat the corona. As Hugh S. Hudson showed in 1991, the hypothetical heating flares have to be an entirely different breed from normal solar flares. Extrapolations from existing observations suggested that the low-energy variety does not occur often enough to collectively supply the corona's heat.

X-ray bright points do, for example, occur more frequently at lower energies, but the number does not increase anywhere near fast enough to provide the corona's heat. Yohkoh's sensitive SXT has revealed a large number of transient X-ray events, occurring at the rate of 1,200 events per hour over the whole Sun, with a typical thermal energy content of 10^{19} Joule. Yet, that amounts to only about 0.002 Joule per second per square meter, or way below that required to heat the corona even outside active regions (Table 4.1).

In another perspective, observations of extreme ultraviolet lines, originating in the transition region at temperatures of about 100 thousand degrees, have revealed the presence of localized hot spots that explode and hurl material outwards at speeds of hundreds of thousands of meters per second. Such high-speed jets of matter were discovered in 1983 by Guenter E. Brueckner and John-David F. Bartoe during brief rocket observations of the transition region. During the subsequent decade, Kenneth P. Dere and his colleagues at the Naval Research Laboratory have carried out additional observations of the explosive transition-region jets, from rockets and the Space Shuttle, showing that they often occur in the magnetic network lanes. Jason G. Porter, David A. Falconer, Ronald L. Moore and colleagues reasoned in the mid 1980s that ultraviolet flares in active regions and the magnetic network are frequent and powerful enough to heat the corona.

SOHO's ultraviolet and extreme ultraviolet eyes have now shown that the tiny explosive events are going off all over the Sun and all the time – as many as 20,000 of them per minute, like an endless string of firecrackers. Richard A. Harrison and his colleagues have used SOHO instruments to show that the inner solar atmosphere remains a vigorous and violent place, even when its 11-year magnetic activity cycle is in a slump. During the apparent lull at activity minimum, the whole Sun seems to sparkle in the ultraviolet light of thousands of localized bright spots formed at temperatures of up to a million degrees Kelvin (Fig. 4.12). Since these hot spots seem to be anchored within magnetic loops that move about and interact, magnetic reconnection is a prime candidate for powering the myriad of explosions.

When SOHO looks at the Sun in extreme ultraviolet radiation, it is mottled all over with a granular appearance, like an orange, a stone beach, or a festering rash (Fig. 4.12). Each stone is a continent-sized bubble of hot gas, that flashes on and off in about 10 minutes and reaches temperatures of several hundred thousand to a million degrees. About 3,000 of these brightenings, known as blinkers, are seen erupting all over the Sun, including the darkest and quietest places at the solar poles. A thermal energy of about 10^{18} Joule is dissipated in each flash. So, even though there are a lot of these blinkers, their total thermal energy is comparable to that of X-ray bright points, and still significantly less than that required to heat the corona.

Many scientists have nevertheless continued to search for numerous low-level flares that might heat the corona. They have been called microflares, with micro designating small. Parker dubbed them nanoflares, perhaps be-

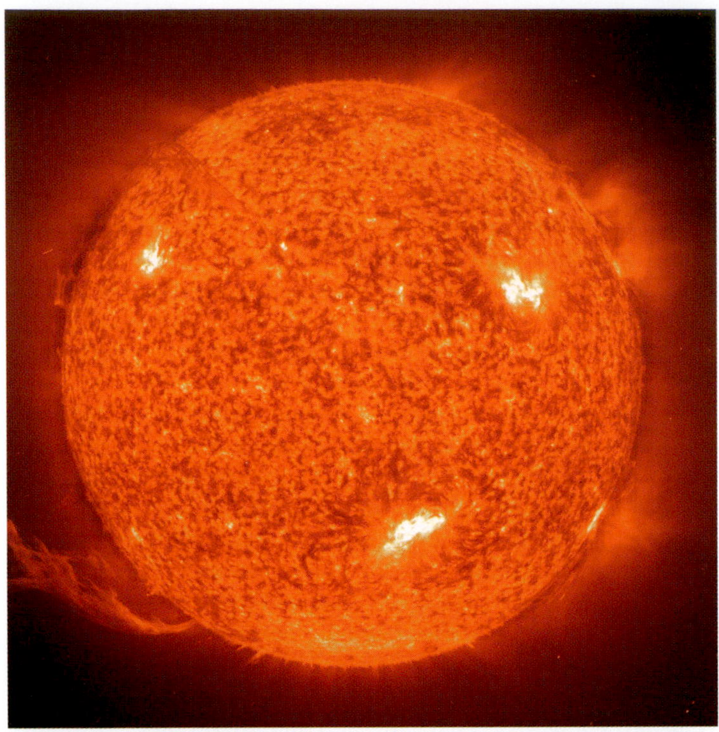

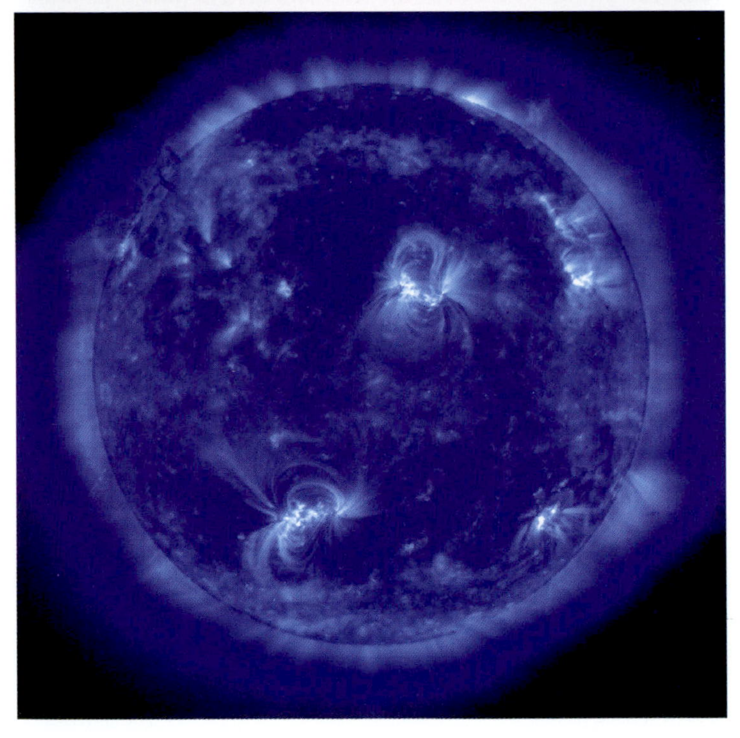

cause billions of them seemed to be required. Microflares have alternatively been described as those with energies of about 10^{19} Joule, or a millionth of that of a powerful flare at 10^{25} Joule. If nanoflares were true to their name, a billion of them would be required to release the same amount of energy as a normal flare.

All observers have agreed that the low-energy flares occur much more frequently than the high energy ones, but there is disagreement over the combined total energy released by the numerous weaker flares. The microflares or nanoflares could dominate the corona energetically, perhaps supplying a substantial portion of the heat of the quiet corona, or the low-energy flares might not occur often enough to provide sufficient power to heat the corona.

In scientific parlance, it all depends on the power law index of the flare frequency distribution. When there is a negative index larger than 2.0 most of the energy is released by the sum of the smallest flares, and when that index is less than 2.0 there is not sufficient combined power for the small ones to do the job.

Arnold O. Benz and Säm Krucker have used Yohkoh and SOHO observations to conclude, in 1998 and 1999, that microflares have a steep power law exponent of -2.5. This suggests that the weak flaring multitudes can rise up to rule the heating process. Furthermore, the brightest coronal microflares lie above enhanced elements of the magnetic network, and the denser ones exhibit greater fluctuations. This suggests that the lower corona is not just heated, but continuously replenished by underlying chromospheric material that has been heated to coronal temperatures. In other words, the energy input to the newly heated material provides a substantial fraction of the observed radiation loss of the brightest network elements.

In one attractive scenario, new magnetic fields emerge from within the Sun at the centers of supergranules, and are carried to their edges by convective motion. The fields then merge together and reconnect, releasing the energy that powers the microflares detected in the overlying low corona.

Other observers have disputed some of the results in 2000. Clare E. Parnell and P. E. Jupp have used observations from the Transition Region and Coronal Explorer (TRACE) satellite to show that an index of -2.5 can be obtained for nanoflares, but that it depends critically on the assumed three-dimensional form of the explosive events. They conclude that there is insufficient energy

Fig. 4.12. Bright Spots. Ultraviolet images from the SOHO EIT instrument show the solar corona at about 1 million degrees Kelvin (*bottom*) and the underlying chromosphere at temperatures of 60,000 to 80,000 degrees Kelvin (*top*). They show bright spots all over the Sun that can contribute to coronal heating. Two intense active regions with numerous magnetic loops are also seen (*top* and *bottom*), as well as a huge eruptive prominence at the solar limb (*top*). The image at the bottom was taken on 11 September 1997 in the extreme ultraviolet resonance lines of eight and nine times ionized iron (Fe IX/X) at 17.1 nanometers. The image at the top was taken on 14 September 1997 in the extreme ultraviolet resonance line of singly ionized helium (He II) at 30.4 nanometers. (Courtesy of the SOHO EIT consortium. SOHO is a project of international cooperation between ESA and NASA)

input from nanoflares to explain the total energy losses of the quiet corona, but do not rule out the possibility that yet weaker, and unobserved, events could heat it. Markus J. Aschwanden and colleagues use TRACE data to obtain an index for nanoflares of -1.8, arguing that the power in nanoflares is insufficient to heat the corona. Yet, the measured values of the index are suspiciously close to the 2.0 dividing line, and all of the heating might not be directly detected by events observed at extreme ultraviolet of X-ray wavelengths.

There is growing evidence that coronal heating is linked to frequent magnetic interaction. SOHO scientists have correlated highly-detailed, time lapse images of the photosphere's magnetism with extreme ultraviolet movies of intensity variations in the low corona. This has provided direct evidence for the transfer of magnetic energy from the visible solar disk toward the corona.

At any one time, SOHO's magnetic images, or magnetograms, reveal a salt-and-pepper sprinkling of about fifty thousand magnetic concentrations. Each one of them has a north and south pole connected by a rising magnetic arch. They together form a complex, tangled web of magnetic fields low in the corona (Fig. 4.13). Uninterrupted by atmospheric turbulence, bad weather, and night, SOHO's magnetograms follow the small magnetic loops continuously, showing that they appear, fragment, drift and disappear in hours.

Observations by Carolus J. Schrijver, Alan M. Title and colleagues, published in 1998, suggest that the Sun's small-scale magnetic fields account for large-scale coronal heating. On the smaller scales, individual bipolar regions rise up into the photosphere and disappear within days, but they are continuously replenished every 40 hours by the emergence of new small bipolar

Fig. 4.13. Magnetic Carpet. Magnetic loops of all sizes rise far into the solar corona, arising from regions of opposite magnetic polarity (*black* and *white*) in the photosphere. Energy released when oppositely directed magnetic fields meet in the corona is one likely cause for making the solar corona so hot. (Courtesy of the SOHO EIT and MDI consortia. SOHO is a project of international cooperation between ESA and NASA)

regions. The frequent magnetic reconnections associated with the cancellation and disappearance of the small magnetic concentrations may be heating the corona.

Yohkoh soft X-ray observations discussed by Eric Priest and colleagues, also in 1998, additionally suggest that the heating of large-scale magnetic loops in the diffuse corona is uniform, and probably explained by frequent breaking and reconnection of the magnetic field lines.

Simultaneous comparisons of SOHO's magnetograms with its high-resolution extreme ultraviolet images indicate that whenever the underlying magnetism converges there is a brightening overhead. Moreover there are tens of thousands of the loops, rising up to form new connections all the time and all over the Sun. This strongly suggests that the inner corona is heated when the numerous, small magnetic loops connect with each other.

SOHO therefore seems to have identified one of the elusive heating mechanisms of the corona. One enthusiast asserts that this magnetic carpet can provide anywhere from 10 to 1,000 times the energy needed to raise the entire corona to its million-degree temperatures. Since the interacting magnetism surges up over the entire solar surface, and not just in active regions, it could also explain why the corona stays hot at a minimum in its 11-year activity cycle when active regions go away.

To sum up, the new findings by the SOHO and Yohkoh spacecraft help explain why the corona stays hot. Magnetic connections in the low corona, producing motions down below, account for a substantial part of the energy needed to heat the high-temperature corona, or at least provide plausible explanations for it. Rapid jets have been discovered that are naturally produced by magnetic reconnection. Intense coronal brightenings, known as X-ray bright points, seem to be heated by magnetic reconnection, and the ultraviolet blinkers may also be attributed to it. Numerous nanoflares , that are most likely produced by magnetic reconnection, provide at least some of the heat of the solar corona. The small-scale magnetism of the solar photosphere is completely replenished every 40 hours, apparently producing rapid magnetic reconnection. Such events could explain the uniform heating of the diffuse corona.

Yet, the mystery is not completely solved. All of the new mechanisms operate in the low corona, near the Sun where the ubiquitous magnetic loops rise and disappear. The outer corona must also be heated, in the extended coronal streamers and above the coronal holes. As we shall next see, these locations coincide with the places where the Sun's winds originate and are accelerated.

Table 4.2. Key events in coronal heating studies*

Date	Event
1939–41	Walter Grotrian and Bengt Edlén identify coronal emission lines with highly ionized elements, indicating that the Sun's outer atmosphere has a temperature of millions of degrees. The conspicuous green line was identified with Fe XIV, an iron atom missing 13 electrons.
1946	Vitaly Ginzburg, David F. Martyn and Joseph L. Pawsey independently confirm the existence of a million-degree corona from observations of the Sun's radio emission.
1946–48	Ronald G. Giovanelli develops a theory of solar flares involving the magnetic fields in the solar atmosphere above sunspots, including electric currents at magnetic neutral points.
1947	Hannes Alfvén argues that convective granulation can generate magnetohydrodynamic waves that can heat the inner corona.
1948–49	Ludwig F. Biermann, Martin Schwarzschild and Evry Schatzman independently reason that sound waves, produced by convective granulation, might transport mechanical energy into the chromosphere and corona. These sound waves would be quickly deformed into shock waves as they pass into regions of decreasing density, dissipating energy and heating the gas.
1960	Thomas Gold and Fred Hoyle show that magnetic energy must power solar flares, and argue that these explosions are triggered when two magnetic loops of opposite sense or direction interact, merge and suddenly dissipate their stored magnetic energy in the corona.
1964	Thomas Gold suggests that the corona is heated by the relaxation of twisted or stressed coronal magnetic fields, driven by turbulent motion in the convective zone, and compares the energy dissipation mechanism to the magnetic interaction theory of solar flares developed by Gold and Fred Hoyle in 1960.
1964	Harry E. Petschek clarifies the process of magnetic field reconnection and shows that it can occur rapidly even in a highly conducting plasma.
1969–71	Magnetic fluctuations are observed in the solar wind from Mariner 5 on its way to Venus, and attributed to large-amplitude Alfvén waves by John W. Belcher, Leverett Davis Jr., and Edward J. Smith.
1970–77	X-ray photographs from rockets, and then from the Skylab mission, are used to show that magnetic fields create a three-fold structure in the inner corona, with its coronal holes, coronal loops and X-ray bright points.
1974	Leon Golub, Allen S. Krieger, J. Kevin Silk, Adrienne F. Timothy and Giuseppe S. Vaiana provide the first detailed studies of X-ray bright points using Skylab data. They give values for their densities, temperatures, sizes, lifetimes and rate of occurrence.
1976	Alan H. Gabriel introduces a two-dimensional model of the chromosphere and corona, and the transition region between them, in which magnetic flux is concentrated at the boundaries of supergranular convection cells that produce a magnetic network.
1976	George A. Doschek, Uri Feldman and J. David Bohlin use Doppler wavelength shifts of ultraviolet lines observed from Skylab to suggest that most observable material in the transition region is falling down into the Sun rather than moving away from it.
1978	John T. Mariska, Uri Feldman and George A. Doschek show that extreme ultraviolet lines observed from Skylab are wider than would be expected from thermal Doppler broadening alone. The non-thermal component corresponds to a velocity of about 5–10 thousand meters per second at the base of the transition region and rises with increasing temperature.

1978	Robert Rosner, Wallace H. Tucker and Giuseppe S. Vaiana use Skylab X-ray data and a theoretical model to derive a scaling law connecting the length, temperature and density of the ubiquitous coronal loops that dominate the structure of the low solar corona.
1978–79	R. Grant Athay and Oran R. White, and independently Elmo C. Bruner, Jr., observe chromosphere oscillations in the ultraviolet lines of ionized carbon from the eighth Orbiting Solar Observatory (OSO 8) spacecraft. These observations indicate that the chromosphere might be heated by sound waves, but that there is not enough power in the sound waves to heat the overlying corona.
1983	Eugene N. Parker argues that the main source of coronal heating is the dynamical dissipation of energy from coronal magnetic fields that have been twisted by sub-photosphere convection.
1983	Rocket observations of the ultraviolet solar spectrum by Guenter E. Brueckner and John-David F. Bartoe reveal high-energy jets in the transition region outside active regions. Kenneth P. Dere and colleagues subsequently provide detailed observations of these compact, short-lived explosive events from rockets, the space shuttle, and satellites, showing that they occur in the solar magnetic network lanes at the boundaries of supergranular convective cells and might play a role in coronal heating.
1984–95	Jason G. Porter, David A. Falconer, Ronald L. Moore and colleagues observe low-lying, ultraviolet microflares in the transition region and low corona, using the Solar Maximum Mission (SMM) satellite and SOHO. These events are located on magnetic neutral lines in active regions and in the magnetic network, and estimates suggest that they are frequent enough and powerful enough to drive the heating of the entire corona.
1986	Eric R. Priest and Terry G. Forbes put the idea of fast magnetic reconnection on a firm foundation, and discover a new family of fast regimes.
1988	Eugene N. Parker interprets the solar X-ray corona in terms of nanoflares, occurring more often and with less intensity than known solar flares. The nanoflares are related to bipolar magnetic fields driven together by underlying convective motion.
1991	Hugh S. Hudson uses extrapolations from existing flare observations to show that similar, but less energetic, microflares cannot heat the corona.
1991–2000	The Soft X-ray Telescope (SXT) on the Yohkoh satellite, launched on 31 August 1991, shows the structured, dynamic nature of the inner corona more clearly than ever before.
1992–96	Kazunari Shibata, Toshifumi Shimizu, Saku Tsuneta and their colleagues use data from Yohkoh's Soft X-ray Telescope (SXT) to show that magnetic reconnection is rather common in the low corona, particularly inside active regions.
1994	Eric R. Priest, Clare E. Parnell, Sara F. Martin and Leon Golub give a model for heating X-ray bright points by magnetic reconnection, and show that this model explains some of the observed bright points.
1997	Davina Innes, Bernd Inhester, William Ian Axford and Klaus Wilhelm obtain SOHO observations that exhibit the bi-directional jets expected from magnetic reconnection.
1997–98	Richard A. Harrison reports SOHO observations of thousands of small-scale brightening events in the extreme-ultraviolet, inactive Sun. These blinkers flash on and off in about 10 minutes and reach temperatures of up to a million degrees.
1997–98	Carolus J. Schrijver, Alan M. Title, and colleagues compare coronal extrapolations of SOHO magnetograms to extreme ultraviolet images from SOHO, suggesting that the low corona brightens above merging magnetic fields.

	Large-scale coronal heating may be produced by frequent magnetic reconnection involving small-scale dipolar regions that can rise and disappear, canceling out and being replaced in days.
1998	Eric R. Priest and his colleagues show that the heating of the large-scale corona is uniform, and so is likely to be due to turbulent reconnection of many small current sheets.
1998–99 2000	Arnold O. Benz and Säm Krucker use SOHO data to demonstrate that numerous low-level, unexpectedly frequent microflares could heat the quiet corona outside active regions. Subsequent observations by Clare E. Parnell and P. E. Jupp and by Markus J. Aschwanden and colleagues using the TRACE satellite suggest that there may not be enough energy in microflares or nanoflares to heat the entire quiet corona, and that such heating remains controversial.

* See the References for complete references to these seminal papers.

5. Winds Across the Void

Overview

The Sun's hot and stormy atmosphere is forever expanding in all directions, filling the solar system with a ceaseless flow – called the solar wind – that contains electrons, protons and other ions and magnetic fields. Early spacecraft measurements showed that there are two kinds of wind, a fast one moving at about 750 thousand meters per second, and a slow one with about half that speed.

The twinkling, or scintillation, of radio sources suggested that a fast wind is streaming out at high solar latitudes. Measurements from the Helios 1 and 2 spacecraft indicated that the electrons, protons and helium nuclei in the solar wind have different temperatures; in the high-speed wind the more massive particles are hotter. The Ulysses spacecraft has made measurements all around the Sun, at a distance comparable to that of the Earth and near a minimum in the Sun's 11-year activity cycle. Ulysses' velocity data conclusively prove that a uniform fast wind pours out at high latitudes near the solar poles, and that a capricious, gusty, slow wind emanates from the Sun's equatorial regions. Polar plumes are not a major source of the high-speed wind. It originates at the boundaries of the magnetic network in coronal holes. SOHO data, also taken near activity minimum, have shown that the fast wind blows at a wide range of latitudes near the base of the corona, and that the slow wind is associated with the narrow stalks of equatorial coronal streamers.

The high-speed wind is accelerated very close to the Sun, within just a few solar radii, and the slow component obtains full speed much further away, often spurting out of helmet streamers in magnetically-driven blobs. SOHO has additionally demonstrated that heavier particles in polar coronal holes move faster than light particles in coronal holes; the oxygen ions have agitation speeds 60 times greater than those of protons. Magnetic waves might preferentially accelerate the heavier ions in coronal holes.

Ulysses has detected magnetic fluctuations, attributed to Alfvén waves, far above the Sun's poles; they may block cosmic rays trying to enter these regions. The Ulysses and SOHO spacecraft have determined the rough shape and content of the heliosphere, a vast Sun-centered bubble extending far into space.

5.1 The Two Solar Winds

The million-degree solar atmosphere, or corona, is expanding radially out from the Sun in all directions, like heat escaping from a chimney on a cold day. The existence of this perpetual solar wind was suggested in 1954 by Ludwig F. Biermann from observations of comet tails (Sect. 2.1). In 1958, Eugene N. Parker used hydrodynamic theory to show how a solar wind could work (Focus 5.1). The existence of a solar wind was confirmed by direct *in situ* observations in the 1960s by Konstantin I. Gringauz using ion traps aboard the Lunik 2 rocket, and by Marcia Neugebauer and Conway W. Snyder with instruments aboard the Mariner 2 spacecraft (Sect. 2.2). Unlike any wind on Earth, the solar wind is an exceedingly rarefied mixture of electrons, protons, heavier ions and magnetic fields; but like terrestrial winds, the solar gale has strong gusts and periods of relative calm.

The Mariner 2 data showed that the wind is ubiquitous and continuous, with both a fast and a slow component. Comparisons of satellite wind measurements and solar X-ray images, obtained from rockets and the Skylab mission in the 1970s, established that at least some of the high-speed solar wind has its origin in extended, low-density regions called coronal holes (Sect. 2.5). The open magnetic fields in coronal holes provide a conduit for the fast wind, like the express lane of a divided highway.

The solar wind has never disappeared during the more than three decades it has been observed with spacecraft. Two winds are detected – a fast, uniform wind blowing at about 750 thousand meters per second, and a variable, gusty slow wind, moving at about half that speed.

As we shall see, these two winds do not blow uniformly from all points of the Sun, but instead depend on heliographic latitude. Near a minimum in the Sun's 11-year cycle of magnetic activity, a steady torrent of high-speed wind rushes out of the open magnetic fields in the Sun's polar regions; a slow, variable wind moves away from regions near the Sun's equator.

Near activity minimum, the Sun can be described as a simple magnet with north and south poles where large, unipolar coronal holes are located. The northern hole is of one magnetic polarity, or direction, and the southern one of opposite polarity. According to a model proposed by Gerald W. Pneuman and Roger A. Kopp in 1971, the negative and positive field lines meet in between the poles, near the solar equator, where a magnetically neutral layer, or current sheet, is dragged out into space by the radially outflowing wind (Fig. 5.1). Near the Sun, the current sheet is rooted in a belt of streamers that seems to meander across the star like the seam of a baseball.

The wind's magnetic sectors, detected by spacecraft in the ecliptic, have been described by a "ballerina model" advocated in the late 1970s by Hannes Alfvén (Sect. 2.4). Since the Sun's magnetic dipole axis is tilted with respect to its rotation axis, the current sheet is warped (Fig. 5.2). As the Sun rotates, the current sheet wobbles up and down, like the edge of a spinning ballerina's skirt, sweeping sectors of opposite magnetic polarity past the Earth.

The distant reaches of the solar wind can be investigated using remote radio sources that fluctuate, or scintillate, in much the same way that stars twinkle when seen through the Earth's wind-blown atmosphere. The radio waves are perturbed when they pass through the solar wind, producing a hazy, blinking and distorted image. It is something like looking at a light from the bottom of a swimming pool.

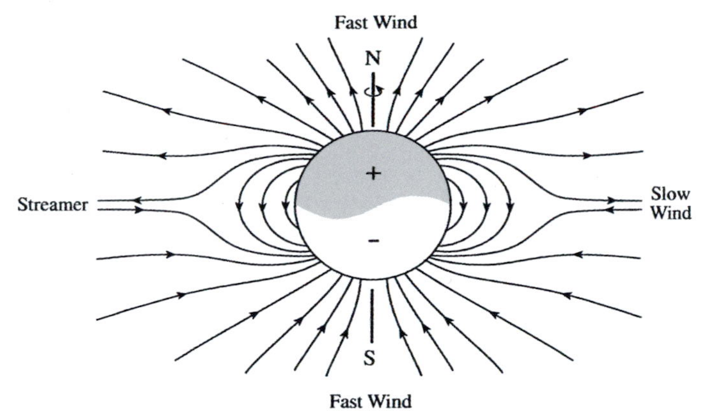

Fig. 5.1. Magnetism at Minimum. Theoretical cross sections of the magnetic fields lines expected in the Sun's corona with a dipole-type geometry and equatorial current sheet near the minimum in the 11-year cycle of solar magnetic activity. The high-speed wind escapes along the open magnetic field lines in the Polar Regions. At the equator, where the slow wind originates, the magnetic field lines have been pulled outward by the solar wind into oppositely directed, parallel magnetic fields separated by a neutral current sheet. Here, the transition from closed to open field lines at the equator occurs at two and a half times the Sun's radius. (Adapted from G. W. Pneuman and R. A. Kopp, 1971)

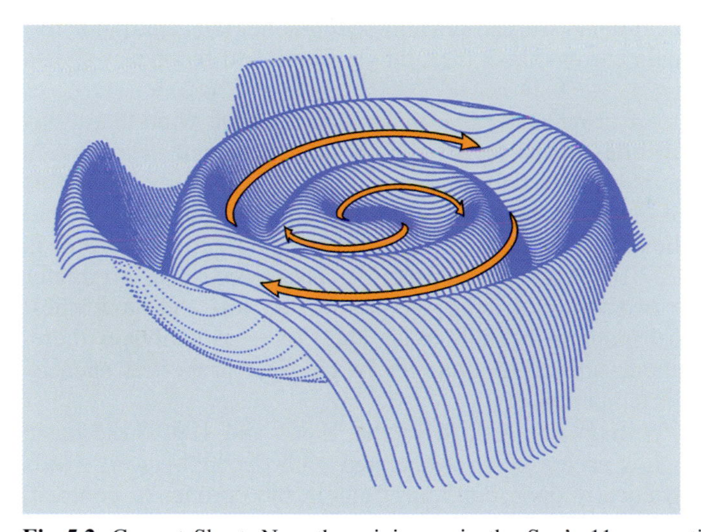

Fig. 5.2. Current Sheet. Near the minimum in the Sun's 11-year activity cycle, the Sun's magnetic field is primarily directed outward from one pole and inward in the other. The oppositely directed magnetic field lines meet near the solar equator, forming a thin, wave-like neutral current sheet that divides magnetic fields directed away from the Sun and those directed toward it. The current sheet rotates with the Sun and sweeps regions of opposite magnetic polarity past the Earth. The expansion of the solar wind combined with the Sun's rotation carries and twists the Sun's magnetic field into a spiral shape indicated by the arrows. It is sometimes known as the "Parker spiral", named after the scientist who first described it. Thus, the neutral current sheet winds out above the Sun's equator, with a spiral shape and a warped structure that resembles the skirt of a ballerina or a whirling dervish. (Courtesy of J. Randy Jokipii, and the Advanced Composition Explorer mission)

The fluctuating radio signals sweep past the Earth like the beacon of a search light, and the velocity of the solar wind can be inferred from the time it takes the sweeping signal to move between two antennas. You could similarly watch the shadow of a wind-blown cloud and determine the cloud's speed by seeing how long it takes the shadow to move from place to place. This interplanetary scintillation technique was used throughout the 1970s and 1980s, primarily by Barney J. Rickett and William A. Coles in the United States and by Masayoshi Kojima and Takakiyo Kakinuma in Japan.

The radio scintillation data showed that the solar wind has an uneven flow. Near the minimum in the 11-year activity cycle, the average wind velocity increases from the solar equator to higher latitudes toward the solar poles where coronal holes are most often observed. The slow wind is then confined to low latitudes near the solar equator. The blinking radio signals also indicated that the slow-speed wind does not blow steadily; it exhibits squalls and calms. In contrast, a uniform, fast wind seemed to spill out of higher latitudes, at least near solar minimum.

The spatial distribution of the winds is very different near the maximum in the Sun's activity cycle. At maximum the slow winds seem to emanate from all over the Sun, the high-speed winds abate, and polar coronal holes shrink or disappear. The global magnetic structure is then changing radically, with a switch in overall magnetic polarity; the north magnetic pole becomes south and *vice versa*. So, the dipolar, warped-current-sheet model falls apart at activity maximum. The main current sheet then moves about and secondary sheets are often present.

Helios 1 and Helios 2 provided *in situ* analysis of the solar wind that tightened constraints on theories for its origin. These twin spacecraft, respectively launched on 10 December 1974 and 15 January 1976, repeatedly looped as close as 0.3 AU from the Sun for years – until March 1986 for Helios 1 and March 1980 for Helios 2. (The mean distance of the Earth from the Sun is 1 AU, or 149,597,870,610 meters to be precise.) They confirmed, in greater detail, that there are two kinds of solar wind flow, the fast and slow ones, with different physical properties (Table 5.1). The fast wind is always there, lasting for years without substantial changes in physical properties such as composition, speed or temperature.

As suggested by William Ian Axford in the 1980s and 1990s, the basic mode of solar wind flow seems to be associated with the high-speed winds emanating from magnetically open configurations in the corona. In contrast, the slow wind, which is filamentary and transient, apparently involves the intermittent release of material from previously closed magnetic regions, and thus may not be treated as an equilibrium flow in a steady state.

Instruments aboard the two Helios spacecraft measured temperatures of the charged particles blowing in the wind, showing significant and unexpected differences. In the high-speed wind, heavier particles have higher temperatures; but it is the other way around in the slow wind, where lighter particles are hotter. In the high-speed wind, the protons are a few times hotter than the electrons, and the helium nuclei, or alpha particles, are even hotter than the protons.

Table 5.1. Average solar wind parameters measured from Helios 1 and 2 between December 1974 and December 1976 normalized to the distance of the Earth's orbit at 1 AU*

Parameter	Fast Wind	Slow Wind
Source	Coronal Holes	Equatorial Steamers
Composition, Temperature and Density	Uniform	Highly Variable
Proton Density, N_p	3.0 million m^{-3}	10.7 million m^{-3}
Proton Speed, V_p	667 thousand $m\,s^{-1}$ (750 thousand $m\,s^{-1}$)**	348 thousand $m\,s^{-1}$
Proton Flux, $F_p = N_p V_p$	$1.99 \times 10^{12}\ m^{-2}\,s^{-1}$	$3.66 \times 10^{12}\ m^{-2}\,s^{-1}$
Proton Temperature, T_p	280,000 degrees K	55,000 degrees K
Electron Temperature, T_e	130,000 degrees K	190,000 degrees K
Helium Temperature, T_α	730,000 degrees K	170,000 degrees K
Helium to Proton Abundance, A*	0.036 (constant)	0.025 (very variable)

* Adapted from Schwenn (1990). Measurements are referred to a distance of 1 AU = 1.496×10^{11} meters. The helium ion to proton abundance $A = N_\alpha / N_p$, where N is the number density and the subscripts α and p respectively denote the helium ions and the protons.

** The Helios 1 and 2 spacecraft traveled near the ecliptic where the slow solar wind dominates the flow, and this led to an underestimate of the velocity of the high-speed component. It has a speed of about 750 thousand meters per second.

The energy transported by the solar wind is dominated by the massive protons. The spacecraft showed that the proton density is high whenever the wind is slow, and that the proton density is low when the wind speed is high. The product of the proton density and velocity, or the proton flux, is about the same in the fast and slow winds, with a value of between 1.5 and 4.0 million million [(1.5 to 4.0) $\times 10^{12}$] protons $m^{-2}\,s^{-1}$ at the Earth's distance from the Sun.

Once the existence of the solar wind had been firmly established, and its details known, theoretical scientists began the difficult task of explaining it. To simplify the equations, a spherically symmetric and steady flow was initially assumed. The starting point was Eugene N. Parker's prophetic speculations, with supersonic flow beyond a critical distance from the Sun and a spiral-shaped twist to the interplanetary magnetic field (Sect. 2.4). Then as

Focus 5.1

Why a Solar Wind Has to Exist

The theoretical concept of a solar wind grew out of consideration of a static, non-expanding, isothermal atmosphere. At a distance, r, in a static, isothermal atmosphere, the balance between the pressure gradient, dp/dr, and the gravitational force can be expressed by:

$$\frac{\mathrm{d}p}{\mathrm{d}r} = -\frac{GM_{\odot}\rho}{r^2} ,$$

where G is the gravitational constant, $M_{\odot}$ is the Sun's mass, and ρ is the corona's mass density. If we take the coronal protons and electrons to have the same temperature, T, the ideal gas law becomes

$$p = nk(T_{\mathrm{e}} + T_{\mathrm{p}}) = 2nkT ,$$

where n is the number density of particles per unit volume, k is Boltzmann's constant, and the subscripts e and p denote electrons and protons. The coronal mass density is given by

$$\rho = n(m_{\mathrm{e}} + m_{\mathrm{p}}) = nm = mp/(2kT) ,$$

and m is the sum of the electron and proton mass, or essentially the proton mass since the electron has a relatively low mass.

By substitution, we obtain the differential equation:

$$\frac{1}{p}\frac{\mathrm{d}p}{\mathrm{d}r} = -\frac{GM_{\odot}m}{2kT}\frac{1}{r^2} .$$

The solution of this equation for the pressure $p(r)$ at distance, r, is:

$$p(r) = p_0 \exp\left\{\frac{GM_{\odot}m}{2kT}\left(\frac{1}{r} - \frac{1}{R_{\odot}}\right)\right\} ,$$

where p_0 is the pressure at the base of the corona and $R_{\odot}$ is the radius of the Sun. As one would expect, the pressure decreases with increasing distance, r, but the difficulty is that it does not decrease fast enough. If we let the distance go to infinity, or $r \rightarrow \infty$, the pressure approaches the value:

$$p(\text{infinity}) = p_0 \exp[-GM_{\odot}m/(2kTR_{\odot})] ,$$

where the coronal temperature can be taken to be $T = 10^6\,^{\circ}\mathrm{K}$, since this temperature falls off very slowly with distance owing to the high thermal conductivity of the corona. The other constants are $G = 6.6726 \times 10^{-11}\,\mathrm{N\,m^2\,kg^{-2}}$, $M_{\odot} = 1.989 \times 10^{30}\,\mathrm{kg}$, $m = m_{\mathrm{p}} = 1.62726 \times 10^{-27}\,\mathrm{kg}$, $k = 1.38066 \times 10^{-23}\,\mathrm{J\,K^{-1}}$, and $R_{\odot} = 6.955 \times 10^8$ meters.

With about 10^{14} electrons per cubic meter in the low corona, $p_0 = 0.001\,\mathrm{N\,m^{-2}}$, and $p(\text{infinity}) = 10^{-13}\,\mathrm{N\,m^{-2}}$ which may be comparable to the pressure thought to exist in the interstellar medium. This means that the static, isothermal corona may not be in equilibrium with the distant interstellar spaces, which is what led Eugene N. Parker to introduce an expanding corona with nonzero flow speeds, v, to obtain the differential equation:

$$\rho v\frac{\mathrm{d}v}{\mathrm{d}r} = -2kT\frac{\mathrm{d}n}{\mathrm{d}r} - \rho\frac{GM_{\odot}}{r^2} .$$

the complexity of the observed solar wind was gradually realized, the models were improved to permit two components (electrons and protons), and a spatial asymmetry between the winds flowing from the equatorial and polar regions of the Sun. Some of these theories are highlighted in Table 5.2 at the end of this chapter. They include novel calculations by William Ian Axford, John W. Belcher, R. E. Hartle, Thomas E. Holzer, Klaus Jockers, Roger A. Kopp, Egil Leer, Gerald W. Pneuman, Peter Sturrock and Edmund J. Weber. The curious reader can consult the original papers provided in the references at the book's end.

It was nevertheless difficult to confront the theoretical models with conclusive observational tests, for spacecraft had only measured the wind parameters within the plane of the Earth's orbit, and far from the regions of wind acceleration and origin. It remained for Ulysses to directly measure the speed and composition of the solar wind at high solar latitudes, and for SOHO to observe regions near the Sun where the solar wind is accelerated and the corona is heated. Both spacecraft have made unique, pioneering measurements, and they have both helped solve the riddles of where the two components of the solar wind originate and what accelerates them to supersonic velocities.

5.2 From Whence do the Winds Blow?

The intrepid explorer, Ulysses, has extended our parochial view of the Sun's domain by venturing above its poles in regions never before visited by any spacecraft. In 1994–95, near a minimum in the Sun's 11-year magnetic activity cycle, Ulysses provided the first measurements of the wind speed all around the Sun, over the full range of heliographic latitudes (Fig. 5.3). The Ulysses velocity data, reported by John L. Phillips and his colleagues, conclusively proved the existence of two basic types of solar wind at activity minimum. There is a fast gale blowing at a smooth, uniform and steady clip at high latitudes, with an average speed of 750 thousand meters per second, and a slower, more gusty breeze confined to low latitudes and moving at about 400 thousand meters per second.

At activity minimum, the slow component of the solar wind is very focused and localized near the plane of the solar equator. The fast solar wind is found everywhere else. It covers a wide latitude range and is not located just near the poles of the Sun. So the three-dimensional structure of the solar wind near activity minimum is dominated by the fast wind that occupies most of the volume in interplanetary space. This has been confirmed by Shadia R. Habbal, Richard Woo and their colleagues by observing the fluctuations, or scintillations, in radio communication signals coming from interplanetary spacecraft as they dip behind the solar wind. The radio scintillations show that the slow wind at activity minimum is associated with the narrow stalks of coronal streamers near the solar equator.

As the winds blow away, they must be replaced by hot gases welling up from below. However, Ulysses always kept its distance, never passing

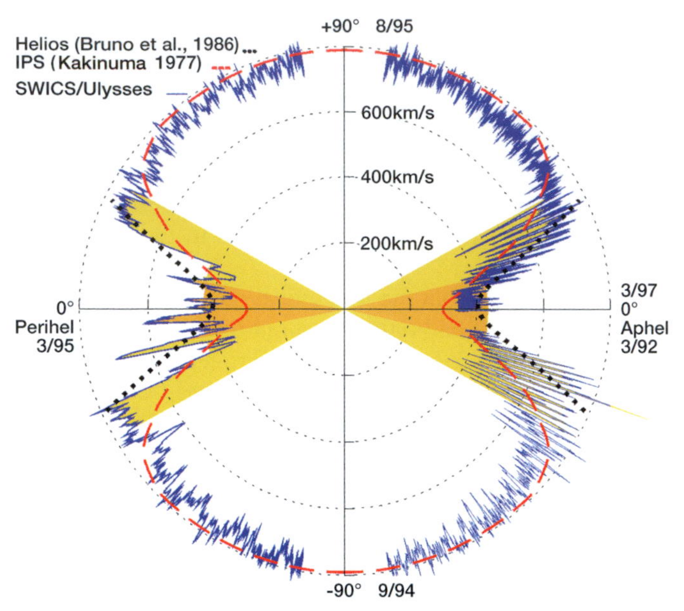

Fig. 5.3. Solar Wind Speeds. A polar diagram of the solar wind speed variation with heliographic latitude, from 0 degrees at the solar equator to ±90 degrees at the Sun's north and south poles (*top* and *bottom* respectively). These one-day averages of the proton speed, derived from Ulysses SWICS observations from March 1992 to March 1997, show that the fast wind dominates the outflow at most latitudes near the minimum in the 11-year cycle of solar activity, when the slow wind is localized near the equatorial regions. The equatorial slow-speed wind exhibits strong fluctuations; the polar high-speed one also fluctuates but at much lower levels. There are fewer fluctuations in the slow wind at the left-center because Ulysses was then moving rapidly over the equatorial regions. No shading represents fast coronal hole streams, while the reddish color denotes the slow solar wind and the streamer belt. These data are compared with previous measurements at lower latitudes by the Helios spacecraft (*dotted line*, R. Bruno, et al., 1986) and by interplanetary scintillations of extragalactic radio sources (*dashed red line*, T. Kakinuma, 1977). (Adapted from J. Woch et al., 1997, and courtesy of Joachim Woch)

closer to the Sun than the Earth does. Scientists therefore had to rely on other instruments to tell exactly where the winds come from. Fortunately, the Ulysses data, as well as simultaneous Yohkoh and SOHO results, were obtained near activity minimum with a particularly simple corona characterized by marked symmetry and stability. There were pronounced coronal holes at the Sun's north and south poles, and its equator was encircled by coronal streamers.

Comparisons of Ulysses' high-latitude passes with Yohkoh soft X-ray images showed that at least some of the high-speed component of the solar wind escapes from holes in the corona near the Sun's poles (Fig. 5.4). The coronal holes mark the base of very long magnetic "loops" which are unable to restrain the coronal gas, so the hot particles rush out unimpeded to form the fast solar wind.

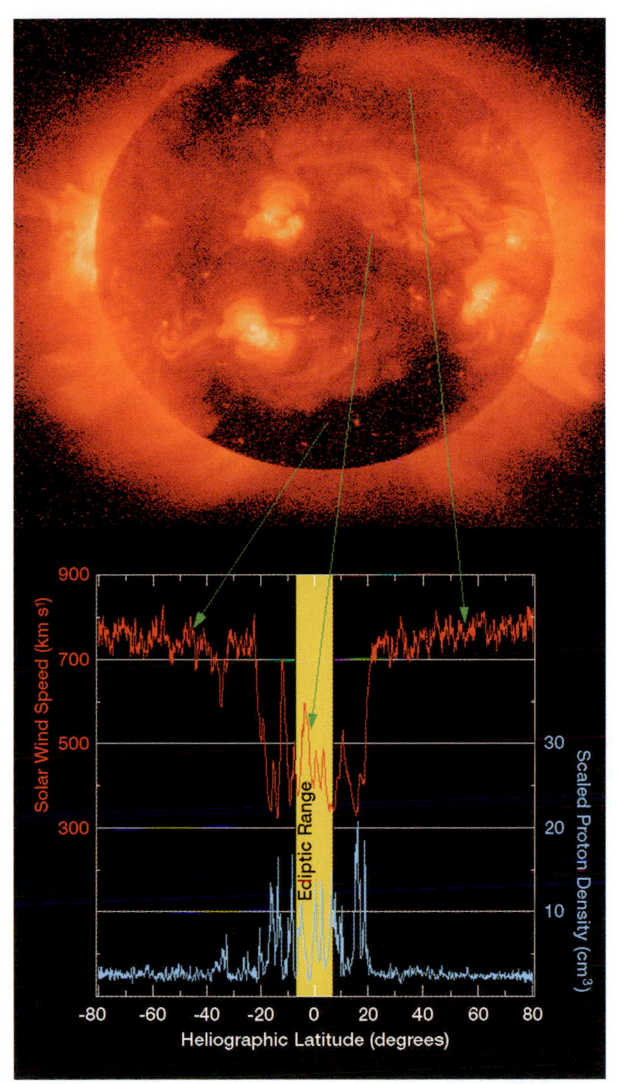

Fig. 5.4. Winds and Holes. An X-ray image taken with the Soft X-ray Telescope (SXT) on Yohkoh (*top*) is compared with measurements of the solar wind proton flow speed and density obtained with the Ulysses SWOOPS instrument (*bottom*). At least some of the high-speed solar wind originates in coronal holes (the dark areas shown at both poles of the X-ray image). At high latitudes the velocity is high and the density is low; near the equator the velocity is low and the density is high. (Adapted from J. L. Phillips et al, 1995, and courtesy of the Ulysses SWOOPS consortium. Ulysses is a project of international cooperation between ESA and NASA)

SOHO has a nested set of three white-light coronagraphs that use an occulting disk to block the photosphere's underlying glare, and obtain an edge-on view of the corona in regions where the solar wind originates. They detect visible sunlight scattered off free electrons in the corona, providing information about their time-varying distribution. This SOHO instrument is known as LASCO, an acronym for Large Angle and Spectrometric Coronagraph . It observes the corona from 1.1 to 30 solar radii from Sun center, looking closer to, and further from, the Sun than all previous space-borne coronagraphs.

Comparisons of Ulysses data with coronagraph images pinpointed the streamers as the birth place of the slow and sporadic wind. At the base of streamers, electrified matter is densely concentrated within magnetic loops rooted in the photosphere. The ragged, slow wind probably remains largely bottled up in this magnetic cage. The part that manages to escape seems to be worn out and varies in strength as the result of the effort.

Farther out in the corona, the streamers narrow and stretch out into stalks bounded by magnetic fields running in and out of the Sun. They coincide with a current sheet that separates oppositely directed, large-scale magnetic fields. The streamer stalks seem to mark the well-springs of the slow solar wind. Another source may be the edges of the coronal holes that are next to the streamers.

SOHO's LASCO has zoomed in to get a close-up view of the places that the slow wind originates. As shown by Neil R. Sheeley, Yi-Ming Wang and their colleagues, the equatorial streamers live up to their name; SOHO detects material flowing along their magnetically open stalks, like leaves floating on a moving stream. Time-lapse sequences of LASCO coronagraph images give the impression of a continual, gusty flow of bright, dense material concentrations, or blobs, coasting along and dribbling from the streamer stalks. So, at least some of the slow wind seems to leak out of the bottled-up part of the streamers in spurt-like blobs, like water working its way through a beaver dam or down a clogged sink.

The outward-moving blobs are emitted from the elongated tips of stretched-out helmet streamers, and may be injected into the slow wind during a disconnection of magnetic fields rooted in the Sun. In some instances, sequential LASCO images reveal a succession of narrow loop-tops that rise to form a bright streamer stalk that continues to lengthen and narrow with time. It eventually gets so stretched out and constricted that it pinches itself off and snaps in two at about four solar radii from Sun center. The inward- and outward-directed magnetic fields on each side of the narrow streamer apparently join together at the place where they touch. The lower parts of the streamer close down and collapse, and the outer disconnected segment is propelled out to form a gust in the slow solar wind.

Soft X-ray movies from Yohkoh have shown that magnetic loops in active regions often expand out into space, and that such expansion contributes to the slow component of the solar wind (Fig. 5.5). Way down, at the very bottom of streamers, magnetized coronal loops are found. The intense, looping magnetic fields trap the million-degree coronal gas that has been closely scru-

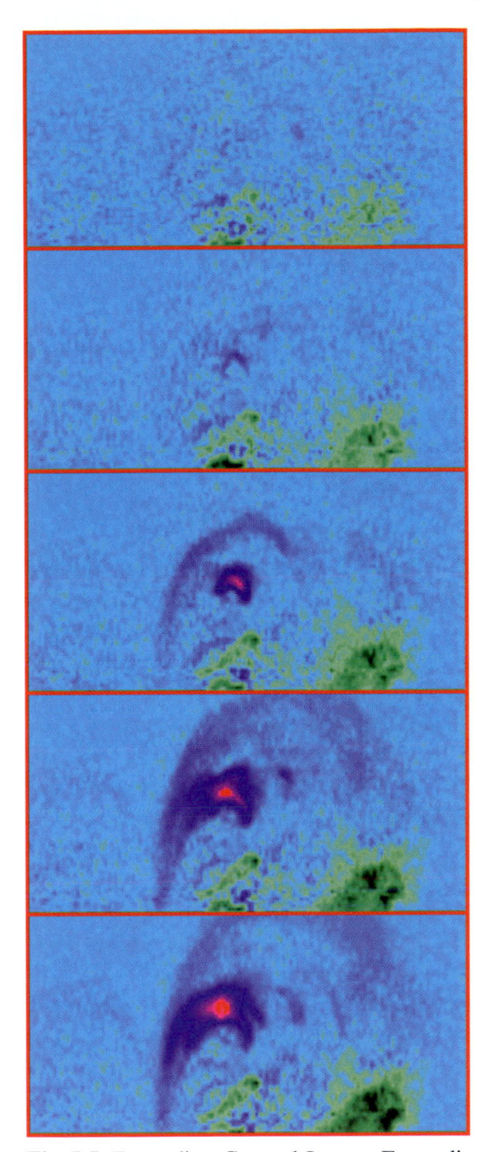

Fig. 5.5. Expanding Coronal Loops. Expanding active-region loops were discovered by the Soft X-ray Telescope (SXT) on Yohkoh. This figure shows difference images that reveal the expansion of the corona in a time sequence, running from top to bottom for about one half an hour on 22 April 1992. The fainter loops near the top are expanding with an apparent velocity of about 40 thousand meters per second. This is an especially clear, and fast, example of the common tendency of magnetic loops in solar active regions to expand outward, producing mass outflow from the Sun into the solar wind. (Courtesy Alan McAllister, NASA, ISAS, the Lockheed-Martin Solar and Astrophysics Laboratory, the National Astronomical Observatory of Japan, and the University of Tokyo, also see Y. Uchida et al., 1992)

tinized by the Soft X-ray Telescope (SXT) aboard Yohkoh. As first shown by Yutaka Uchida and his colleagues in 1992, the SXT data indicate that the active region corona is expanding, in some cases almost continuously, contrary to the commonly accepted models invoking static equilibrium. When viewed at the Sun's limb or edge, coronal loops are seen rising upwards at speeds of some tens of thousands of meters per second, maintaining their basic shape during the expansion (Fig. 5.5).

Images obtained with SOHO's coronagraphs have supported the idea that expanding loops may provide the energy and mass for the slow solar wind. These instruments show small loops emerging into the equatorial regions all the time. They expand into the global magnetic field at the current sheet, and establish new magnetic connections with it. The reconnection could happen again and again, becoming the engine that drives and perhaps accelerates the slow solar wind. The small expanding loops also carry material with them, perhaps accounting for the running blobs detected further out in the low-latitude, slow-speed solar wind.

For more than thirty years, scientists have known that much, if not all, of the high-speed solar wind comes from the open magnetic fields in coronal holes (Sect. 2.5 and Fig. 5.4), but the detailed structure of these regions has only recently been investigated by SOHO. Its Solar Ultraviolet Measurements of Emitted Radiation, or SUMER for short, spectrometer has investigated whether polar plumes help generate the high-speed solar wind. These tall, conspicuous bright features, in the otherwise dark coronal holes, extend out millions of kilometers into space like long ropes (Fig. 5.6). The feathery plumes rise out of Earth-sized magnetic regions in the photosphere, where they blink on and off in ultraviolet light. They seem to be rooted where small dipoles are in contact with unipolar flux, and may be energized by reconnection between the two types of magnetic features.

It was once thought that material might be preferentially squirted out of the open, nozzle-like magnetic channels, but careful SOHO measurements suggest that the high-speed wind is coming out of the entire coronal hole, with no substantial difference in speed between plumes and adjacent places. If the high-speed wind originated in polar plumes, then the magnetic fields in the wind would have to come from the plumes, and therefore packed within a much smaller area at their source. The magnetic fields in the plumes would then have to be much stronger than those detected in interplanetary space, but such intense magnetism has not yet been observed in the plumes. Since the interplume regions occupy most of the polar hole area, they are probably the main source of the polar high-speed wind. So appearances are once again deceiving. The brightest, most-obvious places are not providing most of the action in coronal holes.

By examining ultraviolet lines emitted at the base of the corona in coronal holes, Donald Hassler and his colleagues have used SUMER to show that the high-speed outflow is concentrated at the boundaries of the magnetic network formed by underlying supergranular convection cells (Fig. 5.7). These edges are places where the magnetic fields are concentrated into inverted magnetic funnels that open up into the overlying corona (Fig. 4.6, Sect. 4.3). The

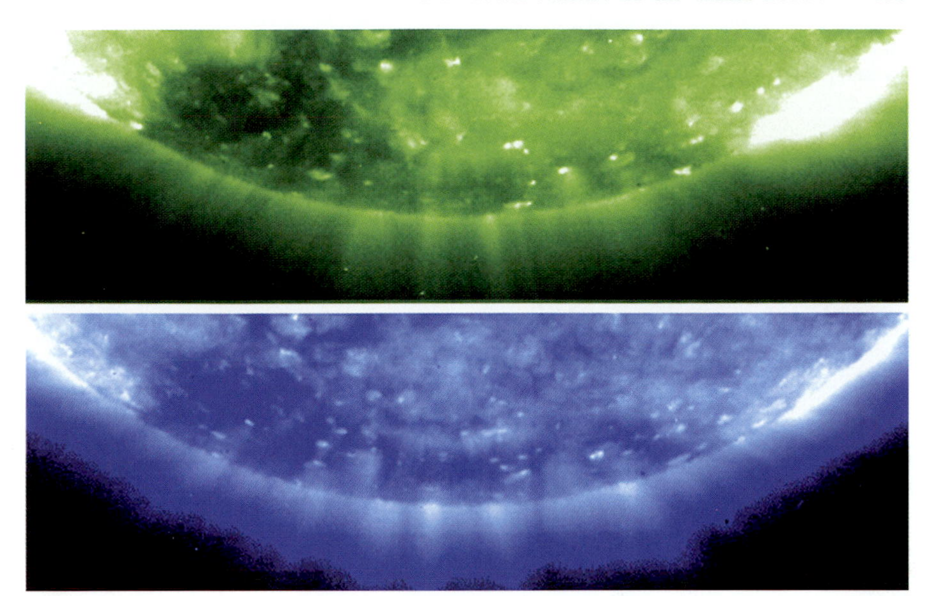

Fig. 5.6. Polar Plumes. Close-up, extreme ultraviolet views of the south polar coronal hole (*large dark region*) obtained with the SOHO EIT instrument on 8 May 1996. The fast component of the solar wind emanates from these coronal holes. Plumes can be seen emerging from tiny bright spots, but they are not the main source of the high-speed winds. These images were taken in the extreme ultraviolet light of highly ionized iron – the Fe XII emission line at 19.5 nanometers (*top*), formed at a temperature of about 1.5 million degrees Kelvin, and the Fe IX/X emission lines near 17.1 nanometers (*bottom*), formed at a temperature of about 1.0 million degrees Kelvin. (Courtesy of the SOHO EIT consortium. SOHO is a project of international collaboration between ESA and NASA)

strongest high-speed flows gush out of the crack-like edges of the network, like grass or weeds growing in the dirt where paving stones meet. Thus, SOHO has for the first time discovered the exact sources of the fastest winds.

The high-speed wind is not only found above polar coronal holes. At roughly the Earth's distance from the Sun, Ulysses found fast solar winds at lower latitudes nearer the solar equator and outside the radial projection of the coronal hole edges (Fig. 5.3). One explanation is that the magnetic field lines emerging from coronal holes splay and bend outward, like the petals of an opening flower, allowing the fast wind to spread out as it escapes. Streamers detected in the white-light view of the corona suggest that global solar magnetism does bend over within the low corona (Fig. 5.8), and polar plumes near the boundaries of coronal holes do turn toward the solar equator. However, the fast wind will also shoot radially out of coronal holes that extend down to low latitudes, even crossing the equator, and directly feed low-latitude regions.

Although much, if not all, of the fast solar wind gushes out of coronal holes, they might not necessarily be the only source of the high-speed wind. When scientists look close enough to the photosphere, one does not see a big

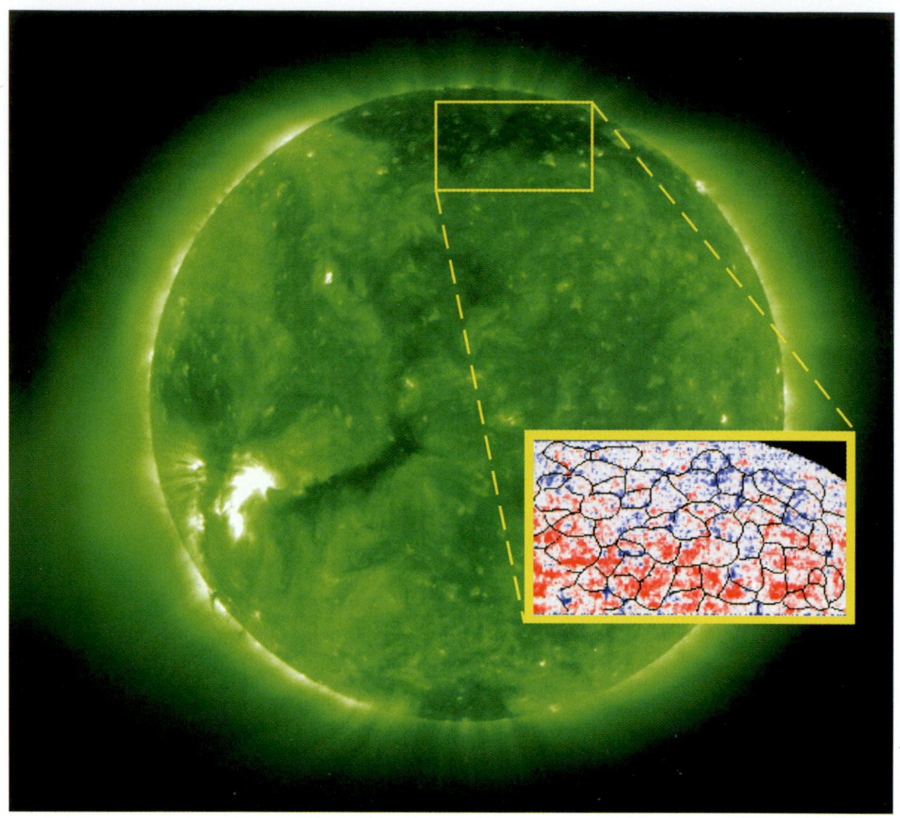

Fig. 5.7. Source of the High-Speed Solar Wind. This full-disk image of the Sun shows coronal gas at 1.5 million degrees Kelvin, shining in the extreme ultraviolet light of ionized iron (the Fe XII line at 19.5 nanometers). Bright regions indicate hot, dense material confined within strong magnetic loops, while the dark polar regions (*top* and *bottom*) imply an open magnetic field geometry. These polar coronal holes are the source of much of the high-speed solar wind. The inset provides a close-up, Doppler velocity map of the million-degree gas at the base of the corona where the fast solar wind originates, taken in the extreme ultraviolet light of ionized neon (the Ne VIII line at 77.0 nanometers). Dark blue represents an outflow, or blueshift, at a velocity of 10,000 meters per second; it marks the beginning of the high-speed solar wind. Dark red indicates a downflow at the same speed. Superposed are the edges of the "honey-comb" shaped pattern of the magnetic network, where the strongest outward flows (*dark blue*) are found. The relationship between the outflow velocities and the network suggests that the high-speed wind emanates from the boundaries and boundary intersections of the magnetic network. These observations were taken on 22 September 1996 with the Extreme ultraviolet Imaging Telescope (EIT – full disk) and the Solar Ultraviolet Measurements of Emitted Radiation (SUMER – velocity inset) spectrometer on SOHO. (Adapted from D. M. Hassler, et al., 1998; courtesy of the SOHO EIT consortium and the SOHO SUMER consortium. SOHO is a project of international collaboration between ESA and NASA)

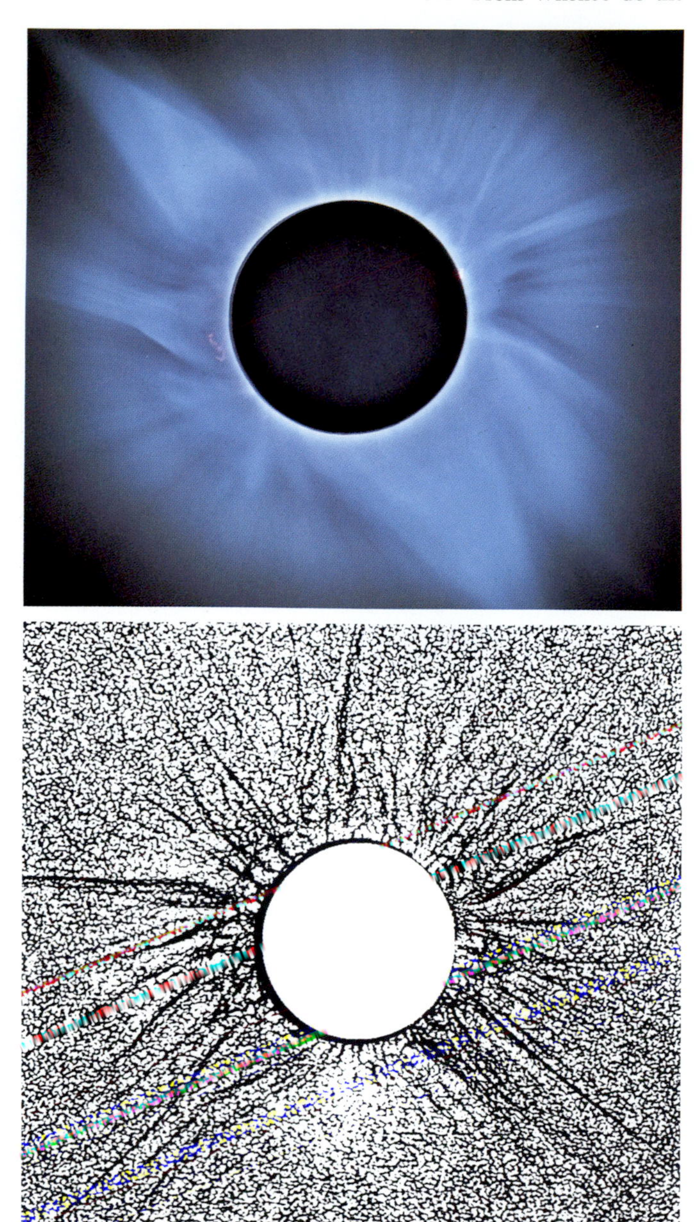

Fig. 5.8. Coronal Structures. The Sun's corona photographed in white light during the total solar eclipse on 11 July 1991 (*top*); it extended several solar diameters and had numerous fine rays as well as larger helmet streamers. A radially-filtered, digitally processed image of the total solar eclipse on 22 July 1990 (*bottom*) is filled with straight, thread-like rays extending as far as six solar radii, or 4 billion (4×10^9) meters from the center of the Sun. (Photograph by Shigemi Numazawa, president of the Japan Planetarium Laboratory, Nigata, Japan – *top*, and Serge Koutchmy, Institut d'Astrophysique de Paris, CNRS – *bottom*)

magnetic change as a function of latitude, and SOHO instruments found that the fast wind can blow at a wide range of latitudes within the inner corona. This suggests that some of the fast wind could be squirted out along straight, open magnetic channels all over the Sun and not just near the poles. Scintillations of spacecraft radio communication signals indicate that the corona might be filled with such straight, thread-like filamentary structures, and they have also been inferred from eclipse images of the corona (Fig. 5.8).

So, much of the fast wind might bend down from coronal polar holes, or rush straight out of equatorial ones, but some of it could flow along radial magnetic tubes originating at mid-latitudes. There is an ongoing controversy about the various explanations, with the scientific majority currently favoring coronal holes as the main source of the fast solar wind. Only time and future experiments will conclusively demonstrate the exact role of radial, low-latitude tubes in expelling the high-speed winds.

5.3 Some Like It Hot

What forces propel the solar wind to its high velocities? Energy is transported from the solar interior into the corona, and deposited there to heat the gas to temperatures of millions of degrees Kelvin. The hot corona then expands and becomes the source of the solar wind. So, the basic solar-wind mystery is the elusive heat source for the corona, discussed in Sects. 4.2, 4.3 and 4.4.

The corona's expansion will begin slowly near the Sun, where the solar gravity is the strongest, and then continuously accelerate out into space, gaining speed with distance and reaching supersonic velocities. But the wind can't go on accelerating forever, for there is a limit to the amount of energy being pumped into it. The solar wind therefore breaks away from the Sun, and is set free to cruise along at a roughly constant speed, called its asymptotic or terminal velocity.

More than one mechanism is required to explain the two types of solar wind, the slow and fast ones, that are accelerated in very different ways. The slow wind naturally reaches terminal velocities of a few hundred thousand meters per second as the million-degree corona expands away from the Sun. Additional energy must be deposited in the low corona to give the fast wind an extra boost and double its speed. In technical terms, the fast wind has a velocity and mass flux density that are too high to be explained by heat transport and classical thermal conduction alone.

The composition of material originating from the sources of the slow and fast winds differs, providing clues to physical conditions near the place of their origin (Fig. 5.9). As ions move out into the increasingly rarefied gases, collisions become infrequent and the ions decouple from other particles, retaining an unchanging temperature. Johannes Geiss, George Gloeckler and their colleagues used Ulysses measurements of these thermometers to show

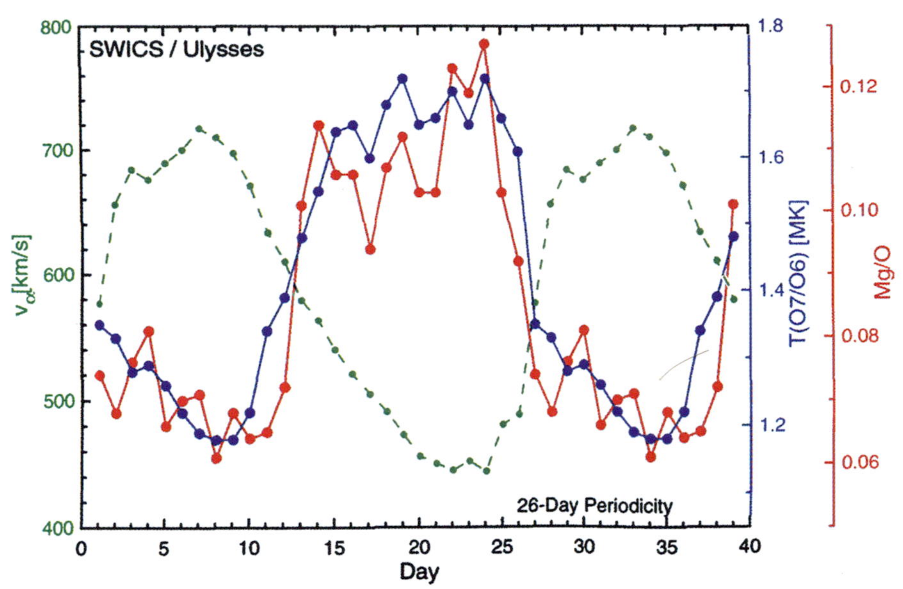

Fig. 5.9. Composition and Temperature Differences of Fast and Slow Streams. Three parameters are shown over an interval of one and a half solar rotations. They are the solar wind speed, *V* (*dashed line*); the abundance ratio, Mg/O (*red line*), of two heavy ions in the solar wind – magnesium, Mg, and oxygen, O; and the so-called "freezing-in" temperature, T(O7/O6), derived from the relative abundance of oxygen ions – seven, O7, and six, O6, times ionized oxygen atoms. The solar wind speed shows alternating high-speed streams (from high latitudes) and low-speed streams (from low latitude). The Mg/O abundance is larger in slow streams than in the fast ones. The temperature where the oxygen ions are created is high in the low-speed wind, over 1.6 million degrees Kelvin (denoted MK), indicating a hot coronal source, while the temperature of the high-speed coronal wind source is relatively low, about 1.2 million degrees. [Adapted from J. Geiss, G. Gloeckler and R. Von Steiger (1995) and courtesy of the Ulysses SWICS consortium. Ulysses is a project of international collaboration between ESA and NASA]

that the fast polar wind originates in a relatively low-temperature region in the corona.

The temperatures of the coronal electrons can be determined from their X-ray emission, showing that the higher corona is hotter than the lower corona (Fig. 5.10). The data show that the corona does not become fully heated until a height of between 0.2 and 0.5 solar radii, or between 140 and 348 million meters. Moreover, at the same height, the electron temperatures in coronal steamers are hotter than those in coronal holes. The electron temperatures in coronal holes appear to be several hundred thousand degrees cooler than the temperatures of electrons in lower latitude streamers.

Time-lapse sequences of SOHO's LASCO images suggest that the slow wind accelerates at a leisurely pace, and takes a long time to get up to speed (Fig. 5.11). Elongated structures formed near the tips of helmet streamers at a few solar radii from Sun center have to move out to 20 or 30 solar radii to accelerate to speeds of 300 thousand meters per second. In contrast, the

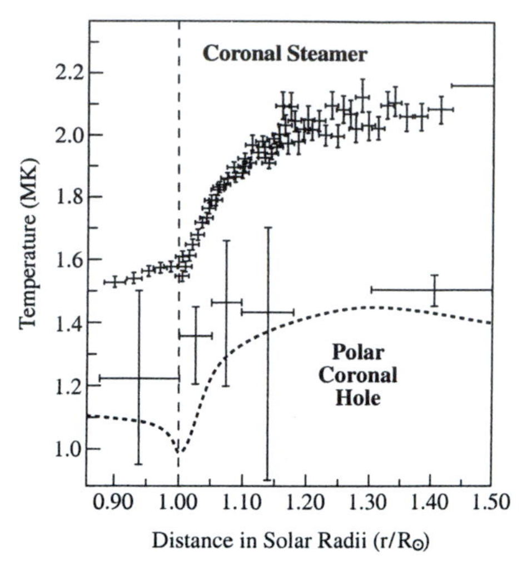

Fig. 5.10. Heating Up Close. The increases in temperature with height in a coronal streamer (*top*) and a coronal hole (*bottom*) have been inferred from data taken with the Soft X-ray Telescope on Yohkoh. Here the temperature is given in millions of degrees Kelvin, or MK, and the distance, *r*, from Sun center is specified in units of the Sun's radius. At a given height in the low corona, the temperature of coronal streamers is hotter than the temperature of coronal holes. Both regions seem to be fully heated by between 1.3 and 1.5 solar radii from Sun center. [Adapted from C. R. Foley, J. L. Culhane and L. W. Acton (1997) for the coronal hole, where the dashed-dotted line denotes the temperatures expected from a model given by G. L. Withbroe (1988) and the most distant point is from the Ulysses SWICS instrument and Ko et al. (1997). Yohkoh is a project of international cooperation between ISAS and NASA]

fast wind accelerates quickly, like a racing horse breaking away from a starting gate. Radio scintillation measurements indicate that the polar wind reaches terminal speeds of 750 thousand meters per second within just 10 solar radii or less (Fig. 5.12), so the high-speed wind is accelerated relatively close to the Sun.

In addition to LASCO, another instrument on SOHO directly examines the regions where the corona is heated and the winds originate. It is known as UVCS for UltraViolet Coronagraph Spectrometer. Like LASCO, the UVCS uses an occulting disk to block the photosphere's light and view the low corona at the Sun's edges. UVCS measures ultraviolet spectral features, or lines, determining temperatures and velocities within the source regions of the solar wind from 1.2 to 10 solar radii from Sun center. It has, for example, used the Doppler shifts of the spectral lines to show that the high-speed solar wind, emerging from coronal holes, accelerates to supersonic velocity within just 2.5 solar radii from Sun center.

Spectral line profiles, obtained by John L. Kohl, Giancarlo Noci and their colleagues with SOHO's UVCS at a few solar radii from Sun center, have revealed a surprising, marked difference in the speeds at which hydrogen and oxygen ions move within coronal holes. Above two solar radii from the

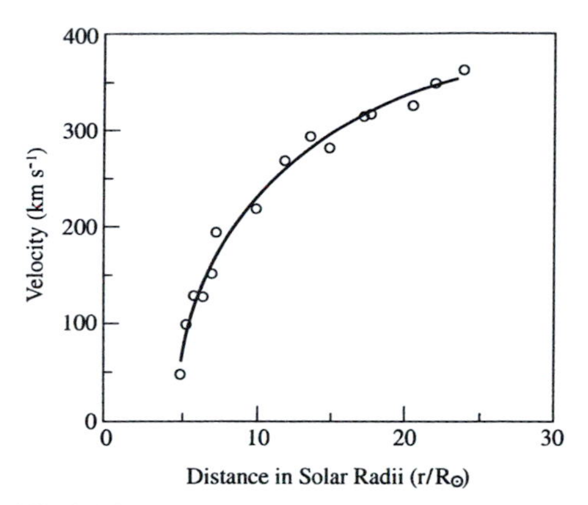

Fig. 5.11. Breaking Out Slowly. Time-lapse sequences of white-light images obtained with the SOHO LASCO instrument show prominent features, sometimes called blobs, that move radially out in equatorial regions from helmet streamers. Their speed typically doubles from 150 thousand meters per second near 5 solar radii from Sun center to about twice that speed near 25 solar radii, with a nearly constant acceleration of about 4 meters per square second through most of this distance. This data was taken on 30 April 1996, at a time near a minimum in the 11-year solar activity cycle. The solid line denotes the best fit to the data using an exponential function starting at a distance of 4.5 solar radii from Sun center and with an asymptotic speed of 418.7 thousand meters per second at a distance of about 15 solar radii from Sun center. [Adapted from N. R. Sheeley Jr., et al. (1997), and courtesy of the SOHO LASCO consortium. SOHO is a project of international collaboration between ESA and NASA]

Sun's center, oxygen has the higher agitation speed, approaching 500 thousand meters per second in the holes, while hydrogen moves at about half this speed (Fig. 5.13). In contrast, within equatorial regions where the slow-speed wind begins, the lighter hydrogen moves faster than the oxygen, as one would expect for a gas with thermal equilibrium among different types of particles.

As incredible as it might seem, the proton velocities are inferred from those of atomic hydrogen – not the protons or hydrogen ions. The hydrogen atoms emit the brightest ultraviolet radiation from the corona, the so-called Lyman α transition at 121.6 nanometers. You might expect that all the electrons would be torn from the hydrogen atoms by the intense radiation or during frequent collisions in the high-temperature gas, leaving the atomic nuclei, or protons, behind. However, the protons also recombine with electrons to make a relatively small number of hydrogen atoms, each consisting of one proton and one electron. An equilibrium is therefore set up between the creation and destruction of hydrogen atoms, and calculations indicate that there is always about one hydrogen atom present in the solar corona for every 500 thousand protons. Since there are an awful lot of protons, the number of hydrogen atoms in the corona is enough to produce exceedingly bright ultraviolet radiation.

In any event, the main SOHO UVCS result is that the heavier oxygen ions in polar coronal holes move faster and seem to be hotter than the protons there. If the particle agitation speeds measure temperatures, then the particles are heated to sizzling temperatures in proportion to their mass and the square of their thermal velocity (Focus 5.2). The temperatures of the protons would be about 7 million degrees Kelvin hot, and the oxygen ions would be at searing temperatures of hundreds of millions of degrees Kelvin. So, the oxygen ions would be more than ten times hotter than the center of the Sun.

The apparent temperature differences indicate that the material in coronal holes is not in thermal equilibrium. The oxygen ions and the protons seem to live on their own, isolated from each other and cut off from their neighbors. They are not near enough to each other, and do not have enough time to jostle together and smooth out their velocity or temperature differences.

That would also explain fast-wind observations by other spacecraft, including Ulysses and Helios 1 and 2, that showed that the temperatures of the heavy ions exceed those of the protons far out in the fast winds, and that the protons are hotter than the electrons there (Table 5.1). In contrast, frequent collisions within coronal streamers, where the density is greater, would adjust particle temperatures to similar values, while also wiping out any memory of the initial acceleration mechanism and erasing signatures of it.

Focus 5.2

Temperature, Mass and Motion

The kinetic temperature, T, of a particle is obtained by equating the thermal energy of the particle to its kinetic energy of motion, or by:

$$\text{Thermal Energy} = \frac{3}{2}kT = \frac{1}{2}mV_{\text{thermal}}^2 = \text{Kinetic Energy}\ ,$$

or equivalently:

$$T = \frac{mV_{\text{thermal}}^2}{3k}\ ,$$

where m is the mass of the particle and Boltzmann's constant $k = 1.38066 \times 10^{-23}\,\text{J K}^{-1}$.

For protons in coronal holes, $m = m_{\text{p}} = 1.67 \times 10^{-27}$ kilograms, and the measured velocities of $V = 250$ thousand meters per second imply a temperature of $T = 2.5 \times 10^6$ degrees, or 2.5 million degrees Kelvin if the velocities are thermal. Oxygen ions are 16 times heavier than the protons and moving twice as fast, so they would be 64 times hotter. However, the particles are not in equilibrium, so our basic assumptions are incorrect. In thermal equilibrium, the particles interact with each other often enough to have the same uniform temperature, which is not the observed situation in coronal holes. Moreover, at a given temperature the thermal velocity will decrease with increasing particle mass, and that is also not observed. So, there has to be some non-thermal process that preferentially accelerates the heavier particles in coronal holes (Focus 5.3).

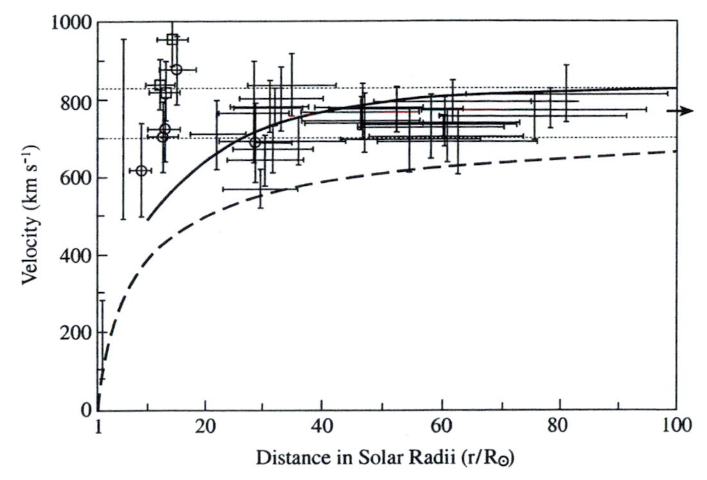

Fig. 5.12. Rapid Acceleration. Acceleration Interferometric observations of interplanetary radio scintillations indicate that the fast solar wind accelerates to its high velocity very close to the Sun, within at least 10 solar radii. The vertical bar on each data point is the 90 percent confidence limit, and the horizontal bar indicates the distance range over which the scintillation estimate is averaged. Measurements with the Very Long Baseline Array, or VLBA for short, are marked with circles and squares. The upper and lower bounds of the Ulysses measurements are plotted as horizontal dotted lines, and the mean Ulysses fast-wind speed is marked with an arrow at 100 solar radii. The flow speed from a wave-assisted acceleration model is plotted as a dashed line, and the apparent scintillation velocity calculated from this model is plotted as a heavy solid line. The point nearest the Sun is estimated from Spartan-201 coronagraph measurements. (Adapted from R. R. Grall, et al., 1996)

Of course, the peculiar thing is that the heavier particles in coronal holes are moving faster. That does not seem to make sense. It would be something like preferentially exciting all the overweight people at a party, and causing them to dance wildly about when all their slim companions are waltzing at a slow tempo.

Waves of magnetism could preferentially accelerate and heat the heavier ions in coronal holes. The ponderous magnetic waves remind us of the waves in a stormy ocean that push heavy logs to shore. The lighter shells twist and spiral about in the pounding surf, rarely reaching the beach. That is why the heaviest debris is sometimes found left on the beach after high tide.

In one explanation for the preferential acceleration of heavy particles in coronal holes , developed by James F. Mc Kenzie, William Ian Axford, Eckart Marsch and others, magnetic waves pump energy into the heavy ions by driving their gyrations around the magnetic field lines. The charged particles move around the magnetic fields in coronal holes in the same way that charged particles move in laboratory particle accelerators or cyclotrons. According to the cyclotron resonance theory, the heavier ions gyrate with lower frequencies where the magnetic waves are most intense (Focus 5.3). The heavier ions therefore consume more magnetic-wave energy, and are pumped up to higher temperatures and accelerated to higher speeds.

As the waves move along the magnetic fields, they will produce rapid gyrations in the direction perpendicular to the fields and little extra motion along them, something like a hula hoop that moves in and out from your hips but not up and down them. UVCS measurements confirm that the oxygen ions are moving at high speed and temperature across the field lines, and at much lower speeds and temperatures along them. That is, the oxygen ion velocity measured in the direction perpendicular to the magnetic field is greater than that in the parallel direction, and a similar velocity anisotropy is found for the protons. So, the polar corona and fast solar wind might be heated by waves which favor the perpendicular temperature of heavier ions. This brings us to magnetized waves blowing further out in the winds.

Focus 5.3

Particles Gyrating About Magnetic Fields

If a magnetic field of strength, $\boldsymbol{B}$, acts on a particle with charge eZ, for electron charge e, and velocity, $\boldsymbol{V}$, the particle experiences a force, $\boldsymbol{F}$, called the Lorentz force:

$$\boldsymbol{F} = eZ(\boldsymbol{V} \times \boldsymbol{B}) ,$$

and from Newton's law for a particle of mass, m, and momentum, $m\boldsymbol{V}$, we have

$$m\frac{\mathrm{d}\boldsymbol{V}}{\mathrm{d}t} = \boldsymbol{F} = eZ(\boldsymbol{V} \times \boldsymbol{B}) ,$$

where we have assumed that gravitational forces are negligible.

The motion of the charged particle is a circle, and it does not change the particle's kinetic energy. If $V_\perp$ denotes the component of velocity perpendicular to the magnetic field, we can rewrite our equation as:

$$\frac{\mathrm{d}^2 V_\perp}{\mathrm{d}t^2} = \frac{eZB}{m}\frac{\mathrm{d}V_\perp}{\mathrm{d}t} = \left(\frac{eZB}{m}\right)^2 V_\perp ,$$

which describes circular motion with a frequency ν_g given by (Focus 2.2):

$$\nu_\mathrm{g} = eZB/(2\pi m) = 2.8 \times 10^{10}(Zm_\mathrm{e}/m)B \text{ Hertz} ,$$

where m_e denotes the electron mass, the magnetic field strength B is in units of Tesla, and one Hertz is equivalent to one cycle per second.

The radius of gyration, R_g, given in Focus 2.2, is obtained from

$$R_\mathrm{g} = \frac{V_\perp}{2\pi\nu_\mathrm{g}} = \frac{mV_\perp}{eZB} .$$

In the context of the acceleration of particles by waves in coronal holes, there is more power in the lower frequencies, and heavier particles gyrate at these lower frequencies.

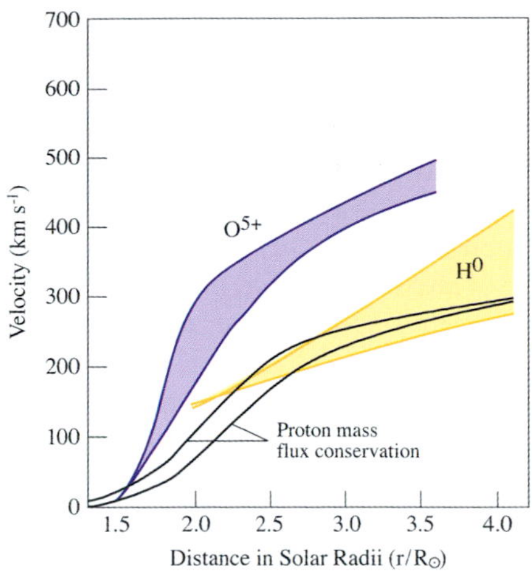

Fig. 5.13. Heavier Ions Move Faster In Coronal Holes. Empirical particle outflow speeds at different distances over the solar poles, derived from SOHO UVCS observations of hydrogen atoms (H I or H^0 – Lyman alpha) and ionized oxygen (O VI or O^{5+}) in late 1996 and early 1997. Here the distances are given in units of the solar radius. The yellow region denotes the range of hydrogen speed able to reproduce the observational data, and the blue region denotes the corresponding range of speed for ionized oxygen. The black lines denote the proton outflow speed derived from mass flux conservation; for a time-steady flow, the product of the density, speed and flow-tube area should be constant. The density and polar-tube information is derived from UVCS white light data, and *in situ* mass flux data from Ulysses define the constant. These data show that the heavier oxygen ions move out of coronal holes at faster speeds than the lighter protons, and that the oxygen ions attain supersonic velocities within 2.5 solar radii from the Sun center. (Adapted from J. L. Kohl et al., 1998, and courtesy of the SOHO UVCS consortium. SOHO is a project of international collaboration between ESA and NASA)

5.4 Magnetized Waves, Exotic Particles and the Edge of the Solar System

Magnetic waves, detected by Ulysses far above the Sun's poles, may also help pump up the flow of the solar wind. The spacecraft's magnetometers detected the continuous presence of magnetic fluctuations in the fast, high-latitude wind, with periods of 10 to 20 hours. They were changes in the direction of the magnetic field, and not variations in its strength. Andre Balogh, Edward J. Smith, Bruce T. Tsurutani, and their colleagues attribute the fluctuations to large-amplitude Alfvén waves, similar in properties to those seen in the fast solar wind near the ecliptic by Helios 1 and 2, and previous spacecraft. When these Alfvén waves are added to the heat-driven wind, they can provide an extra boost that pushes the polar winds to higher speeds. These waves may also explain significant, high-latitude distortions of the spiral shape of the solar magnetic field. The spiral field extending into interplanetary space above the Sun's poles appears to be underwound when compared to expectations using the equatorial rotation rate at the Sun's underlying photosphere.

The magnetic waves rippling through the Sun's polar regions probably block incoming high-energy cosmic rays (Sect. 2.3), intruders from outer space. Scientists expected that the charged cosmic-ray particles would be able to penetrate deep within regions above the Sun's poles, where the magnetic fields stretch out radially and smoothly with little twist. Since solar rotation winds up the solar wind's magnetism in the equatorial plane of the Sun, cosmic rays might have more difficulty in penetrating these regions. After all, increased magnetism in the solar wind, associated with a maximum in the 11-year solar cycle, was known to cut off cosmic rays so that fewer of them reach the Earth at activity maximum (Sect. 2.4).

Ulysses' instruments surprised nearly everyone; they did not register substantially more cosmic rays over the poles than near the ecliptic. As suggested by J. Randy Jokipii and Joseph Kóta in 1989, strong magnetic waves in the polar regions can repel the high-energy ions back into space (Fig. 5.14). The incoming cosmic rays meet an opposing force, like a swimmer entering the surf on a distant shore or one trying to swim upstream against the current of a powerful river. To put it in more scientific language, the Alfvén waves are very long, with wavelengths reaching 0.3 AU, so they can resonate with the energetic cosmic rays and oppose their entry into the polar regions.

The magnetic detectors on Ulysses made another interesting observation; the wind's distant magnetic field is uniform at all latitudes. In contrast, the

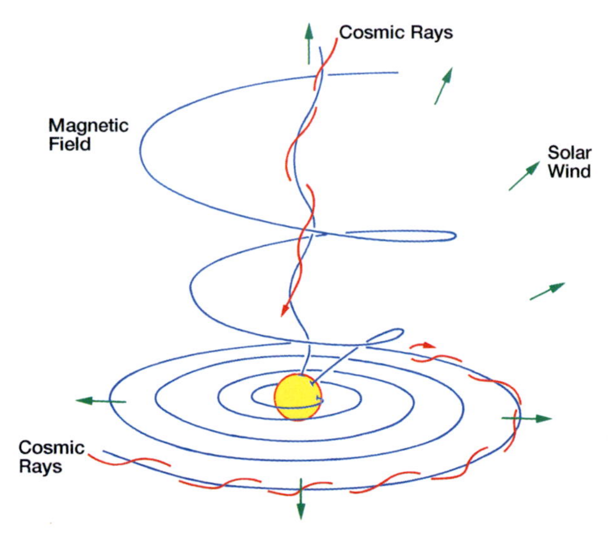

Fig. 5.14. Diffusion of Cosmic Rays. Within the ecliptic, or the plane of the solar equator, the magnetic fields of the solar wind are wound up in a spiral pattern due to the rotation of the Sun. Since the solar rotation velocity is lower at higher latitudes, the magnetic fields are oriented more or less radially above the solar poles. Scientists therefore predicted that the abundance of cosmic-ray ions would increase over the Sun's polar regions. However, Ulysses did not find the expected increase of cosmic rays, apparently because strong magnetic waves above the poles act to repel the high-energy ions back into space

dipole magnetic field near the surface of the Earth is more concentrated over the poles, with magnetic fields that come together at the poles and spread out between them (Sect. 7.1, Fig. 7.1). Edward J. Smith and Andre Balogh found that this kind of intensification does not exist out at Ulysses' distance from the Sun, where there is no concentration of magnetic flux above the solar poles.

As the spacecraft traveled along its orbit, the magnetometers did not detect any variations in the radial component of the magnetic field, indicating that the magnetic flux is independent of heliographic latitude. This is what one would expect from a dipolar magnetic field with a neutral current sheet wrapped around the solar equator, where the streamers appear at activity minimum. The dipole is stretched way out at its middle, resulting in two polar monopoles whose magnetic field lines do not cross the equatorial regions. Magnetic pressure may be driving a non-radial expansion of the solar wind near the Sun, pushing magnetic fields from the polar regions to lower latitudes, like strands of seaweed in ocean currents. This would redistribute the magnetism and produce a uniform radial field.

There are all kinds of particles in the space near Earth, originating at the Sun, in other dying stars, or in space itself. Most of them flow out of the Sun with the solar wind; these have relatively low energies associated with the production of the wind, unless briefly accelerated during transient explosions in the corona. The cosmic rays are distinguished by both their low density and very high energies (Sect. 2.3). They are most likely hurled into space when an entire star explodes. Other types of rapidly-moving particles are energized in interplanetary space.

Some ions found in the solar wind move far too swiftly to have been accelerated at the wind's source. When the fast winds overtake the more sluggish flows in their path, they produce shocks that locally accelerate ions to high speeds, like ocean waves taking a surfer for a ride. Such shocks, originating at places where the slow and fast winds interact, have been observed many times with Ulysses.

Other heavy nuclear particles, such as helium, nitrogen and oxygen, come from interstellar space, but they are ionized and accelerated within the solar system. Because the compositions of these particles are peculiar, when compared to other types of low-energy cosmic rays, scientists dubbed them anomalous cosmic rays. They were first observed in the early 1970s, and their unusual composition provided the vital clue that enabled Lennard A. Fisk and his colleagues to deduce their origin.

The anomalous material begins its life as electrically neutral, uncharged interstellar atoms, and then drifts into the solar system at low velocities, where ultraviolet sunlight tears off just one electron from each atom and thereby ionizes the atom. The singly ionized atoms have increased mobility in the interplanetary magnetic field, compared with other cosmic rays that have lost most of their electrons by more violent processes, often associated with the death throes of stars.

The newborn ions are then picked up by the magnetic fields entrained in the outflowing solar wind, and taken for a ride out to the edge of the solar system. There they can rebound at higher velocities due to shocks, something like a tennis ball bouncing off a wall. Ulysses discovered a vast population of the pick-up ions, many for the first time, showing that they are also accelerated where the fast and slow winds interact. These particles have additionally been used to determine the abundance of elements in the interstellar space that they come from.

Where does it all end? The relentless solar wind streams out in all directions, rushing past the planets and clearing out a large cavity in interstellar space. This huge Sun-centered bubble is called the heliosphere, from the Greek word *Helios* for the Sun (Fig. 5.15). Within the heliosphere, physical conditions are dominated by the magnetized and electrified solar wind.

Contemporary spacecraft have measured the shape and content of the heliosphere. One instrument on board SOHO, the Solar Wind ANisotropies (SWAN), examines interstellar hydrogen atoms sweeping through our solar

Focus 5.4

Edge of the Solar System

The solar wind carves out a cavity in the interstellar medium known as the heliosphere. The radius of the heliosphere can be estimated by determining the standoff distance, or stagnation point, in which the ram pressure, P_W, of the solar wind falls to a value comparable to the interstellar pressure, P_I. As the wind flows outward, its velocity remains nearly constant, while its density decreases as the inverse square of the distance. The dynamic pressure of the solar wind therefore also falls off as the square of the distance, and we can use the solar wind properties at the Earth's distance of 1 AU to infer the pressure, P_{WS}, at the stagnation point distance, R_S. Equating this to the interstellar pressure we have:

$$P_{WS} = P_{1\,AU} \times \left(\frac{1\,AU}{R_S}\right)^2 = (m_p N_{1\,AU} V_{1\,AU}^2) \times \left(\frac{1\,AU}{R_S}\right)^2 = P_I \,,$$

where the proton's mass $m_p = 1.67 \times 10^{-27}$ kg, the number density of the solar wind near the Earth is about $N_{1\,AU} = 5$ million protons per cubic meter and the velocity there is about $V_{1\,AU} = 400$ thousand meters per second.

To determine the distance to the edge of the solar system, R_S, we also need to know the interstellar pressure, and that is the sum of the thermal pressure, the dynamic pressure, and the magnetic pressure in the local interstellar medium. It is about $P_I = (1.3 \pm 0.2) \times 10^{-13}\ N\,m^{-2}$. The estimate obtained from the equation is $R_S = 100$ AU or more, far beyond the orbits of the known outer planets. However, the estimates by different authors give a broad range for the distance to the edge of the solar system, depending on the uncertain values of various components of the interstellar pressure.

system from elsewhere. The Sun's ultraviolet radiation illuminates this hydrogen, much the way that a street lamp lights a foggy mist at night. Since solar wind particles tear the hydrogen atoms apart, their ultraviolet glow outlines the asymmetric shape of the Sun's winds and also establishes its flux.

So far SWAN's measurements indicate that the solar wind is more intense in the equatorial plane of the Sun than over the north or south poles, which is consistent with Ulysses' measurements of the latitudinal variations of wind speed and density. Both SWAN and Ulysses find that 2.2 million million (2.2×10^{12}) protons zip through every square meter of the solar wind every second out at 1 AU, the distance of the Earth's orbit. That amounts to a quarter million tons of protons every second, obtained by multiplying the proton flux by the area at the Earth's distance and the mass of the proton, and noting that one ton = 1,000 kilograms. Moreover, since roughly the same flux was detected by the Helios spacecraft in 1974–75, the amount of this loss apparently remains essentially constant for decades.

Since the solar wind thins out as it expands into a greater volume, it is eventually no longer powerful enough to repel interstellar matter. The Sun's wind meets the ionized matter coursing between the stars at a turbulent boundary called the heliopause (Fig. 5.15). There is a celestial standoff out there between the solar wind and interstellar forces, like two gunfighters facing off at sundown.

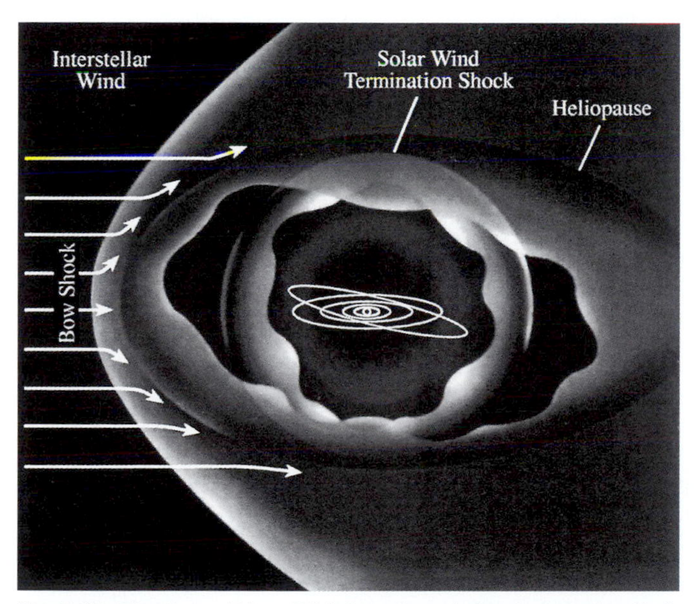

Fig. 5.15. The Outer Edge. With its solar wind going out in all directions, the Sun blows a huge bubble in space called the heliosphere. The heliopause is the name for the blurred boundary between the heliosphere and the interstellar gas outside the solar system. Interstellar winds mold the heliosphere into a non-spherical shape, creating a bow shock where they first encounter it. The orbits of the planets are shown near the center of the drawing.

The size of the heliosphere has been inferred from the twin Voyager spacecraft, cruising far beyond the outermost planets. At the time of writing they are more than 21 years old and approaching 71 AU – they don't make them like they used to. Strong shock waves, associated with intense explosions on the Sun, have plowed into the cold interstellar gas at the heliopause, generating a hiss of radio noise detected by the remote Voyagers. Thirteen months before the spacecraft detected the radio hiss, unusually intense eruptions on the Sun generated one of the largest interplanetary disturbances ever observed. From the measured speed of the disturbance, and the time it took to travel to the heliopause and generate the radio signals, the outer edge of our solar system has been located somewhere between 110 and 160 AU, or roughly a hundred times further from the Sun than the Earth. That is where the solar system ends, in a gigantic distant wall of compressed gas that fences off our Sun from the rest of the cosmos.

Table 5.2. Key events in studies of the solar wind*

Date	Event
1869	At the solar eclipse of 7 August 1869, Charles A. Young, and, independently, William Harkness, discover a single, bright, green emission line in the spectrum of the solar corona. This conspicuous feature remained unidentified with any known terrestrial element for more than half a century, but it was eventually associated with highly ionized iron atoms missing 13 electrons (Fe XIV), indicating that the corona has a million-degree temperature (see 1939–41).
1889–90	Frank H. Bigelow argues that the structure of the corona detected during solar eclipses provides strong evidence for large-scale solar magnetic or electric fields. He correctly speculated that polar rays delineate open field lines along which material escapes from the Sun, and that equatorial elongations of the corona mark closed magnetic field lines.
1896–1913	Kristian Birkeland argues that polar auroras and geomagnetic storms are due to beams of electrons.
1908	George Ellery Hale measures intense magnetic fields in sunspots, thousand of times stronger than the Earth's magnetism.
1919	Frederick Alexander Lindemann (later Lord Cherwell) suggests that an electrically neutral plasma ejection from the Sun is responsible for non-recurrent geomagnetic storms.
1939–41	Walter Grotrian and Bengt Edlén identify coronal emission lines with highly ionized elements, indicating that the Sun's outer atmosphere has a temperature of millions of degrees Kelvin. The conspicuous green emission line was identified with Fe XIV, an iron atom missing 13 electrons.
1944–56	Herman Bondi, Fred Hoyle and William H. Mc Crea develop a theory for spherically symmetric accretion of interstellar matter by a star, including a critical solution for transonic accretion flow; it is applicable to the solar wind, with flow away from, instead of into, the star.
1946	Vitaly L. Ginzburg, David F. Martyn and Joseph L. Pawsey independently confirm the existence of a million-degree solar corona from observations of the Sun's radio radiation.
1948–49	Soft X-rays from the Sun are first detected on 5 August 1948 with a V-2 rocket experiment performed by the U. S. Naval Research Laboratory, reported by T. R. Burnight in 1949. Subsequent sounding rocket observations

	by the NRL scientists revealed that the Sun is a significant emitter of X-rays and that the X-ray emission is related to solar activity.
1950–54	Scott E. Forbush demonstrates the inverse correlation between the intensity of cosmic rays arriving at Earth and the number of sunspots over two 11 year solar activity cycles.
1951–52	Herbert Friedman and his colleagues at the U. S. Naval Research Laboratory use instruments aboard sounding rockets to show that the Sun emits enough X-ray and ultraviolet radiation to create the ionosphere.
1951–57	Ludwig F. Biermann argues that a continuous flow of solar corpuscles is required to push comet ion tails into straight paths away from the Sun, correctly inferring solar wind speeds of between 0.5 and 1.0 million meters per second.
1955	Leverett Davis Jr. argues that solar corpuscular emission will carve out a cavity in the interstellar medium, now known as the heliosphere, accounting for some observed properties of low-energy cosmic rays.
1955	Horace W. Babcock and Harold D. Babcock use magnetograms taken over a two-year period, 1952 to 1954, to show that the Sun has a general dipolar magnetic field of about 10^{-4} Tesla, or 1 Gauss, in strength usually limited to heliographic latitudes greater than ± 55 degrees. Bipolar regions are found at lower latitudes. They argued that occasional extended unipolar areas of only one outstanding magnetic polarity might be related to 27-day recurrent terrestrial magnetic storms.
1956	Peter Meyer, Eugene N. Parker and John A. Simpson argue that enhanced interplanetary magnetism at the peak of the solar activity cycle deflects cosmic rays from their Earth-bound paths.
1957	Max Waldmeier observes intense coronal emission lines, calling attention to seemingly vacant places that he called coronal holes.
1957–59	Sydney Chapman shows that a hot, static corona should extend to the Earth's orbit and beyond.
1957–67	Rocket observations by Richard Tousey and colleagues at the U. S. Naval Research Laboratory indicate that the brightest line in the ultraviolet spectrum of the solar corona is the Lyman alpha transition of hydrogen atoms.
1958	Eugene N. Parker suggests that a perpetual supersonic flow of electric corpuscles, that he called the solar wind, naturally results from the expansion of a very hot corona. He also demonstrates that the solar magnetic field will be pulled into interplanetary space, modulating the amount of cosmic rays reaching Earth and attaining a spiral shape in the plane of the Sun's equator due to the combined effects of the Sun's rotation and radial wind flow.
1960–61	Konstantin I. Gringauz reports that the Soviet spacecraft, Lunik 2, launched on 12 September 1959, has measured high-speed ions in interplanetary space outside the Earth's magnetic field, with a flux of 2 million million (2×10^{12}) ions (protons) per square meter per second.
1962–67	Mariner 2 was launched on 7 August 1962. Using data obtained during Mariner's voyage to Venus, Marcia Neugebauer and Conway W. Snyder demonstrate that a low-speed solar wind plasma is continuously emitted by the Sun, and discover high-speed wind streams that recur with a 27-day period within the orbital plane of the planets.
1964–66	Magnetometers aboard NASA's Interplanetary Monitoring Platform 1, launched on 27 November 1963, are used by Norman F. Ness and John M. Wilcox to measure the strength and direction of the interplanetary magnetic field. They show that the interplanetary magnetic field is pulled into a spiral shape by the combined effects of the Sun's rotation and radial wind flow. They also discover large-scale magnetic sectors in interplanetary space that point toward or away from the Sun.

1966 Peter A. Sturrock and R. E. Hartle propose a two-component (electrons and protons) model of the solar wind driven by electron heat conduction from a hot corona. However, the wind is too slow and the protons are too cool when compared with observations of the fast solar wind, normalized to the distance of the Earth's orbit.

1967 Edmund J. Weber and Leverett Davis Jr. consider the effects of solar rotation and magnetic fields on a steady solar-wind flow in the equatorial plane. They show that co-rotation of the wind and Sun exists out to the Alfvén critical point, or radial distance, where the Alfvén Mach number is one.

1969–71 Magnetic fluctuations are observed in the solar wind from Mariner 5 on its way to Venus, and attributed to large-amplitude Alfvén waves by John W. Belcher, Leverett Davis Jr., and Edward J. Smith.

1970 Klaus Jockers and others demonstrate that totally collisionless (exospheric) models of the solar wind do not work.

1970 Thomas E. Holzer and William Ian Axford review the theory of steady, radial, spherically symmetric flow, introducing heating of the corona to several million degrees (with thermal velocities above the Sun's escape velocity) and showing that rapid wind acceleration occurs close to the Sun. They additionally pointed out that ionization states can be used to determine coronal temperatures.

1971 Alan H. Gabriel explains the coronal Lyman alpha line in terms of resonant scattering of ultraviolet light generated below the corona.

1971 Gerald W. Pneuman and Roger A. Kopp propose a dipolar magnetic model for the Sun, in which the solar wind drags the Sun's magnetic field into a neutral current sheet of oppositely directed magnetism near the equator.

1972 Johannes Geiss and Hubert Reeves publish measurements of the solar wind helium abundance using foil collectors left by American astronauts on the Moon.

1973 Allen S. Krieger, Adrienne F. Timothy and Edmond C. Roelof compare an X-ray photograph, obtained during a rocket flight on 24 November 1970, with satellite measurements of the solar wind to show that coronal holes are the source of recurrent high-speed streams in the solar wind.

1973 Giuseppe S. Vaiana and his colleagues use solar X-ray observations taken from rockets during the preceding decade to identify the three-fold magnetic structure of the solar corona – coronal holes, coronal loops and X-ray bright points.

1973–77 X-ray photographs of the Sun taken with the Apollo Telescope Mount on the manned, orbiting Skylab satellite, launched on 14 May 1973, fully confirm coronal holes, the ubiquitous coronal loops and X-ray bright points. Detailed comparisons of Skylab X-ray photographs and measurements of the solar wind, made from Interplanetary Monitoring Platforms (IMPs) 6, 7 and 8, confirm that solar coronal holes are the source of the high-velocity solar wind streams as well as 27-day-recurrent geomagnetic disturbances. John Wilcox had previously suggested that the fast streams might originate in magnetically open, unipolar regions on the Sun, but the Skylab X-ray photographs definitely identified the place.

1974–78 Hannes Alfvén, and independently Lief Svalgaard and John M. Wilcox, interpret the magnetic structure of the solar wind, at activity minimum, in terms of a warped neutral current sheet dividing the solar wind into two hemispheres of opposite magnetic polarity.

1974–86 The Helios 1 and 2 spacecraft, respectively launched on 10 December 1974 and on 15 January 1976, measure the solar wind parameters as close as 0.3 AU from the Sun for a whole 11-year solar cycle. They confirmed the existence of two kinds of solar wind flow. There is a steady, uniform high-speed wind and a varying, slow-speed wind.

1976	Alan H. Gabriel introduces a two-dimensional model of the chromosphere and corona, and the transition region between them, in which magnetic flux is concentrated at the boundaries of supergranular convection cells that produce a magnetic network.
1977–91	Interplanetary scintillations of extragalactic radio sources are used by Barney J. Rickett and William A. Coles, and independently by Masayoshi Kojima and Takakiyo Kakinuma, to investigate the solar wind speed outside the plane of the ecliptic. Their data show that near the minimum in the 11-year solar activity cycle, the slow wind is confined to low latitudes, while the fast wind emanates from high latitudes. Near activity maximum, the slow wind is dominant over the whole range of observable latitudes.
1977–80	Helmuth R. Rosenbauer, Wolfgang K. H. Schmidt and their colleagues use measurements from Helios 1 and 2 and the first International Sun-Earth Explorer (ISEE-1) spacecraft to show that helium and other heavy ions are hotter and move faster than protons in the high-speed wind. In addition, the electrons are cooler than the protons in this fast component of the solar wind.
1978	Edward J. Smith, Bruce T. Tsurutani and Ronald L. Rosenberg use observations from Pioneer 11 to show that the solar wind becomes unipolar, or obtains a single magnetic polarity, at high heliographic latitudes near 16 degrees.
1982–83	Gary J. Rottman, Frank Q. Orrall and James A. Klimchuk obtain rocket observations of extreme ultraviolet resonance lines formed in the low corona and transition regions, showing that the lines are systematically shifted to shorter wavelengths in large polar coronal holes with well developed, low-latitude extensions. Outflow velocities of between 7 to 8 thousand meters per second are inferred from these Doppler shifts.
1986	James F. Dowdy, Jr., Douglas M. Rabin and Ronald L. Moore show that narrow magnetic funnels open up into the base of the corona, emerging from only a fraction of the magnetic network.
1989	J. Randy Jokipii and Joseph Kóta argue that Alfvén waves streaming out of the Sun's polar regions may block incoming cosmic rays.
1992	Yutaka Uchida and his colleagues use data from the Soft X-ray Telescope aboard Yohkoh to show that the active-region corona is continuously expanding, perhaps as the source of the slow solar wind. This idea was subsequently supported by SOHO LASCO coronagraph images that show expanding coronal loops near the Sun's equatorial regions.
1994–95	The Ulysses spacecraft measures the velocity of the solar wind over the full range of solar latitudes. The spacecraft moved above the poles of the Sun at a distance of about 2 AU, near a minimum in the Sun's 11-year activity cycle. John L. Phillips and his colleagues use correlation with other data to show that the slow wind is narrowly confined to low latitudes above an equatorial steamer belt. They also showed that at least some of the fast wind is emitted from polar coronal holes, and that the fast wind extends to lower latitudes than the radial extension of coronal holes.
1995	Andre Balogh, Edward J. Smith and their colleagues show that the distant radial component of the magnetic field detected, and normalized to 1 AU, by Ulysses does not vary with solar latitude.
1995–96	Ulysses' measurements of cosmic rays, obtained by John A. Simpson and colleagues, do not show substantially more cosmic rays above the Sun's poles than near the ecliptic. This may be explained by Alfvén waves observed in the polar fast wind by Edward J. Smith, Bruce T. Tsurutani and colleagues with Ulysses; the magnetic waves repel incoming cosmic rays above the polar regions.

1995–97 James F. Mc Kenzie, William Ian Axford, Eckart Marsch and others develop fast solar wind models in which magnetic waves heat the corona and preferentially accelerate heavier ions.

1995–97 Data from the Ulysses, Yohkoh and SOHO spacecraft independently show that the polar fast wind originates in a relatively low electron temperature region (coronal holes) when compared with the electron temperature of the source of the slow wind.

1997 Eckart Marsch and Chuan-Yi Tu provide a theoretical model in which the solar wind is heated and accelerated by Alfvén waves in magnetic funnels opening into the corona from the chromosphere magnetic network.

1997–98 John L. Kohl, Giancarlo Noci and their colleagues use SOHO UVCS measurements to show that oxygen ions flowing out of coronal holes have extremely high energies corresponding to temperatures of over 200 million degrees Kelvin, and accelerate to supersonic outflow velocities within 2.5 solar radii of Sun center. The outflow velocities of the oxygen ions are faster than protons in coronal holes, and electrons in these regions move at even slower speeds.

1997–98 Neil R. Sheeley, Yi-Ming Wang and their colleagues use time-lapse SOHO LASCO coronagraph sequences to show that one component of the slow wind may be emitted far out in coronal streamers, and that it does not accelerate to terminal velocity until 20 or 30 solar radii from Sun center.

1998 Klaus Wilhelm and colleagues use the SUMER instrument on SOHO to determine the physical parameters above a polar coronal hole, showing that polar plumes are not a major source of fast solar wind streams.

1998 Donald M. Hassler and his colleagues use the SOHO SUMER spectrometer to show that the high-speed, solar-wind outflow velocity, observed in the low corona in a polar coronal hole, is spatially correlated with the boundaries of the magnetic network seen in the underlying chromosphere.

---

* See the References for complete references to these seminal papers. An AU is the mean distance of the Earth from the Sun, or about 146 billion (1.46×10^{11}) meters.

6. Our Violent Sun

Overview

The relatively calm solar atmosphere can be torn asunder by sudden, brief and intense outbursts called solar flares. They are the most powerful explosions in the solar system, releasing energies of up to 20 million 100 megaton terrestrial nuclear bombs (amounting to 10^{25} Joule or 10^{32} erg) in 100 to 1,000 seconds.

Solar flares generate high-speed electrons that emit intense radiation at invisible radio and X-ray wavelengths.

Flare-associated protons and heavier ions can be beamed down into the lower solar atmosphere, producing nuclear reactions with the emission of gamma-ray spectral lines, meson decay gamma rays and neutrons that move nearly at the speed of light.

Another type of explosive solar activity, called coronal mass ejections, or CMEs for short, is observed using white-light coronagraphs aboard spacecraft, including the Large Angle Spectrometric COronagraph (LASCO) aboard the SOlar and Heliospheric Observatory (SOHO). CMEs expand away from the Sun at speeds of hundreds of thousands of meters per second, becoming larger than the Sun and removing up to fifty billion tons (5×10^{13} kilograms) of coronal material. The depleted regions of the corona are detected as reductions in the soft X-ray emission observed by the Yohkoh spacecraft.

Explosive solar flares and CMEs can be ignited when current-carrying magnetic loops come together and coalesce in a process called magnetic reconnection. During this coronal merging process, the stressed magnetic fields partially annihilate each other, release energy stored in them, and reconnect into less-energetic, more stable configurations. The Soft X-ray Telescope, or SXT, aboard Yohkoh finds a soft X-ray, cusp-type geometry in the low corona, seen edge-on at the apex of long-lived, gradual (hours) flaring loops that can be associated with coronal mass ejections. This morphology was predicted by magnetic reconnection theory. Yohkoh's Hard X-ray Telescope, or HXT, has discovered non-thermal, loop-top hard X-ray sources located just above short-lived, impulsive (minutes) flaring loops detected by the SXT, suggesting that very energetic electrons can also be accelerated near the reconnection site.

Impulsive solar flares eject protons and electrons into interplanetary space with energies that are thousands and even millions of times greater than those usually present in the solar wind. These particles can be directed by the interplanetary magnetic field to collision with the Earth. Fast coronal mass ejections generate strong interplanetary shock waves that can also hit the Earth; they are associated with long-duration soft X-ray flares and intense long-lived proton and electron events.

6.1 Brightening in the Chromosphere

Transient, explosive perturbations of the solar atmosphere, now called solar flares, occur in solar active regions, the places in, around and above sunspots. Exceptionally powerful ones can be detected in the combined colors of sunlight, or in white light. They were first observed and carefully recorded by two Englishmen, Richard C. Carrington and Richard Hodgson, who noticed a fleeting brightening near a complex group of sunspots on 1 September 1859, lasting just a few minutes. These white-light flares, as they are now called, create only a minor perturbation in the steady luminous output of the photosphere, so they are rarely seen.

Routine visual observations of explosive solar disturbances became possible after George Ellery Hale's development of the spectrohelioscope in the late 1920s. This instrument allowed the Sun to be imaged in the monochromatic radiation of hydrogen atoms – the red hydrogen alpha spectral feature, designated Hα, at 656.3 nanometers. Light at this wavelength originates just above the photosphere, in the chromospheric layer of the solar atmosphere.

When viewed in hydrogen alpha, solar flares appear as sudden brightenings, lasting from a few minutes to an hour, usually in strong, complex magnetic regions. They emit a sudden flash of red light followed by a slower decay, somewhat like suddenly igniting a fire in a pool of gasoline.

The power of a solar flare in red hydrogen light is relatively insignificant when compared to the rest of the Sun's visible radiation. At maximum emission, a large solar flare emits only about 10^{19} watts in Hα light, or about 10^{22} Joule of energy during its flash. A power of one watt is equivalent to the emission of one Joule of energy every second. The power emitted by the same area as a flare in visible sunlight is thousands of times greater, and the total light coming from the Sun is tens of million of times more powerful (3.86×10^{26} watts). As we shall see, most of the awesome energy liberated during solar flares is released in a totally different form, as high-speed electrons and invisible radiation, at the rate of up to a million times that of the Hα emission.

For more than half a century, astronomers throughout the world have used hydrogen alpha observations in a vigilant patrol for chromospheric flares, like hunters waiting for the sudden flush of game birds, providing detailed knowledge of their size, shape, location and temporal evolution. The Hα flares are nearly always located close to sunspots and comparable in area to them, often occupying less than one ten thousandth (0.01 percent) of the Sun's visible disk. Flares occur more often when sunspots are most numerous, and a magnetically complex sunspot group is most likely to emit a flare. The greatest flare frequency is found in groups of sunspots in the so-called delta, or δ, configuration having two or more sunspot umbrae of opposite polarities within a common penumbra. All of this suggested long ago, in the 1940s and 1950s, that flares are energized by the powerful magnetism associated with sunspots.

The Hα eruptions do not occur directly above sunspots, but are instead located between regions of opposite magnetic polarity, near the line or place marking magnetic neutrality. They often appear on each side of the magnetic

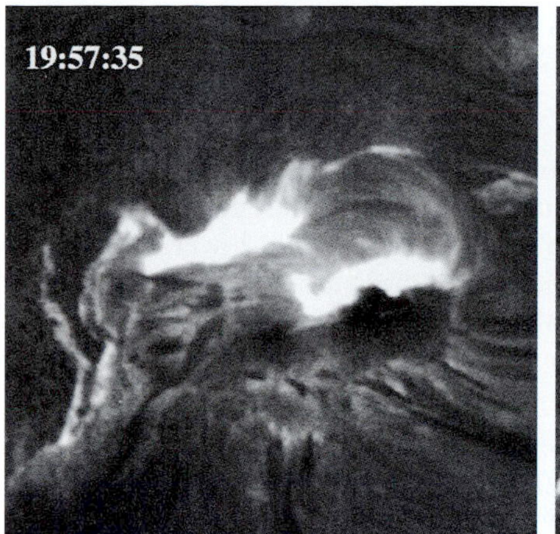

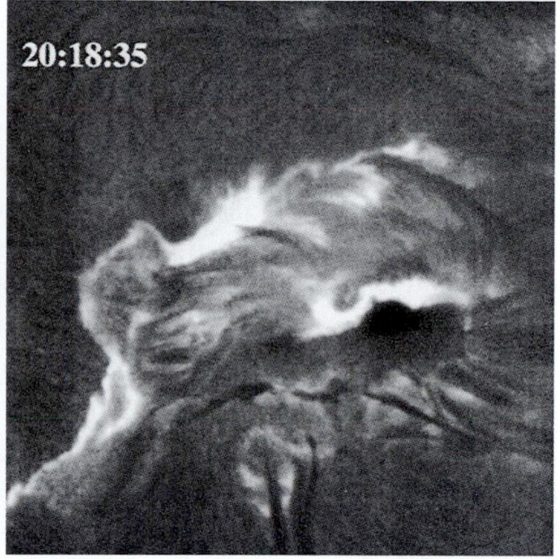

Fig. 6.1. Chromospheric Brightening. A large solar flare observed in the red light of hydrogen alpha (Hα), showing two extended, parallel flare ribbons in the chromosphere. Each image is 200 million meters in width, subtending an angle of 300 arc seconds or about one sixth of the angular extent of the Sun. These photographs were taken at the Big Bear Solar Observatory on 29 April 1998. (Courtesy of Haimin Wang)

neutral line as two extended, parallel ribbons (Fig. 6.1), like the double yellow line at the center of a highway. The two ribbons move apart as the flare progresses, and the space between them is filled with higher and higher shining loops while the lower ones fade away.

Hydrogen alpha images of the thin chromospheric slice of the Sun's atmosphere provide a two-dimensional, flatland picture without information about what is happening above or below it. Observations at radio and X-ray wavelengths provide a three-dimensional perspective, leading to a more complete understanding of the physical processes responsible for solar flares.

6.2 Solar Radio Bursts

The radio emission of a solar flare is often called a radio burst to emphasize its brief, energetic and eruptive characteristics. During such outbursts, the Sun's radio emission can increase up to a million times normal intensity in just a few seconds, so they can outshine the entire Sun at radio wavelengths. Solar radio bursts are very effective probes of the physical state of the flaring solar atmosphere, providing an important diagnostic tool for magnetic and temperature structures and detecting signatures of electrons accelerated to very high speed, approaching that of light. The expulsion of these energetic electrons has been confirmed by direct *in situ* measurements in interplanetary space (Sect. 6.7).

During World War II, amateur radio operators reported a curious noise or hiss in their receivers, occurring only in daytime, and British radar systems, operating at meter wavelengths, were occasionally jammed by severe radio interference coming from the Sun. This solar radio noise was investigated soon after the war, when sudden, intense radio outbursts were detected.

As first shown by Ruby Payne-Scott, D. E. Yabsley and John G. Bolton in 1947, the bursts do not occur simultaneously at different radio frequencies or wavelengths, but instead drift to later arrival times at lower frequencies and longer wavelengths. This was explained by a disturbance that travels out through the progressively more rarefied layers of the solar atmosphere, making the local electrons in the corona vibrate at their natural frequency of oscillation, called the plasma frequency (Focus 6.1).

In the 1950s, J. Paul Wild's group in Australia used a swept-frequency receiver to distinguish at least two types of meter-wavelength radio bursts (Table 6.1 and Fig. 6.3). Designated as type II and type III bursts, they both show a drift from higher to lower frequencies, but at different rates. Solar radio astronomers usually measure this frequency in units of MHz, where one MHz is equal to a million, or 10^6, Hz. A frequency of 300 MHz corresponds to a wavelength of 1 meter.

Ground-based instruments observe the dynamic spectra, or frequency drift, for solar radio bursts between 10 MHz and 8,000 MHz. Solar radiation is reflected by the Earth's ionosphere at frequencies below 10 MHz, or wavelengths longer than 30 meters, and cannot reach the ground. Terrestrial long-wavelength radio communication utilizes this reflective capability of the ionosphere to get around the curvature of the Earth. Instruments have been sent into space, beyond the ionosphere, to detect the frequency drift of solar radio bursts from as low as 0.01 MHz, or 10 kHz, to 10 MHz, first in 1963 from the Alouette-I satellite followed by the first Radio Astronomy Explorer (RAE-1) launched in 1968.

The most common bursts detected at meter wavelengths are the fast-drift type III bursts that provide evidence for the ejection of very fast particles from the Sun. These radio bursts last for only a few minutes at the very onset of solar flares and extend over a wide range of radio frequencies (Fig. 6.3). The rapid frequency drift corresponds to an outward velocity of about half the speed of light (Focus 6.1).

Electrons have to be accelerated to very high energies to move this fast. An electron's energy is often specified in units of kilo electron volts, or keV for short. One keV is equivalent to one thousand electron volts, and to an energy of 1.6×10^{-16} Joule. An electron volt is the energy acquired by an electron when it is accelerated through a potential difference of one volt. The electrons responsible for type III bursts have been accelerated to energies of about 100 keV.

A type III radio burst emits non-thermal radiation, and cannot be due to the thermal radiation of a hot gas. Thermal radiation is emitted by a collection of particles that collide with each other and exchange energy frequently, giving a distribution of particle energy that can be characterized by a single temperature. Impulsive solar flares do not have enough time to achieve this equilibrium, and the flaring electrons are far too energetic for a thermal

Table 6.1. Types of solar radio bursts*

Type I Bursts (Noise Storms)	Long-lived (hours to days) sources of radio emission with brightness temperatures from 10 million to a billion (10^7 to 10^9) degrees Kelvin. Although noise storms are the most common type of activity observed on the Sun at meter wavelengths, they are not associated with solar flares. Noise storms are attributed to electrons accelerated to modest energies of a few keV within large-scale magnetic loops that connect active regions to more distant areas of the Sun.
Type II Bursts	Meter-wavelength type II bursts have been observed at frequencies between 0.1 and 100 MHz. A slow drift to lower frequencies at a rate of about 1 MHz per second suggests an outward motion at about a million meters per second and has been attributed to shock waves.
Type III Bursts	The most common flare-associated radio bursts at meter wavelengths, observed from 0.1 to 1,000 MHz. Type III bursts are characterized by a fast drift from high to low frequency, at a rate of up to 100 MHz per second. They are attributed to beams of electrons thrown out from the Sun with kinetic energies of 10 to 100 keV, and velocities of up to half the velocity of light, or 150 million meters per second. The U-type bursts are a variant of type III bursts that first decrease in frequency and then increase again, indicating motion along closed magnetic field lines.
Type IV Bursts	Broad-band continuum radiation lasting for up to one hour after impulsive flare onset. The radiation from a Type IV burst is partly circularly polarized, and has been attributed to synchrotron emission from energetic electrons trapped within magnetic clouds that travel out into space with velocities from several hundred thousand to a million meters per second.
Centimeter Bursts	Impulsive continuum radiation at centimeter wavelengths that lasts just a few minutes. These microwave bursts are attributed to the gyrosynchrotron radiation of high-speed electrons accelerated to energies of 100 to 1,000 keV. The site of acceleration is located above the tops of coronal loops.
Millisecond Bursts	Radio flares can include literally thousands of spikes, each lasting a few milliseconds, suggesting sizes less than a million meters across and brightness temperatures of up to a million billion (10^{15}) degrees, requiring a coherent radiation mechanism.

* A frequency of one MHz corresponds to a million Hz, or 10^6 Hz. An energy of one keV corresponds to 1.6×10^{-16} Joule or 1.6×10^{-9} erg.

process to work. To make electrons travel at half the speed of light, a thermal gas would have to be heated to implausible temperatures of 1.5 billion degrees, or about 100 times hotter than the center of the Sun. The non-thermal radio bursts are instead accelerated by other processes.

For one special variant of type III bursts, the frequency first decreases and then increases, suggesting disturbances that travel up and down in the corona. They are called U-type bursts since they have the shape of an inverted U in spectral recordings of burst frequency as a function of time. This shape has been attributed to electron trajectories along closed magnetic field lines. The U-type of fast-drift bursts occurs less commonly than the normal type III bursts.

The radio emission from the slow-drift type II bursts consists of two bands that drift to lower frequencies at a leisurely pace, corresponding to an outward velocity of about a million meters per second. They are excited by shock waves set up at the time of a solar explosion and moving out into space. Spatially resolved radio interferometric observations in the 1950s confirmed the outward motion of type II bursts at these speeds, and also indicated that they are very large, with angular extents that can become comparable to that of the Sun.

The associated shocks are also detected with visual telescopes tuned to the red hydrogen alpha, or Hα, line. They are then seen moving across the chromosphere like ripples created when a stone drops in a pond. Such disturbances are named Moreton waves after their discovery by Gail E. Moreton and her colleagues using time-lapse Hα photography in 1960–61.

Other radio outbursts associated with solar flares, and designated type IV bursts, were described by André Boischot in 1957, using radio data obtained at the Nancay Observatory in France. The best studied component is the so-called moving type IV observed at frequencies from roughly 10 MHz to 100 MHz.

Boischot and Jean-Francoise Denisse attributed the moving type IV bursts to energetic electrons trapped in a magnetic cloud that is propelled outward into the solar atmosphere. Radio telescopes have tracked such radiating clouds of gas as they carry their high-speed electrons and entrained magnetic fields to distances of at least a billion meters. They indicate that the broad-band continuum radiation moves outward at a speed of 100 thousand to a million meters per second. Like the non-thermal electrons that produce type III bursts, the electrons giving rise to type IV bursts are traveling at speeds far exceeding those of thermal motions at any plausible temperature.

Focus 6.1

Exciting Plasma Oscillations in the Corona

At the high million-degree temperature of the solar corona, electrons are stripped from the gaseous atoms by innumerable collisions, leaving electrons and ions that are free to move about. The electrons have a negative charge, and since ions are atoms that are missing one or more electrons, they are positively charged. An un-ionized atom is electrically neutral without charge. In the solar corona, the negative charge of the electrons equals the positive charge of the protons, so the mixture of electrons and ions, called a plasma, has no net charge. The entire Sun, including its outer atmosphere, is nothing but a giant, hot ball of plasma.

When a flare-associated disturbance, such as an electron beam or a shock wave, moves though the coronal plasma, the local electrons are displaced with respect to the ions, which are more massive than the electrons. The electrical attraction between the electrons and ions pulls the electrons back in the opposite direction, but they overshoot the equilibrium position. The light, free electrons therefore oscillate back and forth when a moving disturbance passes through the corona.

The natural frequency of oscillation, called the plasma frequency, depends on the local electron concentration, with a higher plasma frequency at greater coronal electron densities. The exact expression for the square of the plasma frequency, ν_p^2, is:

$$\nu_p^2 = e^2 N_e / (4\pi^2 \varepsilon_0 m_e) = 81 N_e \, \mathrm{Hz}^2 ,$$

where the electron density, denoted by N_e, is in units of electrons per cubic meter, the electron charge $e = 1.60 \times 10^{-19}$ Coulomb, the electron mass $m_e = 9.11 \times 10^{-31}$ kilograms, and the permittivity of free space $\varepsilon_0 = 10^{-9}/(36\pi)$ farad m^{-1}. Low in the solar corona, where $N_e \approx 10^{14}$ m^{-3}, the plasma frequency is $\nu_p \approx 9 \times 10^7$ Hz $= 90$ MHz, where one MHz is a million Hz or a million cycles per second. Thus, a solar flare can set the corona oscillating at radio frequencies, and this plasma frequency decreases with diminishing electron density at greater distances from the Sun.

As the explosive disturbance moves out through the progressively more rarefied layers of the corona, it excites radiation at lower and lower radio frequencies. Ground-based radio telescopes can identify these radio signals by changing the frequency to which their receiver is tuned (Fig. 6.3). As an example, the radiation at a frequency of 200 MHz, or at 1.5 meters wavelength, might arrive about a minute before the 20 MHz (15 meters) outburst. The product of frequency and wavelength is equal to the velocity of light $c = 2.9979 \times 10^8$ meters per second.

With an electron density model of the solar atmosphere, such as that shown on the next page, the emission frequency can be related to height, and combined with the time delays between frequencies to obtain the outward velocity of the moving disturbance. This electron density model is based on eclipse or coronagraph measurements at different times and different places in the corona. One can alternatively use the relationship between the drift rate, $d\nu/dt$, near frequency ν, and the disturbance velocity, V:

$$d\nu/dt = (N^{-1} dN/ds)\nu V/2 ,$$

where the gradient of density, N, with height, s, is $(N^{-1} dN/ds) = 10^{-8}$ per meter for the solar corona at a temperature of 2 million degrees.

For type III radio bursts, a velocity between 0.2 and $0.8c$ is determined, with an average velocity of $0.4c$. Beams of high-speed electrons, thrown out from solar flares, apparently excite progressively lower plasma frequencies. The kinetic energy of the flare-associated electrons is 10 to 100 keV. One keV is equivalent to an energy of 1.6×10^{-16} Joule, and the rest mass energy of the electron, $m_e c^2 = 511$ keV, where m_e is the mass of the electron and c is the velocity of light.

As the hot corona expands into space, creating the solar wind, it fills an ever greater volume and thins out, becoming more and more tenuous at greater distances from the Sun. There are only about 10 million electrons per cubic meter in interplanetary space near the Earth's orbit, corresponding to

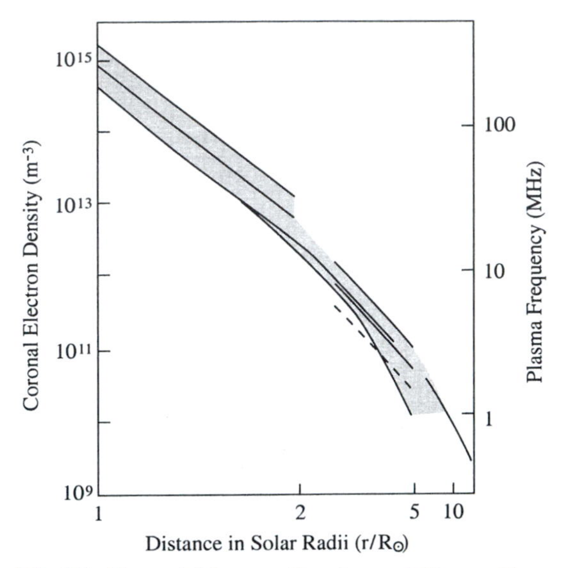

Fig. 6.2. Coronal Electron Density and Plasma Frequency. The solid lines are measurements for coronal condensations and the backround corona; the dashed line is for coronal holes. (Adapted from Mc Lean, 1985)

a plasma frequency of about 28 thousand Hz, or 28 kHz. Radio experiments aboard spacecraft have followed the high-speed electrons of type III bursts as they have moved out into these more rarefied regions, mapping out the spiral structure of the interplanetary magnetic field that guides the electron motion (Sect. 6.7, Fig. 6.20). Radiation at these low frequencies and long wavelengths must be detected from space, for the Earth's ionosphere keeps it from reaching the ground.

The ionosphere is also a plasma, with a plasma frequency of up to 10 MHz. This electrically-charged layer is created by the Sun's X-ray and extreme ultraviolet radiation. The ionosphere is located between 50 thousand and a million meters above the Earth's surface. It reaches a maximum density of almost one million million (10^{12}) electrons per cubic meter at an altitude of a few hundred thousand meters. Radiation with frequencies lower than the ionosphere's plasma frequency is reflected by the ionosphere. This mirroring capability explains how low frequency, long wavelength, radio waves get around the Earth's curvature, enabling world-wide radio communication.

The ionosphere similarly reflects incoming solar radio waves back into space if their frequency is lower than about 10 MHz. Spacecraft lofted above the ionosphere must therefore be used to track high-speed electrons or shock waves at remote distances from the Sun, where the density and plasma frequency are lower. They have monitored the corona's plasma radiation at frequencies from 0.01 to 10 MHz for more than three decades.

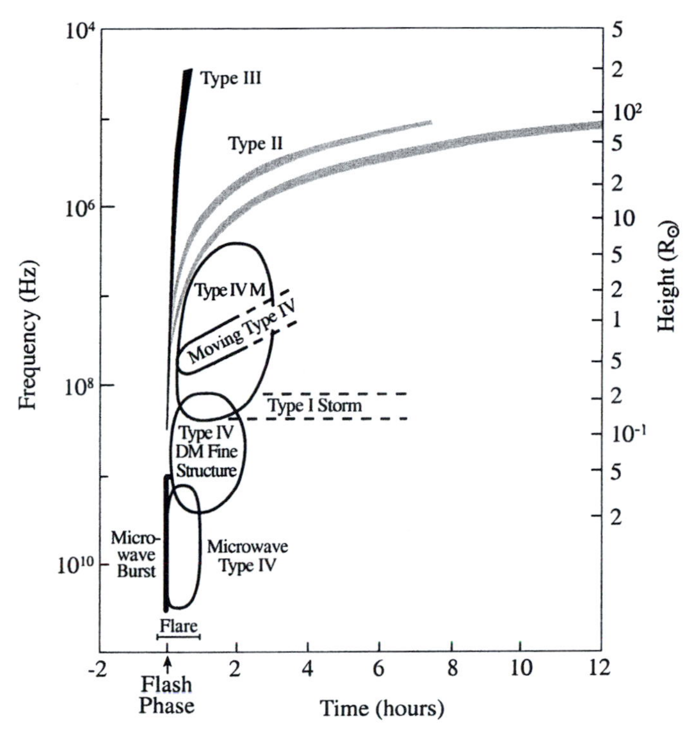

Fig. 6.3. Solar Radio Bursts. Schematic representation of the radio spectrum during and after a large solar flare. It can be associated with several different kinds of intense radio emission, depending on the frequency or wavelength (*left vertical axis*) and time after the explosion (*bottom axis*). In these plots, the impulsive, or flash, phase of the solar flare is indicated at 0 hours; it normally lasts about 10 minutes and is associated with a powerful microwave burst. Dynamic spectra at frequencies of about 10^8 Hz, or 100 MHz, show bursts that drift from high to low frequencies as time goes on, but at different rates depending on the type of burst. Types II and III radio bursts have been respectively attributed to shock waves and electron beams moving outward into the solar atmosphere, or corona, exciting plasma oscillations. The height scale (*right vertical axis*) corresponds to the height, in units of the Sun's radius of 696 million meters, at which the coronal electron density yields a plasma frequency corresponding to the frequency on the left-hand side. At frequencies less than about 10^7 Hz, or 10 MHz, the bursts must be observed from spacecraft, above the ionosphere that deflects incoming radio waves at these low frequencies and long wavelengths. (Adapted from H. Rosenberg, 1976)

The high-speed electrons that emit type III or type IV bursts spiral around the magnetic field lines, moving rapidly at velocities near that of light, and send out radio waves called synchrotron radiation after the man-made synchrotron particle accelerator where it was first observed (Fig. 6.4). Unlike the thermal radiation of a hot gas, the non-thermal synchrotron radiation is most intense at long, invisible radio wavelengths rather than short visible ones. Synchrotron radiation was used to explain the radio emission from our Galaxy

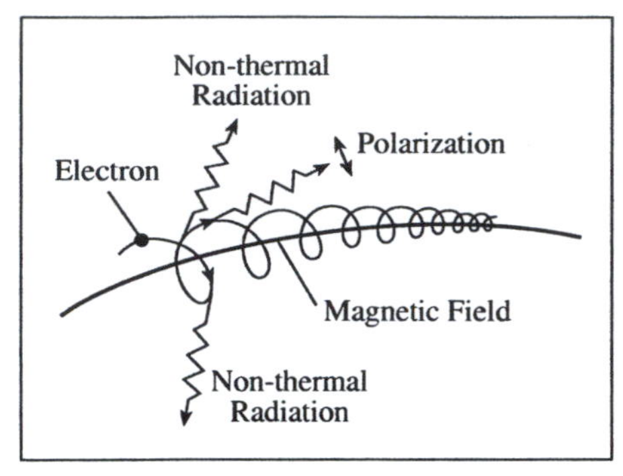

Fig. 6.4. Synchrotron Radiation. An electron cannot cross a magnetic field, but instead circles or gyrates around its lines of force. The electron can also move freely in the direction of the magnetic field, so the electron's trajectory is bent into a spiral, helical path. High-speed electrons moving at velocities near that of light emit a narrow beam of synchrotron radiation that spins abound like the beam of a light house. This emission is sometimes called non-thermal radiation because the electron speeds are much greater than those of thermal motion at any plausible temperature. The name "synchrotron" refers to the man-made, ring-shaped synchrotron particle accelerator where this type of radiation was first observed; a synchronous mechanism keeps the particles in step with the acceleration as they circulate in the ring

and other cosmic radio sources in the 1950s, even before being applied to the radio emission of solar flares.

Radio bursts at meter wavelengths occur at altitudes far above the site where the flare energy is released and particles are accelerated. It took solar physicists surprisingly long to realize that bursts originate much lower in the corona, observed at decimeter wavelengths (1 decimeter = 0.1 meter = 10 centimeter). In the 1980s and 1990s decimeter bursts were systematically studied by Arnold O. Benz and collaborators in Switzerland. They discovered signatures of electron beams at frequencies up to 8,000 MHz, or at wavelengths as short as 3.75 centimeters, moving downward from the acceleration regions into the low corona. A particular kind of rapid bursts around 300 MHz, or 100 cm wavelength, is closely related to metric type III bursts and may originate from the particle acceleration process.

A giant array of radio telescopes, located near Socorro, New Mexico and called the Very Large Array, can zoom in at the very moment of a solar flare, taking snapshot images with just a few seconds exposure (Fig. 6.5). It has pinpointed the location of the impulsive decimeter radiation and the electrons that produce it. These radio bursts are triggered low in the Sun's atmosphere, unleashing their vast power in a relatively small area within solar active regions.

The high-speed electrons are often accelerated just above the apex of magnetic arches, called coronal loops, that link underlying sunspots of opposite magnetic polarity (Fig. 6.5). The acceleration process is most likely related to the magnetic reconnection of oppositely directed magnetic fields near the loop tops (Sect. 6.6). Some of the energetic electrons are confined within the closed magnetic structures, and are forced to follow the magnetic fields down into the chromosphere. Other high-speed electrons break free of their magnetic cage, moving outward into interplanetary space along open magnetic field lines and exciting the meter-wavelength type III bursts.

Radio spikes were the first decimetric bursts to be discovered; they are emitted in a narrow range, or band, of frequencies. In some extreme instances, solar flares include literally thousands of radio spikes, each with a rapid rise

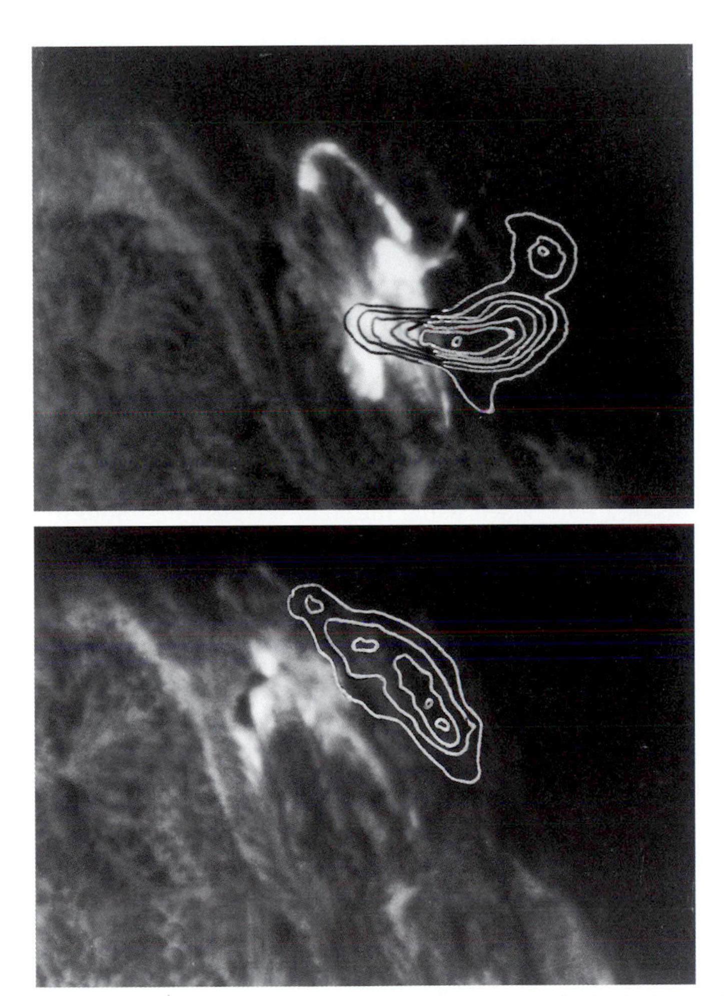

Fig. 6.5. Explosions in the Corona. Solar flares can be ignited near the tops of magnetic loops that reside within solar active regions and are anchored in underlying sunspots. Electrons are rapidly accelerated just above the loop tops during the early stages of the flare, emitting the powerful loop-like radio signals mapped here with white contours. The underlying hydrogen-alpha (Hα) emission is shown in the accompanying photographs. These 10-second snapshot radio maps were obtained with the Very Large Array (VLA) at a wavelength of 20 centimeters, or a frequency of 1,420 MHz. The angular extent of each 20-cm flaring loop is about one minute of arc, or one thirtieth of the angular width of the visible solar disk. (Adapted from R. F. Willson and K. R. Lang, 1984)

as short as a few milliseconds and similarly brief duration. Since nothing can move faster than the velocity of light, it sets an upper limit on any travel time. Rapid changes over time scales of a few milliseconds therefore restrict the size of the radiating volume to relatively small spatial scales of less than a few million meters across. The radio spikes must have brightness temperatures as high as a million billion (10^{15}) degrees if their intense radiation is emitted from such a small source.

The brightness temperature is that temperature the source would have if it was emitting thermal radiation from a hot gas. Yet, such high temperatures are impossible to achieve by heating any gas, and have only been realized in the earliest, big-bang stages of the expanding Universe where material particles, as we know them, did not exit. The high brightness temperatures of millisecond radio spikes are explained by coherent radiation of ordinary particles working together in an organized manner, rather than by the random, incoherent motions of heated gas particles

6.3 X-ray Flares

What does a solar flare look like? It depends on how you look at it. A solar flare produces copious radiation across the full electromagnetic spectrum, and one sees varied aspects of a flare when using different wavelengths. Both the spatial and temporal behavior of a solar flare depend on the perspective you choose, and each view focuses in on a different aspect of the flaring mechanism.

In this section, we will discuss X-rays from solar flares. The wavelength of an X-ray is on the order of, or smaller than, the size of atoms, or between 10^{-11} and 10^{-9} meters. Although X-rays can penetrate small amounts of material substances, including your skin and muscles, solar X-rays are totally absorbed in the atmosphere. This radiation is therefore now observed from satellites orbiting the Earth above our air. Laurence E. Peterson and John Randolph Winckler were the first to detect it, using a balloon during a solar flare on 10 March 1958. Their detectors measured radiation with energies between 200 and 500 keV, lasting less than a minute and coincident in time with a solar radio burst and an Hα flare. One keV is equivalent to an energy of 1.6×10^{-16} Joule.

Researchers describe X-rays by the energy they carry. There are soft X-rays with relatively low energy and modest penetrating power. The hard X-rays have higher energy and greater penetrating power. As a metaphor, one thinks of the large, pliant softballs and the compact, firm hardballs, used in the two kinds of American baseball games.

The energy of the X-ray radiation is a measure of the energy of the electrons that produce it. The high-energy, hard X-rays are produced by non-thermal electrons accelerated to nearly the velocity of light. Like the solar radio bursts, the hard X-ray radiation of solar flares tells us about the acceleration, propagation and confinement of very energetic electrons. The soft X-rays describe the thermal radiation of hot electrons of lower energy.

Like energetic charged particles, the energy of flaring X-rays is often specified in kilo electron volts, denoted keV. Soft X-rays have energies between 1 and 10 keV, and hard X-rays lie between 10 and 100 keV. Gamma rays are even more energetic than X-rays, exceeding 100 keV in energy. The wavelength of radiation is inversely proportional to its energy, so hard X-rays are shorter than soft X-rays, and the wavelengths of gamma rays are still smaller. X-rays with a given energy are produced by electrons with roughly the same amount of energy.

Pioneering rocket observations in the late 1950s and early 1960s, by Herbert Friedman and his colleagues at the Naval Research Laboratory, demonstrated that the X-ray radiation emitted during solar flares can outshine the entire Sun at these wavelengths (see Fig. 6.6 for a recent example). Our knowledge of solar X-ray flares, and the instruments used to detect them, was then improved by NASA's Orbiting Solar Observatory (OSO) series of satellites, beginning with OSO 1 launched in 1962 and continuing through Skylab launched in 1973.

The soft X-rays emitted during solar flares gradually build up in strength and peak a few minutes after the impulsive emission, so the soft X-rays are a delayed effect of the main flare explosion (Fig. 6.7). Moreover, the slow, smooth rise of the soft X-ray afterglow resembles the time integral of the rapid, impulsive microwave bursts detected at centimeter wavelengths. This relationship was reported by Werner M. Neupert in 1968, using soft X-ray observations with OSO 3, and is therefore now known as the Neupert effect.

The Apollo Telescope Mount aboard the manned, orbiting observatory, Skylab, was used to scrutinize the Sun at X-ray, ultraviolet and visible wavelengths for nearly nine months after launch on 14 May 1973, capturing thousands of images on photographic films. From Skylab came the first clear X-ray pictures of the ubiquitous magnetic loops that hold the million-degree coronal gas in place and dominate the structure of the corona (Sect. 2.5). The instruments aboard Skylab additionally resolved soft X-ray flares for the first time, showing that they are also located in magnetic loops within the low corona and can involve extensive magnetic restructuring.

The temperatures of the flaring soft X-ray loops are about ten times hotter than the coronal loops found on the quiescent, or non-flaring, Sun. During solar flares, Skylab and earlier X-ray spectrometers detected spectral lines emitted by iron atoms with almost all of their twenty-six electrons stripped away. Such a degree of ionization can only occur at temperatures of about ten million (10^7) degrees, so the soft-X-ray flares are about as hot as the center of the Sun.

Hotter gases radiate most intensely at shorter wavelengths, and the wavelength at which the brightness is a maximum varies inversely with the temperature. The exact relation, known as the Wien displacement law, indicates that a gas will be brightest at soft X-ray wavelengths (about 10^{-10} meters) when it is heated to 10 million (10^7) degrees Kelvin.

The soft X-rays emitted during solar flares are thermal radiation, released by virtue of the intense heat and dependent upon the random thermal motions of very hot electrons. At such high temperatures, the electrons are set free from atoms and move off at high speed, leaving the ions (primarily protons) behind. When an electron moves through the surrounding material, it is attracted to the oppositely charged protons and emits a kind of thermal radiation called bremsstrahlung (Focus 6.2).

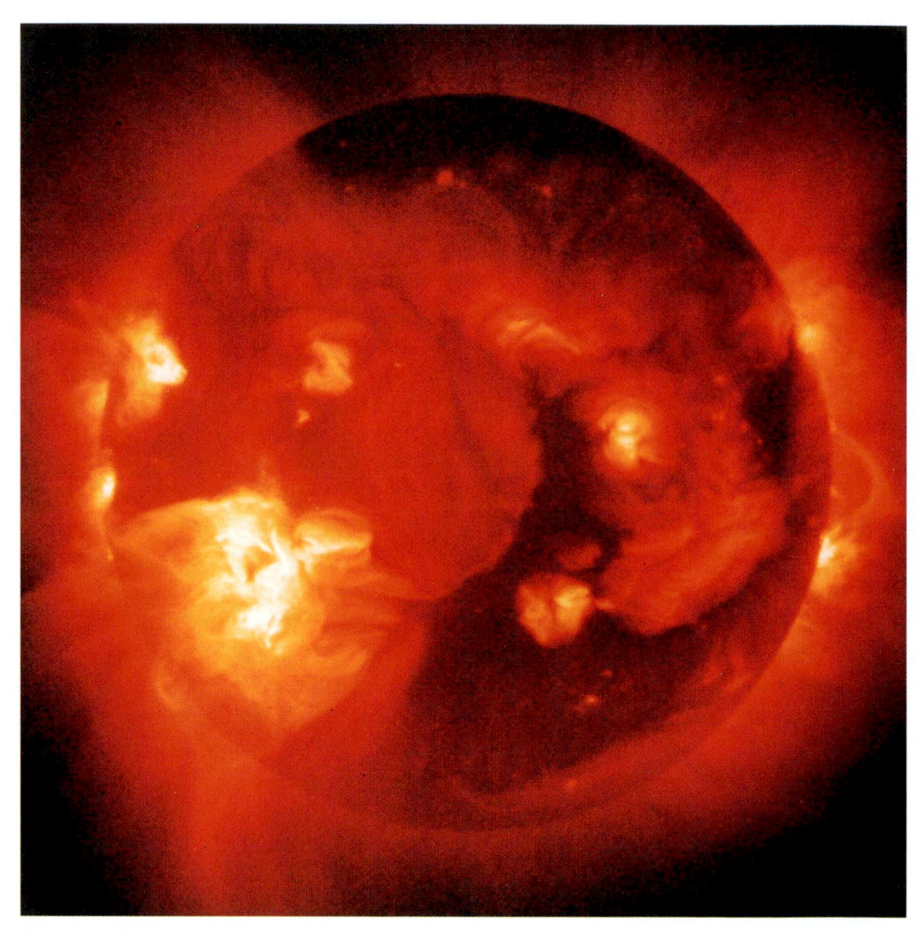

Fig. 6.6. Yohkoh Catches a Flare. A solar flare can result in soft X-ray radiation that outshines the entire Sun at these wavelengths. Bright soft X-ray loops are found in the lower left of this X-ray image, taken a few hours after the impulsive phase of a powerful solar flare. Less luminous coronal loops are found in quiescent, or non-flaring, active regions on other parts of the Sun, and dark coronal holes are also present at both poles (*top* and *bottom*). (This image was taken with the Soft X-ray Telescope (SXT) aboard the Yohkoh satellite on 25 October 1991, Courtesy of Keith Strong, NASA, ISAS, the Lockheed-Martin Solar and Astrophysics Laboratory, the National Astronomical Observatory of Japan, and the University of Tokyo)

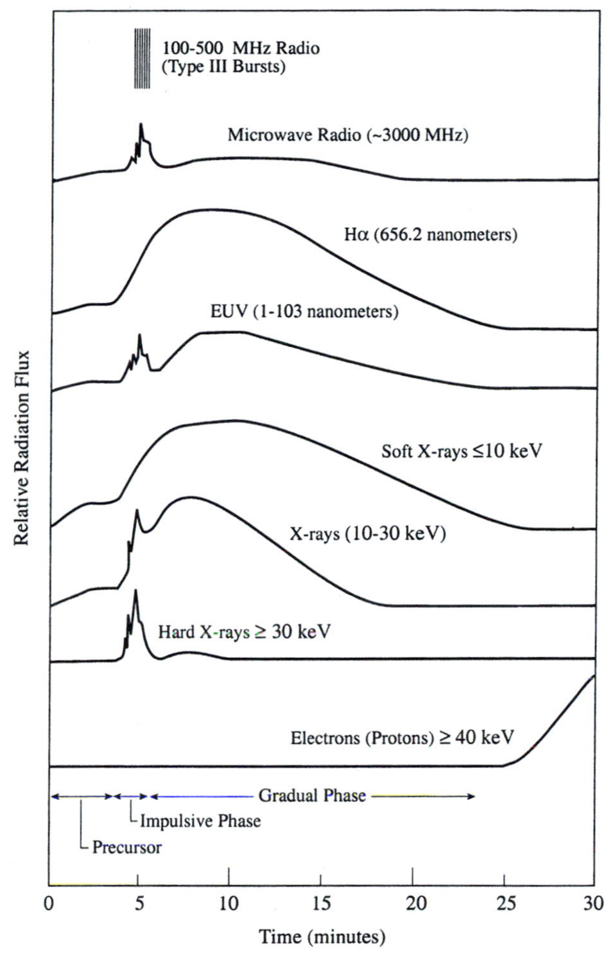

Fig. 6.7. Different Perspectives. What you see during a solar flare depends on how you look at it. This is illustrated in these time profiles of the same flare observed at different wavelengths. The precursor, impulsive and gradual phases of the solar flare are denoted at the bottom. During the early impulsive phase of a solar flare, electrons accelerated to high energies and very rapid speeds emit radio bursts and hard X-rays. The radio emission is at frequencies from 100 to 3,000 MHz, or at wavelengths between 3 and 0.1 meters; the hard X-rays have photon energies greater than 30 keV. The subsequent gradual phase is detected with soft X-rays, at energies of about 10 keV or less, as an aftereffect of the impulsive radiation. The soft X-rays are the thermal radiation, or bremsstrahlung, of a gas heated to temperatures of tens of millions of degrees. (Adapted from S. R. Kane, 1974)

NASA's Solar Maximum Mission (SMM) satellite, launched on 14 February 1980, obtained a new perspective on solar flares by examining spectral lines emitted during these violent events (Table 6.2). The spectral line data provided unique information on physical parameters throughout the flaring solar atmosphere, and contributed key pieces to our understanding of the

Table 6.2. Prominent solar emission lines at wavelengths below 200.0 nanometers*

Wavelength (nm)	Element	Wavelength (nm)	Element	Wavelength (nm)	Element
0.17780	Fe XXVI	13.283	Fe XXIII	97.702	C III
0.18499	Fe XXV	19.204	Fe XXIV	103.191	O VI
0.31769	Ca XIX	28.416	Fe XV	121.567	H I
0.50385	S XV	30.378	He II	139.375	Si IV
0.917	Mg XI	33.541	Fe XVI	140.116	O IV
1.346	Ne IX	49.941	Si XII	154.820	C IV
1.42	Fe XVIII	62.494	Mg X	164.04	He II
1.5012	Fe XVII	62.973	O V	189.203	Si III
1.897	O VIII	77.041	Ne VIII	190.873	C III
2.1602	O VII				

* The symbols Ca, He, Fe, Mg, Ne, O, Si and S denote, respectively, calcium, helium, iron, magnesium, neon, oxygen, silicon and sulfur. Subtract one from the Roman numeral to obtain the number of missing electrons. Thus, the ion Fe XXVI is the iron atom missing 25 electrons. One nm, or one nanometer, is equivalent to one billionth, or 10^{-9}, meters. Historically, solar astronomers have also used the Angstrom unit of wavelength, where one Angstrom is equal to 0.1 nm. High-energy astrophysicists commonly express wavelength in units of energy, most often in keV or kilo electron volts, where one nm $\approx$ 1 keV $\approx 10^{-16}$ Joule.

mechanisms of solar flares. Similar information was obtained at about the same time using a soft X-ray spectrometer aboard the P78-1 spacecraft, launched on 24 February 1979 as part of the U.S. Department of Defense Space Test Program. There were lower resolution X-ray spectrometers on the Orbiting Solar Observatories (OSO) 3, 4, 5 and 6 during the previous decade.

Precise SMM and P78-1 measurements of soft X-ray spectral lines, at about 1 nanometer, or 1 nm and 10^{-9} meters, in wavelength, revealed a Doppler shift to shorter wavelengths, suggesting an upward motion of the hot thermal gas at velocities as high as 400 thousand meters per second. This rise in heated material was explained by a theory called chromospheric evaporation, although it has nothing to do with the evaporation of any liquid. Initially cool chromospheric material is heated by down flowing, or precipitating, flare electrons, and then expands upwards into the low density corona along magnetic loops that shine brightly in soft X-rays after filling.

The SMM and P78-1 data showed that this up welling fills the flaring loops with plasma heated to temperatures of up to 40 million (4.0×10^7) degrees Kelvin. Densities of 10^{17} to 10^{18} electrons per cubic meter have been derived from density-sensitive line ratios. The soft X-ray spectrometer aboard the Japanese solar spacecraft Hinotori, meaning fire-bird and launched on 21 February 1981, also used temperature-sensitive line ratios to show that flares produce plasmas as hot as 30 to 40 million degrees. It independently confirmed the upflow of the heated gas detected at soft X-ray wavelengths during the early phases of solar flares.

In 1980 the SMM obtained pioneering images of the hard X-rays emitted during the impulsive phase of solar flares, permitting the first detailed study of flaring hard X-ray sources (Table 6.3). The two-component, or double, hard X-ray sources, observed at energies up to 30 keV, suggested that this emission corresponds to the footpoints of a magnetic loop. Such coronal loops had been detected nearly a decade earlier at lower soft X-ray energies, of about 6 keV, from the Skylab satellite and rockets.

In fact, SMM showed that hard X-rays, emitted during the impulsive phase of solar flares, can be concentrated at the footpoints of the coronal loops detected at soft X-ray wavelengths, and that the hard X-ray double sources are nearly co-spatial with the hydrogen alpha (Hα) emission in the chromosphere.

The hard X-rays are apparently generated by energetic, non-thermal electrons hurled down the two legs of a coronal loop into the low corona and dense chromosphere. This would explain the double hard X-ray structure, and also account for their simultaneous brightening within the 10 second time resolution of the SMM instrument. High-speed, non-thermal electrons accelerated during impulsive solar flares were discovered and substantiated by their radio emission, whose time profiles at centimeter wavelengths match the hard X-ray time profile (Sect. 6.2).

The hard X-rays marking the flare onset are due to bremsstrahlung, but they are produced by electrons that have much higher energies than those emitting the thermal soft X-ray bremsstrahlung. These energetic, high-speed, non-thermal electrons are believed to be accelerated above the tops of coronal loops, and to radiate energy by non-thermal bremsstrahlung as they are beamed down along the looping magnetic channels into the low corona and chromosphere (Focus 6.2). Only about 0.00001 or 10^{-5}, of the electron's en-

Table 6.3. Instruments aboard the Solar Maximum Mission

Experiment	Energy or Wavelength Range*
Gamma Ray Spectrometer (GRS)	0.1 to 17 MeV and 10 to 160 MeV
Hard X-ray Burst Spectrometer (HXRBS)	25 to 500 keV
Hard X-ray Imaging Spectrometer (HXIS) (Angular resolution 8 to 32 arc seconds, effective time resolution $\geq$ 10 seconds)	3.5 to 30 keV (six channels)
X-ray Polychromator (XRP, with BCS and FCS) (BCS = Bent Crystal Spectrometer) (FCS = Flat Crystal Spectrometer)	0.176 to 0.323 nm and 0.144 to 2.24 nm
UV Spectrometer and Polarimeter (UVSP)	110 to 330 nm
Coronagraph and Polarimeter (C/P)	443.5 to 658.3 nm
Active Cavity Radiometer Irradiance Monitor (ACRIM)	Ultraviolet to Infrared

* A wavelength of one nm, or one nanometer, is equivalent to a billionth (10^{-9}) of a meter, and an energy of about 1 keV. Solar astronomers also often use the Angstrom unit of wavelength, where one Angstrom is equal to 0.1 nm, and an energy of about 10 keV. One keV is equivalent to 1.602×10^{-16} Joule or 1.602×10^{-9} erg, and 1 MeV = 1,000 keV.

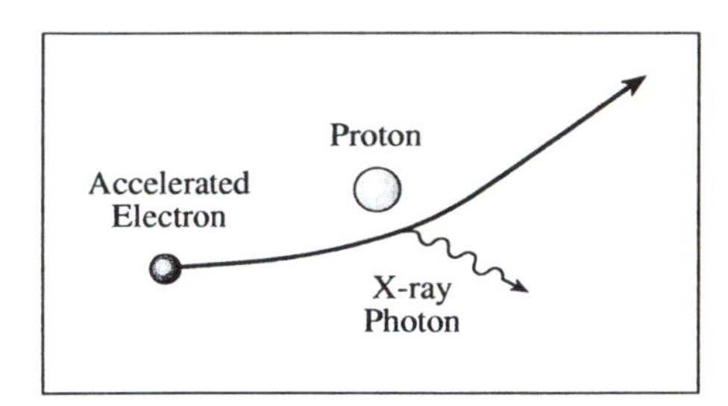

Fig. 6.8. Bremsstrahlung. A fast electron emits bremsstrahlung, or braking radiation, when passing near a proton

radio wavelength radiation by a different synchrotron process that involves magnetic fields (Sect. 6.2).

The greater the number of electrons, the stronger the thermal bremsstrahlung. To be precise, the thermal bremsstrahlung power, P, emitted from a plasma increases with the square of the electron density, N_e, and the volume of the radiating source, V. It also depends upon the temperature of the electrons, T_e, and for a thermal gas this temperature varies inversely with the wavelength of the radiation. A formula for the bremsstrahlung power is:

$$P = \text{constant} \times N_e^2 V T_e^{1/2} \times \text{Gaunt factor},$$

where the Gaunt factor is also a function of the electron temperature. Scientists can use this expression with measurements of the X-ray power during a flare to determine the density of electrons participating in the radiation, assuming that they completely fill the observed volume. Electron density values of $N_e \approx 10^{17}$ electrons per cubic meter are often obtained.

There are two kinds of bremsstrahlung emitted at X-ray wavelengths during solar flares. They are called thermal bremsstrahlung and non-thermal bremsstrahlung, which distinguishes both the method of electron acceleration and the energy of the X-rays. The thermal, soft X-ray bremsstrahlung is produced when electrons are heated to high temperatures of about 10 million degrees, moving at speeds of about 0.05 times the velocity of light and emitting soft X-rays when they encounter protons. Electrons that are accelerated to very high, relativistic speeds, near the velocity of light, emit non-thermal, hard X-ray bremsstrahlung when interacting with ambient protons. Thermal soft X-ray bremsstrahlung is radiated at longer wavelengths and lower energies than hard X-ray, non-thermal bremsstrahlung.

In scientific parlance, the spectrum of thermal bremsstrahlung has an exponential shape while the non-thermal hard X-ray spectrum follows a less steep, power-law drop at increasing energy. For the power law situation, the number of non-thermal electrons with energy, E, varies as E^{-P}, where the power-law index P is a small positive number. Observation of non-thermal bremsstrahlung provides a way to study the accelerated electrons and specify this index.

With Yohkoh, the double-source, loop-footpoint structure of impulsive hard X-ray flares was confirmed with unprecedented clarity. It established a double-source structure for the hard X-ray emission of roughly half the flares observed in the purely non-thermal energy range above 30 keV – see Figs. 6.9 and 6.10 for the time profiles and images of an example. The other half of the flares detected with Yohkoh were either single sources, that could

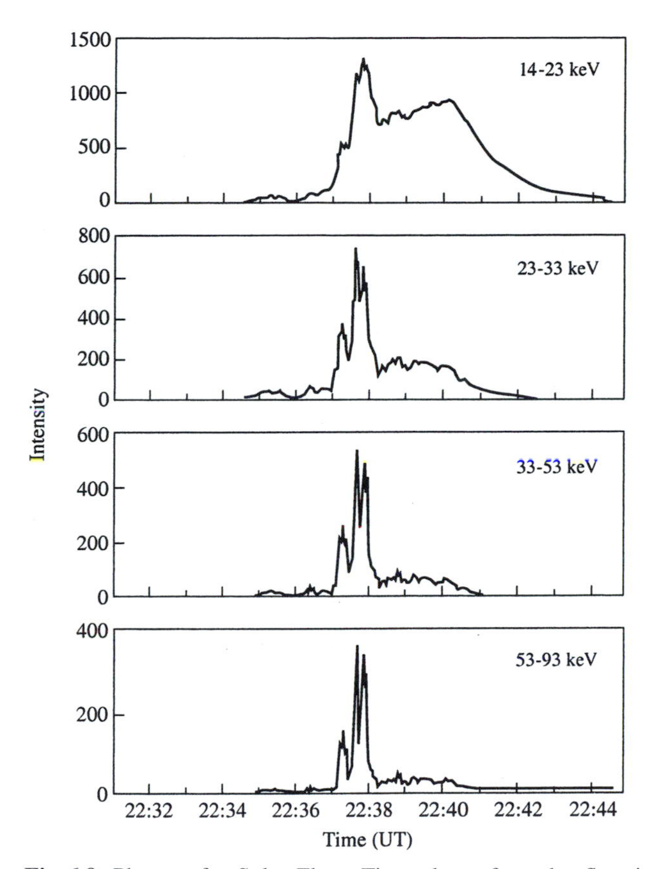

Fig. 6.9. Phases of a Solar Flare. Time plots of a solar flare in four different hard X-ray energy bands on 15 November 1991. The vertical intensity scale provides the average counting rates, derived from 0.5 second integration, of the 64 imaging elements of the Hard X-ray Telescope (HXT) aboard Yohkoh. The time profile in the most-energetic, hard X-ray energies, above 30 keV, is characterized by an impulsive feature that lasts for about one minute at the onset of a flare. This impulsive phase coincides with the acceleration of high-speed electrons that emit non-thermal bremsstrahlung at hard X-ray wavelengths and non-thermal synchrotron radiation at centimeter radio wavelengths. The less-energetic emission shown here, below 30 keV, can be composed of two components, an impulsive component followed by a gradual one. The latter component builds up slowly and becomes most intense during the gradual decay phase of solar flares when thermal radiation dominates. At even lower soft X-ray energies (about 10 keV), the gradual phase dominates the flare emission, and the time profiles of the soft X-ray flux roughly matches the time integral of the radio or hard X-ray profile. (Adapted from T. Sakao, 1994)

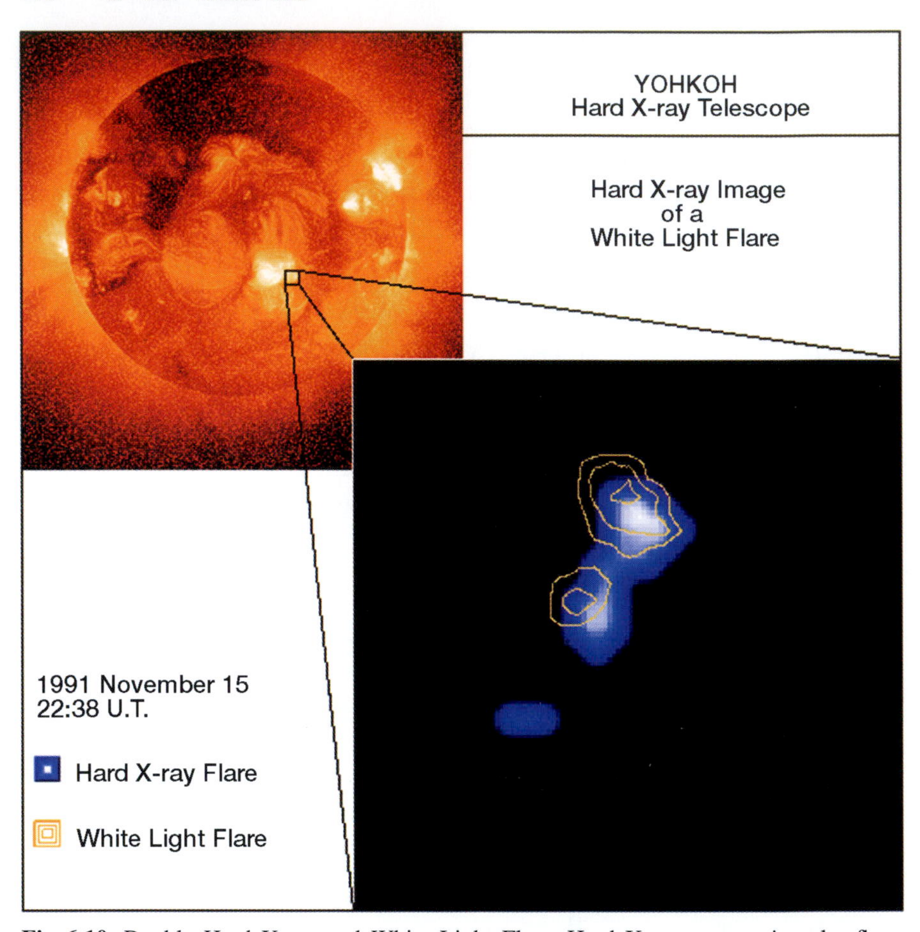

Fig. 6.10. Double Hard X-ray and White Light Flare. Hard X-ray sources in solar flares often occur in simultaneous pairs that are aligned with the photosphere footpoints of a flaring magnetic loop detected at soft X-ray wavelengths. These footpoints can also be the sites of white-light flare emission. The time profiles of this flare, detected on 15 November 1991, show that the increase of white-light emission matches almost exactly that of the hard X-ray flux. This and the simultaneity of hard X-ray emission from the two footpoints establish that non-thermal electrons transport the impulsive-phase energy along the flare loops. The fainter hard X-ray source is located in a stronger magnetic field region than the brighter hard X-ray source, suggesting that intense magnetism can block the electron movement to lower levels. These soft X-ray, hard X-ray and white-light images of the solar flare were taken with telescopes aboard the Yohkoh mission. (Courtesy of NASA, ISAS, the Lockheed-Martin Solar and Astrophysics Laboratory, the National Astronomical Observatory of Japan, and the University of Tokyo)

be double ones that are too small to be resolved, or multiple sources that could be an ensemble of double sources. As subsequently discussed (Sect. 6.6), a third hard X-ray source is sometimes detected near the apex of the magnetic loop joining the other two; this loop-top region marks the primary energy release site and the location of electron acceleration due to magnetic interaction.

Two white-light emission patches were also detected by Yohkoh during at least one flare, at the same time and place as the hard X-ray sources (Fig. 6.10). This shows that the rarely-seen white light flares can also be produced by the downward impact of non-thermal electrons, and demonstrates their penetration deep into the chromosphere.

When the two hard X-ray sources (above 30 keV) are seen by Yohkoh, they are located on both sides of the line that separates regions of opposite magnetic polarity (the magnetic neutral line) in the photosphere, strongly suggesting that the hard X-rays are emitted from the footpoints of a flaring magnetic loop. The double hard X-ray sources occur and vary nearly simultaneously in time, within 0.1 seconds or less, a result that excludes transport mechanisms other than high-energy electrons in coronal loops. Moreover, electron time-of-flight measurements with the Compton Gamma Ray Observatory (CGRO) satellite, published by Markus J. Aschwanden and colleagues in 1996, confirm the existence of a coronal acceleration site for flares observed with both CGRO and Yohkoh.

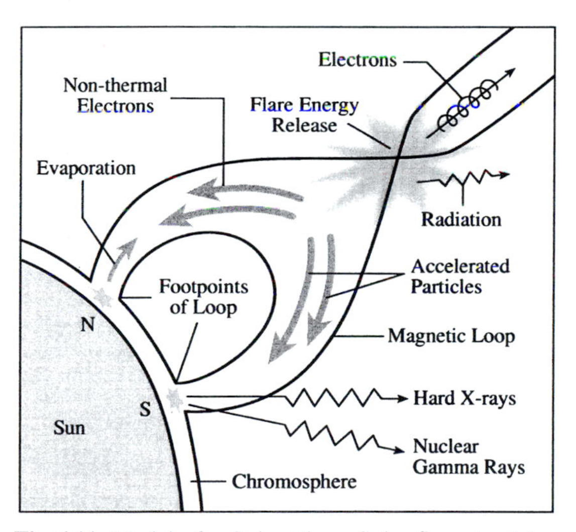

Fig. 6.11. Model of a Solar Flare. Solar flares tend to occur at or just above concentrated magnetic loops extending from the photosphere into the low corona. Stored magnetic energy is released from a magnetic interaction site above the top of the loop shown here. Great numbers of electrons are accelerated to high speed, generating a burst of radio energy as well as impulsive loop-top hard X-ray emission. Some of these non-thermal electrons are channeled down the loop and strike the chromosphere at nearly the speed of light, which creates hard X-rays by electron–ion bremsstrahlung at the loop footpoints. When beams of accelerated protons enter the dense, lower atmosphere, they also produce nuclear reactions that result in gamma-ray spectral lines and energetic neutrons. Material in the chromosphere is heated very quickly and rises into the loop, accompanied by a slow, gradual increase in soft X-ray radiation. This upwelling of heated material is called chromospheric evaporation. (Adapted from S. Masuda, 1994, K. R. Lang, 1995 and M. J. Ashwanden, et al., 1996, 1998)

The Yohkoh spectrometer showed that the upflow of heated material, detected as soft X-ray emission lines that are Doppler shifted to shorter wavelengths, coincides in time with the impulsive phase of flares detected as hard X-ray bursts and microwave bursts. Yohkoh's Soft X-ray Telescope, or SXT, additionally demonstrated that the initial site of chromospheric evaporation coincides with the hard X-ray bursts at the footpoints of magnetic loops. Non-thermal electrons are apparently sent down along the closed magnetic loops into the initially low-temperature chromosphere, and the heated material subsequently expands upward to create the gradual soft X-ray phase seen in the corona.

All of these results confirm the existence of powerful electron acceleration near the tops of magnetic loops, with the non-thermal electrons streaming down to the footpoints at nearly the speed of light (Fig. 6.11). A catastrophic instability triggers the primary energy release of a flare above the apex of magnetized coronal loops. In less than a few seconds, electrons are accelerated there, and beamed into the lower, denser reaches of the solar chromosphere. As the non-thermal electrons spiral down along arching magnetic channels, they generate intense synchrotron radio emission. Further down, the high-speed electrons emit non-thermal hard X-ray bremsstrahlung via interactions with the ambient protons in the low corona, and sometimes produce white-light flares when they strike the chromosphere. The chromosphere at the loop footpoints is heated very rapidly (in seconds) by the accelerated particles that slam into it, and cannot get rid of the excess energy. The high-temperature material is driven upward by the large pressure gradients and "evaporates" along the guiding magnetic field. Relatively long-lived (tens of minutes) soft X-ray radiation is then emitted by thermal bremsstrahlung as the flaring loop is filled with the hot, rising material.

6.4 Gamma Rays from Solar Flares

Protons and heavier ions are accelerated to high speed during solar flares, and beamed down into the Sun where they generate gamma rays, the most energetic kind of radiation detected from solar flares. The protons slam into the dense, lower atmosphere, like a bullet hitting a concrete wall, shattering abundant heavy nuclei in a process called spallation. The lighter nuclear fragments are left in an excited state, but promptly calm down and get rid of their excess energy by emitting gamma rays (Focus 6.3). Other abundant nuclei are directly excited by collision with the flare-accelerated protons, and emit gamma rays when relaxing to their former unexcited state (Fig. 6.12). Thus, nuclear reactions occur on the visible disk of the Sun, as well as deep down in its energy-generating core (Focus 6.3).

The gamma rays from excited nuclei are emitted only at specific, well-defined energies between 0.4 and 7.1 MeV (Table 6.5). One MeV is equivalent to a thousand keV and a million electron volts, so the gamma rays are ten to one hundred times more energetic than the hard X-rays and soft X-rays detected during solar flares. Like X-rays, the gamma rays are totally absorbed in our atmosphere and must be observed from space.

Focus 6.3

Nuclear Reactions on the Sun

Nuclear reactions during solar flares produce gamma ray lines, emitted at energies between 0.4 and 7.1 MeV. They result from the interaction of flare-accelerated protons or helium nuclei, having energies between 1 and 100 MeV, with nuclei in the dense atmosphere below the acceleration site.

When protons with energies above 300 MeV interact with the abundant hydrogen in the solar atmosphere, they can produce short-lived fundamental particles, called mesons, whose decay leads to gamma-ray emission. The decay of neutral mesons produces a broad gamma-ray peak at 70 MeV, but the decay of charged mesons leads to bremsstrahlung giving a continuum of gamma rays with energies extending to several MeV. Neutrons with energies above 1,000 MeV are also produced.

Narrow gamma-ray lines (≤ 100 keV in width) have been observed on the Sun from deuterium formation, electron-positron annihilation, and excited carbon, nitrogen, oxygen and heavier nuclei (Table 6.5). These reactions are often written using letters to denote the nuclei, a Greek letter γ to denote gamma ray radiation, and an arrow $\rightarrow$ to specify the reaction; nuclei on the left side of the arrow react to form products given on the right side of the arrow. The letter p is used to denote a proton, the nucleus of a hydrogen atom, and the Greek letter α signifies an alpha particle, which is the nucleus of the helium atom.

As an example, the collision of a flare-associated proton, p, or alpha particle, α, with a heavy nucleus in the dense solar atmosphere may result in a spallation reaction that causes the nucleus to break up into lighter fragments that are left in excited states denoted by an asterisk $*$. They subsequently de-excite to emit gamma-ray lines. Important examples are the production of excited carbon, $^{12}C^*$, and oxygen, $^{16}O^*$, by flaring protons, p, that break up oxygen, ^{16}O, or neon, ^{20}Ne, nuclei by the reactions:

$$p + {}^{16}O \rightarrow {}^{12}C^* + \alpha$$

and

$$p + {}^{20}Ne \rightarrow {}^{16}O^* + \alpha \, ,$$

with de-excitation and emission of a gamma-ray line, γ, of energy, $h\nu$, by:

$${}^{12}C^* \rightarrow {}^{12}C + \gamma \quad (h\nu = 4.438\,\mathrm{MeV})$$

and

$${}^{16}O^* \rightarrow {}^{16}O + \gamma \quad (h\nu = 6.129\,\mathrm{MeV}) \, .$$

Energetic neutrons can be torn out of the nuclei of atoms during bombardment by flare-accelerated ions. Many of these neutrons are eventually captured by hydrogen nuclei (protons) in the photosphere, to produce deuterons and the delayed emission of a gamma ray line at 2.223 MeV. It is typically the strongest gamma ray line in solar flare spectra.

Flare particles even produce antimatter on the Sun, in the form of positrons, e^+, that annihilate with their material counterparts, the electrons denoted by e^-, producing gamma-ray signatures at 0.511 MeV. It is the second strongest gamma ray line from solar flares. Both the neutron capture and positron annihilation gamma ray lines were first observed by Edward L. Chupp and his colleagues in 1972, from the seventh Orbiting Solar Observatory, or OSO 7, and then detected in profusion in the 1980s by the same group using the Gamma Ray Spectrometer (GRS) aboard the Solar Maximum Mission (SMM) satellite. Energetic neutrons have also been directly measured in space near Earth from the SMM and the Compton Gamma Ray Observatory (Sect. 6.7). In the most energetic flares showing meson decay gamma-ray emission, the associated relativistic neutrons can reach the Earth and produce a signal in ground level neutron monitors.

Another important SMM discovery was that the hard X-ray and gamma ray time profiles of solar flares are time coincident within the accuracy of measurement, demonstrating the simultaneous acceleration of relativistic electrons and energetic ions in solar flares to within a few seconds. Measurements of the high-energy electrons are obtained from the hard X-rays, produced by non-thermal bremsstrahlung during solar flares, and the flare-accelerated ions are observed using gamma rays. In large solar flares, the onset of the impulsive phase emission is simultaneous for radiation with energies from about 40 keV (hard X-rays) to about 40 MeV (gamma rays). This provides important constraints to theories for how particles are accelerated in solar flares.

6.5 Magnetic Bubbles

Just about every day, a vast mass is expelled from the Sun's low corona and travels at high speed into interplanetary space (Fig. 6.13). These sporadic, transient eruptions are known as coronal mass ejections, often designated as CMEs.

Nearly everything we know about coronal mass ejections has been learned in just a few decades. They could not be clearly identified until special telescopes known as coronagraphs were flown in space in the early 1970s. These instruments have a small occulting disk to mask the Sun's face and block out the photosphere's direct sunlight. The bright solar glare previously hid the corona from view, except during rare and brief total eclipses of the Sun.

Coronagraphs image the faint visible sunlight scattered by free electrons in the corona, shining in white light with all the colors combined. Like any atmosphere, the corona is denser at the lower altitudes near the solar disk; it thins out and becomes fainter at greater distances where the gas fills a larger volume and there is less scattered light. Coronagraphs detect the bright solar corona located relatively near the Sun where the corona has its greatest density.

The Earth's bright sky obscures the corona's light and restricts detailed ground-based coronagraph observations to within only about two solar radii from the center of the Sun. The best coronagraph images, with the finest detail, are obtained from satellites in the dark airless sky. They have observed the corona out to thirty solar radii from Sun center.

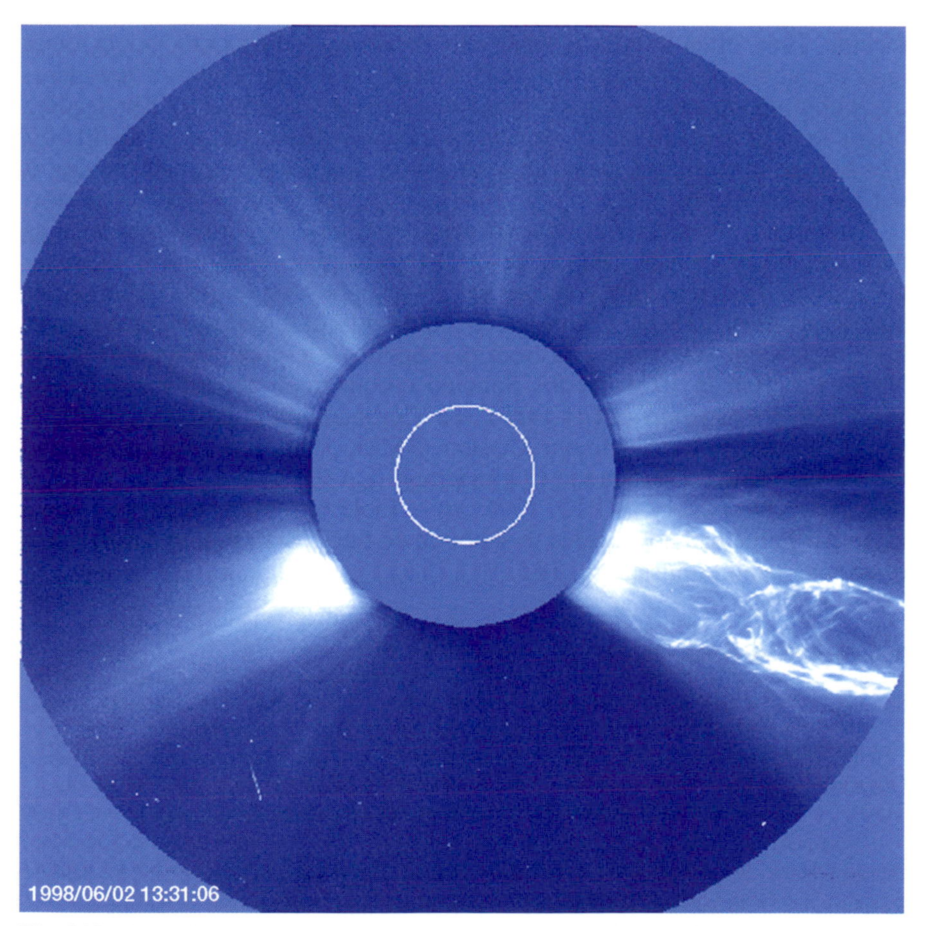

Fig. 6.13. Coronal Mass Ejection. A twisted, helical-shaped coronal mass ejection spins off from the Sun in this image taken by the LASCO C2 coronagraph aboard SOHO on 2 June 1998. The dark disk blocks the Sun, so that the LASCO instrument can observe the structures of the corona in visible light. The white circle represents the size and position of the visible solar disk that we see with our eyes. (Courtesy of the SOHO LASCO consortium. SOHO is a project of international collaboration between ESA and NASA)

upper portions of the magnetic loops are sometimes carried out by the highly-ionized material, while remaining attached and rooted to the Sun at both ends. In other situations, the expelled material stretches the magnetic field until it snaps, taking the coiled magnetism with it and lifting off into space like a hot-

Focus 6.4

Mass, Mass Flux, Energy and Time Delay of Coronal Mass Ejections

Coronal mass ejections are detected as localized increases in the brightness of white-light coronagraph images. Integration of the brightness increase, that depends only on the electron density, N_e, permits evaluation of the total mass, M, of the ejection. For a sphere of radius, R, we have:

$$M = 4\pi R^3 N_e m_p / 3 ,$$

where the proton mass $m_p = 1.67 \times 10^{-27}$ kilograms. The corona is a fully ionized, predominantly (90 percent) hydrogen, plasma, so the number density of protons and electrons are equal, but since the protons are 1,836 times more massive than the electrons, the protons dominate the mass. For a mass ejection with an electron, or proton, density of $N_e = 10^{13}$ electrons per cubic meter (Focus 6.1), that has grown as large as the Sun, with $R = 6.96 \times 10^8$ meters, this expression gives

$$M = 10^{13} \text{ kilograms} = 10 \text{ billion tons} ,$$

where one ton is equivalent to 1,000 kilograms. At the rate of one ejection per day, and 10^{13} kilograms per ejection, this amounts to a mass flow rate of about 10^8 kilograms per second, since there are 86,400 seconds per day.

By way of comparison, the solar wind flux observed in the ecliptic at the orbit of the Earth is about 5×10^{12} protons per square meter per second, or 8.3×10^{-15} kilograms per square meter per second (Table 2.3). If this flux is typical of that over the entire Sun-centered sphere, with an average Sun-Earth distance of $D = 1.5 \times 10^{11}$ meters, we can multiply by the sphere's surface area, $4\pi D^2$, to obtain a solar wind mass flux of about 2×10^9 kilograms per second. Thus, coronal mass ejections contribute roughly five percent of the solar wind mass flux.

The kinetic energy of a coronal mass ejection with a speed of $V =$ 400 thousand meters per second and a mass $M = 10^{13}$ kilograms is:

$$\text{KE} = MV^2/2 \approx 10^{24} \text{ Joule} = 10^{31} \text{ erg} .$$

This is comparable to the energies of large solar flares that lie between 10^{21} and 10^{25} Joule.

At a speed of $V = 400$ thousand meters per second, the time, T, to travel from the Sun to the Earth, at an average distance, D, is:

$$T = D/V = 3.75 \times 10^5 \text{ seconds} = 104 \text{ hours} = 4.34 \text{ days} .$$

air balloon that breaks its tether (Fig. 6.13). Whenever, a big, closed loop of magnetism is unable to hold itself down, a coronal mass ejection takes off.

The Large Angle Spectrometric COronagraph (LASCO), aboard the SOlar and Heliospheric Observatory (SOHO), provides images of coronal mass ejections (CMEs) that are quite similar to those observed previously, but with a larger field of view and better sensitivity (Table 6.6). LASCO observes the corona from 1.1 to 30 solar radii, at least three times more distant than previous coronagraphs. Because it looks further away from the Sun, this coronagraph observes CMEs that sometimes evolve into bigger structures than those seen before, leading to mass estimates up to 50 billion tons, an order of magnitude larger than previously estimated. The unprecedented sensitivity and large dynamic range enable LASCO to detect frequent, smaller and less massive events as well as the large, bright ones. The detected CMEs were therefore several times more frequent than during previous solar activity minima (Table 6.7). There might even be no such thing as a static, unchanging corona, and low-mass ejections could be going off all the time one after the other.

LASCO initially viewed a simple corona at activity minimum – one that was highly symmetrical and stable. During this lull in the Sun's magnetic activity, the Sun's equatorial regions were ringed by helmet streamers that stretched out tens of billions of meters in space, bounded at the top and bottom by oppositely directed magnetic fields. The streamer material seems to flow continuously along their magnetic channels, like a stream of water. At low altitudes, beneath the helmet streamer structure, the corona consists of large-scale magnetic loops that can expand outward (Sect. 5.2, Fig. 5.5). LASCO records dense concentrations moving at greater distances from the Sun through an otherwise unchanging streamer, like seeing logs floating on a river. The low-latitude, slow-speed solar wind may therefore consist of many consecutive, low-mass expulsions, at least during the minimum in the Sun's 11-year magnetic activity cycle (Sect. 5.2).

Unexpectedly wide regions of the Sun seem to convulse when the star releases coronal mass ejections, at least during activity minimum. To almost everyone's astonishment, LASCO found powerful equatorial ejections emitted nearly simultaneously from opposite sides of the Sun. Although coronagraphs only get a side view of the Sun, these ejections could result from global disturbances, extending all the way around the Sun and not just in one or two directions.

When a coronal mass ejection is directed at the Earth, a coronagraph will detect it as a gradually expanding, Sun-centered ring around the Sun. Such halo mass ejections were first observed by Russell A. Howard and colleagues in 1982, using the Solwind coronagraph on the P78-1 satellite. Although coronal mass ejections had previously been observed as magnetic bubbles ejected from one side of the Sun, these events are not likely to collide with the Earth. In contrast, a halo coronal mass ejection may signal a future terrestrial hit, with powerful geomagnetic storms, intense auroras and other threatening consequences (Sect. 7.2). Still, you cannot tell from the halo itself whether the mass ejection is targeted for the Earth or moving away from it in the opposite direction.

Many ejections occur in pre-existing helmet streamers that can persist in the corona for days to months. They bulge and brighten one to several days before erupting, suggesting that more material is accruing there. LASCO even found that the entire streamer belt, that girdles the solar equator during its activity lull, increases in brightness before CMEs. The pressure and tension of this added material could build until the streamer belt blows open. The helmet streamers then reform right after the disruption as if nothing happened, so the large-scale magnetism probably relaxes again to its former state, waiting for the next big one.

Yohkoh's Soft X-ray Telescope (SXT) has observed large-scale dimming of the soft X-ray corona near the time and location of coronal mass ejections. It has detected reduced X-ray emission that marks the scar of giant pieces ripped out of the corona. The expelled mass inferred from this coronal dimming is comparable to that estimated for coronal mass ejections.

Such abrupt depletions of the corona were first recorded by Richard T. Hansen and his colleagues in 1974 using a ground-based white-light coronagraph, but with limited spatial and temporal resolution. The corresponding X-ray effects were noted in Skylab data by David M. Rust and Ernest Hildner in 1978.

About 70 percent of the coronal mass ejections are associated with eruptive prominences or dark filament disappearances (Fig. 6.14). Prominences or filaments are giant, elongated and self-contained features, routinely seen near the solar disk in hydrogen-alpha photographs. They are called prominences when detected at the disk edge, glowing against the dark background and resembling arched viaducts or bridges. The relatively cool (8,000 degrees Kelvin) gas appears as dark, snaking features, called filaments, when projected against the bright chromospheric disk; the dark filaments overly the neutral line that separates regions of opposite magnetic polarity in the underlying photosphere. Prominences or filaments, which are two words for the same thing, can hang almost motionless for weeks or months, apparently suspended by a row of closed magnetic loops like parallel hammocks. Then the supporting magnetism becomes unhinged. Instead of falling down under gravity, the stately structures rise up and break away from the Sun, as though propelled by a loaded spring (Fig. 6.14).

Restraining coronal magnetic fields are apparently removed when a prominence erupts or a dark filament disappears, like opening the top of a jack-in-the-box or taking the lid off a boiling pot of water. The previously closed magnetic structures could be blown open or carried away by the coronal mass ejection, which precedes the prominence eruption. The close proximity of coronal mass ejections and erupting prominences in space and time indicates that they are consequences of a similar instability and restructuring of the large-scale (global) magnetic fields in the corona. As an example, in one model proposed by Eric R. Priest and Terry G. Forbes in 1990, the prominence erupts from twisted magnetic loops, driving magnetic reconnection and creating currents below the rising prominence.

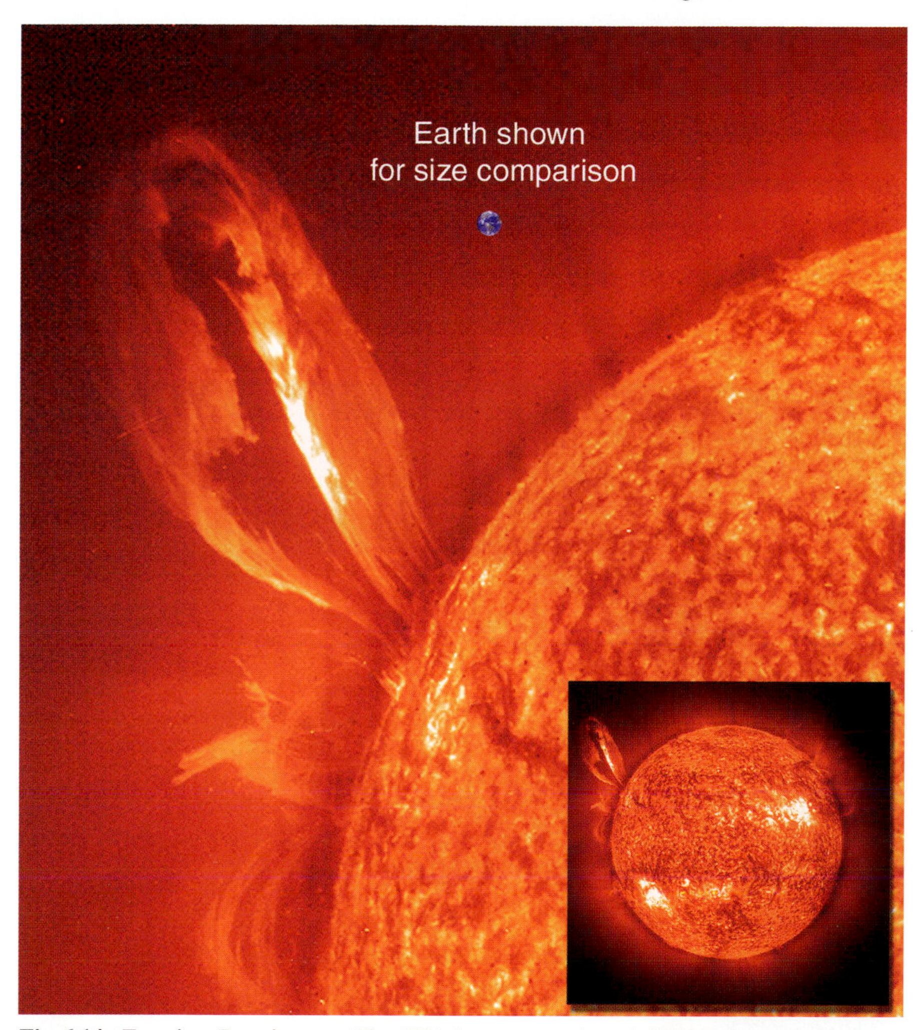

Fig. 6.14. Eruptive Prominence. The EIT instrument aboard SOHO imaged this large erupting prominence in the extreme ultraviolet light of ionized helium (He II at 30.4 nanometers) on 24 July 1999. The comparison image of the Earth shows that the prominence extends over 35 Earths out from the Sun, while the inset full-disk solar image indicates that the eruption looped out for a distance almost equal to the Sun's radius. (Courtesy of the SOHO EIT consortium. SOHO is a project of international collaboration between ESA and NASA)

When a prominence erupts, or a filament disappears, it is replaced by a row of bright X-ray emitting loops (Fig. 6.15), aligned like the bones in your rib cage or the arched trestle in a rose garden. First observed from Skylab, the X-ray loops bridge the magnetic neutral line between opposite polarity regions in the photosphere, stitching together and healing the wound inflicted by emptying that part of the corona. They show that the magnetic backbone of an erupting filament regroups and closes up again, retaining a memory of its former stability.

Yohkoh's SXT has investigated these beautiful, arched, post-ejection structures in detail, and extended their study to weak ones (Fig. 6.15). The loops form progressively from one end of the arcade to the other, and they also rise vertically with time. The enhanced X-ray emission is first seen after the onset of the mass ejection, and remains long after the ejection has departed from the low corona. These long-duration (hours) soft X-ray events, known as gradual flares, are due to global magnetic restructuring that proceeds along the ejection path (Sect. 6.6).

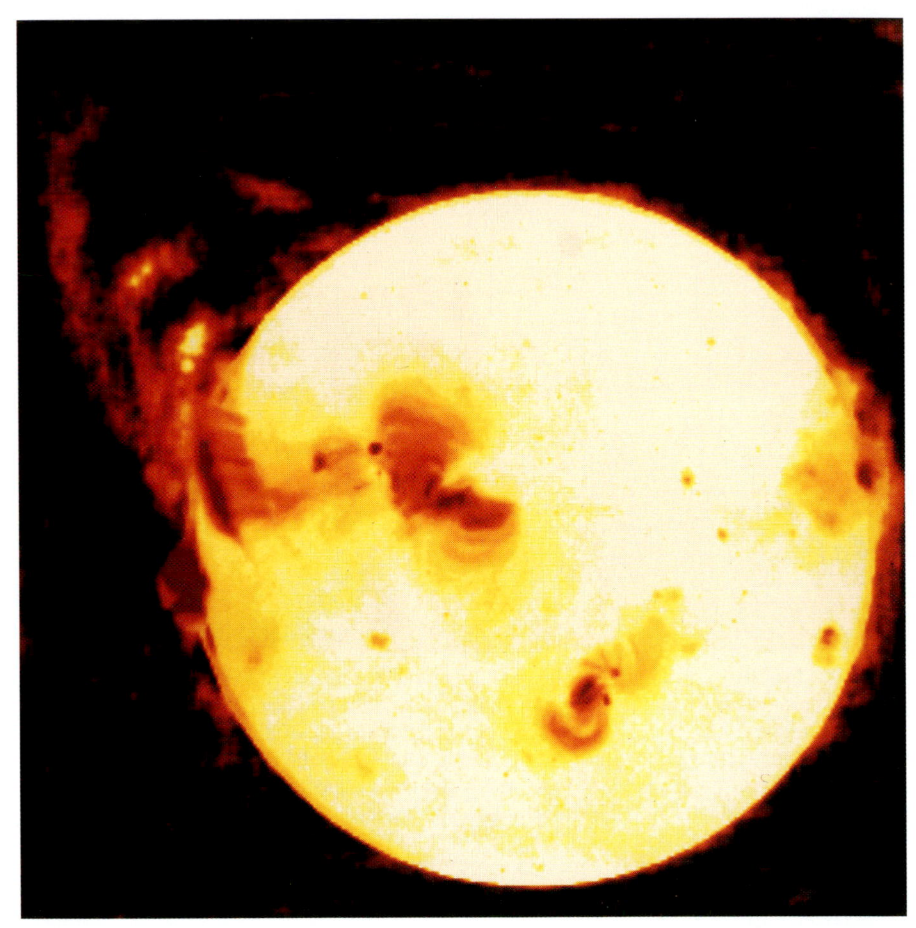

Fig. 6.15. Eruptive Prominence and Arcade Formation. A sequence of three time-spaced radio images show an eruptive prominence above the solar disk at the north-east (*top left*) during a 1.5 hour period. A negative soft X-ray image shows an arcade of loops formed by magnetic reconnection at a slightly later time, as if the Sun was stitching itself back together after the eruption. The radio images are from the Nobeyama Radioheliograph, while the X-ray image was obtained with the Soft X-ray Telescope (SXT) on the Yohkoh satellite. (Courtesy of Shinzo Enome, Nobeyama Radio Observatory)

Roughly 40 percent of coronal mass ejections are accompanied by solar flares that occur at about the same time and place. In fact, the flare-associated, meter-wavelength type II and IV radio bursts suggested the expulsion of mass from the corona long before coronal mass ejections were discovered using white-light coronagraphs. Yet, the physical size of coronal mass ejections is huge compared with flares or even the active regions in which flares occur. Moreover, the accompanying flares can occur at any time before, during or after the departure or "lift-off" of the mass ejection, and the relative locations of the two phenomena also show no systematic ordering.

So, most mass ejections are not initiated by solar flares and most flares are not caused by the ejections. After all, the flares are much more common. Nevertheless, it seems that the two types of solar explosions do involve similar processes, and they can result from the same magnetic activity in the corona.

What causes the Sun's magnetism to suddenly erupt with enough force to drive a large section of the corona out against the restraining force of solar gravity? The triggering mechanism seems to be related to large-scale interactions of the magnetic field in the low solar corona. This magnetism is always slowly evolving. It is continuously emerging from inside the Sun, and disappearing back into it, driven by the 11-year cycle of solar magnetic activity. In fact, both the location and rate of coronal mass ejections vary with this cycle. Near activity minimum coronal mass ejections are largely confined to equatorial regions, but they spread poleward over a broad range of solar latitudes as activity increases – even though sunspots are then migrating in the opposite direction toward the equator. The occurrence rate of coronal mass ejections also follows the state of activity, increasing by more than an order of magnitude from activity minimum to maximum. Coronal mass ejections may therefore be the primary way in which the corona is reconfigured in response to the slow evolution of the Sun's magnetic field. The physical processes involved in storing the magnetic energy in the solar corona and releasing it to either launch coronal mass ejections or ignite solar flares are next discussed.

6.6 Making Solar Explosions Happen

Powerful solar flares involve the explosive release of incredible amounts of energy, sometimes amounting to as much as ten million, billion, billion (10^{25}) Joule in a relatively short time between 100 and 1,000 seconds. This is comparable in strength to 20 million nuclear bombs, each blowing up with an energy of 100 Megatons of TNT.

A substantial fraction of this energy goes into accelerating electrons and ions to high (relativistic) speeds. These high-energy particles are sent down toward the Sun or out into space, and they result in enhanced radio, soft X-ray, hard X-ray and gamma ray radiation. Comparable amounts of energy are released in expelling matter during a coronal mass ejection, or CME, and they are most likely powered by similar processes to those that drive solar flares.

To explain how solar flares happen, we must know where their colossal energy comes from; why that energy is suddenly and rapidly released; and how it is transferred to accelerated particles and the ejection of matter. As stressed by Thomas Gold and Fred Hoyle in 1960, the Sun's magnetic field provides the only plausible source of energy for these powerful outbursts (Focus 6.5). After all, solar flares occur in magnetically active regions where the strongest magnetic fields are concentrated, projecting up into the corona from sunspots. Both solar flares and CMEs also vary in step with

Focus 6.5

Energizing Solar Flares

How much energy is released during a typical solar flare? The total flare energy, E_{T}, expended in producing electrons with energies, E_{e}, of about 30 keV, or 4.8×10^{-15} Joule, in a radius R with an electron density of $N_{\mathrm{e}} \approx 10^{17}$ electrons per cubic meter (Focus 6.2) is:

$$E_{\mathrm{T}} = 4\pi R^3 E_{\mathrm{e}} N_{\mathrm{e}}/3 \approx 2 \times 10^{24} \text{ Joule} ,$$

where the radius of a compact, impulsive flare is $R = 10^7$ meters, subtending an angular radius of 14 seconds of arc when viewed from the Earth.

What fuels these catastrophic eruptions on the Sun, and where does such a large amount of energy come from? The almost universal consensus is that solar flares are powered by magnetic energy. The magnetic energy, E_{M}, for a magnetic field strength, B, in a radius, R, is:

$$E_{\mathrm{M}} = [4\pi/(6\mu_0)] B^2 R^3 = 1.66 \times 10^6 B^2 R^3 \text{ Joule} ,$$

where the permeability of free space is $\mu_0 = 4\pi \times 10^{-7}$ henry m^{-1}, the radius is in meters and the magnetic field strength in Tesla.

Magnetic reconnection in the low solar corona serves as an efficient engine for converting magnetic energy into plasma kinetic energy and thermal energy. Let us suppose that the reconnection site is just above a coronal loop of radius $R = 10^7$ meters, with a comparable size. To provide the flare energy, $E_{\mathrm{T}} = 2 \times 10^{24}$ Joule in a volume of radius $R = 10^7$ meters, a magnetic field of about 0.03 Tesla is required. Solar astronomers often use the c.g.s. unit of Gauss, where 1 Gauss = 10,000 Tesla, so the required magnetic field change in the corona is roughly 300 Gauss.

Some solar flares may have significantly smaller magnetic field strengths near their loop tops in the corona, where the flare energy is thought to be released. In this case, the energy reservoir cannot be supplied by the flare volume seen in soft X-rays, but instead suggests a larger volume above the flare loops. If the radius was just a factor of five larger than our estimate, the volume would be 125 times larger, and the magnetic field strength would be ten times smaller, or about 0.003 Tesla.

the 11-year cycle of magnetic activity, becoming more frequent and violent when sunspots and intense coronal magnetic fields are most commonly observed.

Catastrophic explosions on the Sun must be fueled by magnetic fields looping and threading their way into the low solar corona, above the visible photosphere where sunspots are found. This is because most flares are not seen in the white light of the photosphere; because there is no observed, rapid and significant change in the photosphere magnetic fields during a flare; and because the accelerated electrons giving rise to intense radio emission originate above the photosphere. As shown by Skylab and Yohkoh, such active-region coronal loops emit the most intense X-ray emission on both the flaring and non-flaring Sun.

We can see how excess magnetic energy might be produced in the corona by considering a single magnetic loop within an active region. The coronal loop links sunspots of opposite magnetic polarity or direction, and the line dividing the two polarities is called the magnetic neutral line. If the magnetic fields connect the two sunspots in the shortest, most direct path, and therefore run perpendicular to the magnetic neutral line, scientists say that the magnetic field has a potential configuration. It can be distorted into a non-potential shape when flows at or below the photosphere shear and twist the looping fields. Non-potential magnetic fields have more magnetic energy than the potential ones. This extra energy, called free magnetic energy, builds up from below, due to motions down there, and can be released to power solar flares.

The Sun's magnetism apparently remains closed near the photosphere and opens up further out. In the low corona, strong magnetic fields are tied to the Sun at both ends, trapping hot, dense electrified gas within magnetized loops. Higher up, the relatively weak, outermost loops cannot constrain the outward pressure of the hot gas. The magnetic fields get caught up by the solar wind out there and are pulled into interplanetary space. The overall result consists of closed loops down low, and two extended stalks of oppositely directed magnetic field further out (Fig. 6.16). Such a magnetic geometry was used in one of the first magnetic theories for the origin of solar flares, proposed by Ronald G. Giovanelli in 1946–48.

According to Giovanelli's model, varying magnetic fields generate current sheets at a special place in the corona, called the neutral point. It is located at the interface between the closed and open magnetic structures (Fig. 6.16). The currents build up excess energy at the neutral point until an electrical discharge releases it as a solar flare.

During the subsequent decade, James W. Dungey developed the theory for the growth of currents and subsequent electrical discharge at neutral points of the magnetic field. He noticed that the effect of the discharge would be to quickly "reconnect" the oppositely-directed magnetic lines of force at the neutral point, that marks the place where they touch. This severs the magnetic fields and annihilates some of them.

According to Dungey's model, beams of accelerated particles are shot out from the neutral-point discharge in two directions, up into space and down

but keeping the system near the same critical state. In this analogy, the electric currents and stored magnetic energy in the coronal loop slowly build up until they are on the brink of instability, in a critical condition where further perturbation results in avalanche-like disruptions and magnetic reconnection at or just above the neutral point. Much of the energy that was originally in the magnetic fields is given up to charged particles, accelerating them to high speeds and catapulting them away from the reconnection site.

In other explanations, the explosions are triggered by an instability or loss of equilibrium in a single magnetic loop, like the quick snap of a rubber band that has been twisted too tightly or the sudden flash and crack of a lightening bolt. It has also been likened to the loss of equilibrium during an earthquake. According to this comparison, the moving roots or footpoints of a sheared magnetic loop are analogous to two tectonic plates. As the plates move in opposite directions along a fault line, they grind against each other and build up stress and energy. When the stress is pushed to the limit, the two plates cannot slide further, and the accumulated energy is released as an earthquake. That part of the fault line then lurches back to its original, equilibrium position, waiting for the next earthquake. After an explosive convulsion on the Sun, the magnetic fields similarly regain their composure, fusing together and becoming primed for the next outburst.

All of these theoretical models had a common theme involving free magnetic energy released by magnetic reconnection, but they all remained unsubstantiated by hard evidence. The hypothetical, rapid changes of magnetic fields during solar flares have, for example, never been observed down in the photosphere, and no one has ever measured the predicted depletion of magnetic energy that supposedly spawns eruptive outbursts on the Sun. So, the jury was still out deliberating about the cause of explosions on the Sun. Then, Yohkoh came along to demonstrate that magnetic interactions can indeed strike the match that ignites solar explosions.

Yohkoh's Soft X-ray Telescope, or SXT for short, delineated the ubiquitous coronal loops with a clarity and sensitivity that had not been possible before. Because strong magnetic fields confine the hot and dense coronal gas that emits solar X-rays, the SXT could trace out the changing magnetic topology in the corona. Its rapid, sequential images have demonstrated that the dynamic, ever-changing corona is in a continued state of metamorphosis, forever adjusting to the shifting forces of magnetism.

Sequential SXT images indicate that magnetized loops can become sheared, stretched, tangled and twisted. After a solar flare, the distorted shape can be transformed into a simpler, less-agitated configuration, suggesting the release of free (stored) magnetic energy and relaxation to a lower-energy, more stable, potential-like state with an associated reduction of currents in the corona.

Yohkoh's X-ray movies show sheared coronal loops that converge within active regions or on larger scales, as well as the coalescence of emerging, twisted loops with overlying ones. The magnetic shears and twists suggest the build up of energy that is dissipated at the place where the distorted loops meet.

Yohkoh's SXT has also revealed the probable location of the magnetic reconnection site for the first time (Fig. 6.18). The prototype is the cusp-shaped soft X-ray loop structure detected by Saku Tsuneta and his colleagues during a long-duration (hours) flare seen at the apparent edge or limb of the Sun. The rounded magnetism of a flaring coronal loop has been pulled into a peaked shape at the top, like someone taking a bite out of a candied apple; it formed a few hours after a large-scale coronal eruption, possibly a coronal mass ejection.

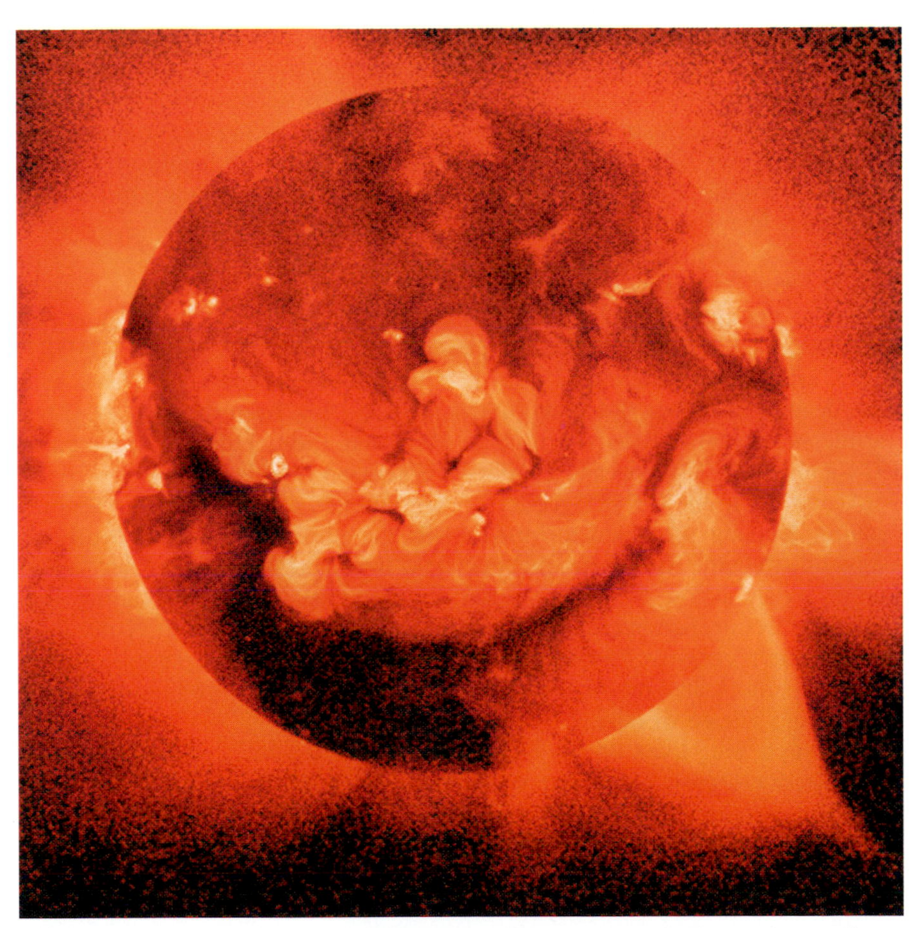

Fig. 6.18. Cusp Geometry. A large helmet-type structure is seen in the south west (*lower right*) of this soft X-ray image obtained on 25 January 1992 following a coronal mass ejection. The cusp, seen edge-on at the top of the arch, is the place where the oppositely directed magnetic fields, threading the two legs of the arch, are stretched out and brought together. Several similar images have been taken with the Soft X-ray Telescope (SXT) aboard Yohkoh, showing that magnetic reconnection is a common method of energizing solar explosions. (Courtesy of Loren W. Acton, NASA, ISAS, the Lockheed-Martin Solar and Astrophysics Laboratory, the National Astronomical Observatory of Japan, and the University of Tokyo)

The sharp, peaked cusp-like structure at the loop top (Fig. 6.18) is attributed to magnetic reconnection along a neutral line, or current sheet, just above the loop. This marks the place where oppositely directed field lines stretch out nearly parallel to each other and are brought into close proximity. Here the magnetism comes together, merges and annihilates part of itself, releasing the energy needed to power a solar explosion.

The cusp geometry has an inverted Y shape that resembles a funnel turned upside down. The rounded, soft X-ray shape seen just below the cusp tip, strongly suggests that accelerated particles pour down the inverted tunnel into the underlying loop, and that magnetic reconnection along the neutral line, or current sheet, just above the loop top is responsible for the energy release.

The apparent loop height and the distance between the two loop footpoints increase as a function of time. This is explained by successive reconnection at a rising point, moving along the outstretched neutral line like a zipper. As the observed soft X-ray loop system grows ever larger, the height of a given underlying magnetic field line is also compressed, so it pulls back, shrinks and withdraws, like a turtle going into its shell.

Yohkoh's SXT has shown that many long-duration flares show cusp-shaped loop structures. The implied magnetic field configuration is quite similar to the classical magnetic reconnection model in which the magnetic structures open and close like a sea anemone. It has been dubbed the CSHKP model after the first letters of the last names of various researchers who have developed it over the years. (Such a model has been invoked by Hugh Carmichael in 1964, Peter Sturrock in 1966 and 1968, Tadashi Hirayama in 1974 and Roger A. Kopp and Gerald W. Pneuman in 1976).

According to the CSHKP model, a coronal mass ejection, with its accompanying erupting prominence, blows open the previously closed, overlying magnetic structure. The open magnetic field lines couple and reconnect again as they pinch below the rising material. Electrons accelerated during this phase of reconnection can explain the subsequent creation of soft X-ray loop arcades, the successive formation of new, higher loops, the separation of the loop footpoints where bright chromosphere, or Hα, ribbons are found, and the long-lasting release of energy.

Although long-lived (hours) flares, sometimes associated with coronal mass ejections, show clear cusp-shaped loop structures suggesting magnetic reconnection, there is no such feature in short-lived (minutes), impulsive flares. Yohkoh has nevertheless gathered evidence for magnetic reconnection during the compact, impulsive flares by observing them at the limb of the Sun at both hard X-ray and soft X-ray wavelengths. To everyone's surprise, the co-aligned images showed a compact, impulsive hard X-ray source well above and outside the corresponding soft X-ray loop structure, in addition to the double-footpoint hard X-ray emission that had been seen before. The prototype, discovered by Satoshi Masuda and his colleagues, is illustrated in Fig. 6.19. Impulsive hard X-ray sources that occur in the corona above the flaring soft X-ray loops have subsequently been found for several impulsive (minutes) flares observed at the Sun's apparent edge.

The expected hard X-ray emission at the loop footpoints is produced when high-speed, non-thermal electrons rain down into dense material in the chromosphere. In contrast, the unexpected, hard X-ray emission above the soft X-ray loops is located in the corona where the density is low. One likely explanation is that the outermost hard X-ray source represents the site where electrons are accelerated to high energy during the impulsive phase. These electrons rapidly move down toward the footpoints, explaining the similarity of the time variations of all three hard X-ray sources.

The coronal hard X-ray source observed during the impulsive flares is located at about the same place as the soft X-ray cusp structure detected during long-duration flares. They could both mark the site of magnetic reconnection, where oppositely-directed, or anti-parallel, magnetic fields meet above the flaring coronal loops.

In summary, a well-developed magnetic theory for solar explosions has received substantial observational verification with Yohkoh's X-ray telescopes. They have shown us how and where these violent outbursts occur, relating them to the interaction of sheared, twisted coronal loops, pinpointing the likely site of magnetic reconnection in the corona, and showing that high-speed particles can be hurled down into the Sun and propelled out into space.

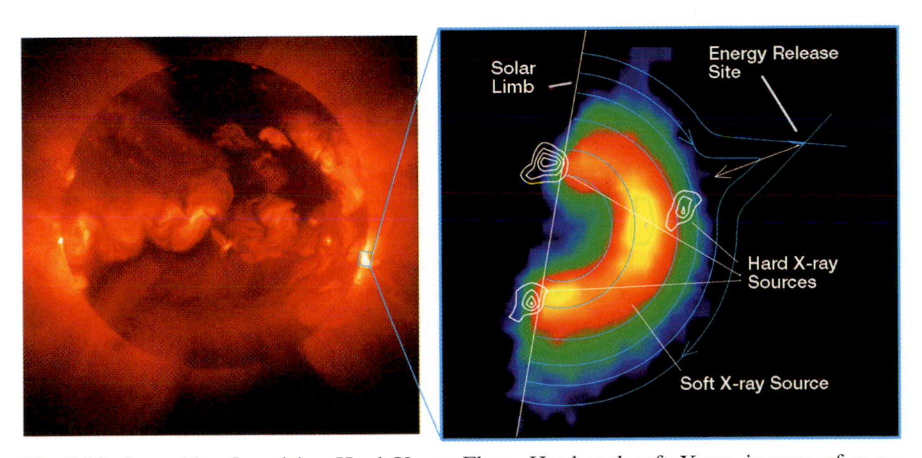

Fig. 6.19. Loop-Top Impulsive Hard X-ray Flare. Hard and soft X-ray images of a solar flare occurring near the solar limb on 13 January 1992. The white contour maps show three impulsive hard X-ray sources from high-energy electrons accelerated during the solar flare, superposed on the loop-like configuration of soft X-rays emitted by plasma heated during the flare gradual or decay phase to temperatures of 10 to 20 million degrees. In addition to the double-footpoint sources, a hard X-ray source exists above the corresponding soft X-ray magnetic loop structure, with an intensity variation similar to those of the other two hard X-ray sources. This indicates that the flare is energized from a site near the magnetic cusp. These simultaneous images were taken with the Hard X-ray Telescope (HXT) and the Soft X-ray Telescope (SXT) aboard the Yohkoh satellite. (Adapted from S. Masuda et al, 1994; Courtesy of NASA, ISAS, the Lockheed-Martin Solar and Astrophysics Laboratory, the National Astronomical Observatory of Japan, and the University of Tokyo)

Such magnetic contortions may provide warning signals of impending solar flares and/or coronal mass ejections that can threaten satellites and humans in space (Sect. 7.2, Figs. 7.9 and 7.10).

6.7 Interplanetary Protons, Electrons and Shock Waves

Energetic subatomic particles can be hurled into space during a solar flare. The brief, transient explosions rapidly release energy stored in unstable magnetic configurations, and convert a significant fraction of that energy into high-speed protons and electrons in less than a second, or about as fast as you can blink your eye.

Flares are composed of the same type of particles as the steady, everflowing solar wind, but with much faster speeds and vastly greater energy. During solar flares, electrified particles are accelerated to speeds of at least 100 times that of the solar wind. The flaring protons and electrons can achieve energies as high as 1,000 million electron volts (1 GeV) and 100 million electron volts (100 MeV), respectively; this is at least a million times more energetic than solar wind particles. Because protons are 1,836 times as massive as electrons, and move with comparable speed, protons are far more energetic than electrons in both flares and the wind,

There are many more particles blowing in the solar wind than emanating from a solar flare (Table 6.8). At the Earth's distance, only about 10 million particles from a typical solar flare cross an area of one square meter each second, which amounts to a flux of 10^7 particles $m^{-2}\,s^{-1}$. About five hundred thousand more solar wind particles pass through the same area in that time (5×10^{12} particles $m^{-2}\,s^{-1}$), but with much lower energy.

We have known about energetic, flare-associated, interplanetary protons for about half a century. In the 1940s, Scott Forbush and his colleagues used cosmic-ray detectors on the ground to record transient increases in the number of energetic charged particles arriving at the Earth after solar flares. Balloon and rocket observations in the late 1950s and early 1960s indicated that these

Table 6.8. Energy and flux of protons arriving at Earth*

Source	Energy (MeV)	Flux (protons $m^{-2}s^{-1}$)
Cosmic Rays	1,000	6×10^2
Solar Flares**	10	1×10^7
Coronal Mass Ejections**	10	3×10^8
Solar Wind	0.001	5×10^{12}

* An energy of $1\,\text{MeV} = 10^6\,\text{eV} = 1.6 \times 10^{-6}\,\text{erg} = 1.6 \times 10^{-13}$ Joule.

** A single particle event associated with a coronal mass ejection usually has a higher flux of energetic protons than a single particle event produced by a solar flare, but coronal mass ejections occur about one hundred times less frequently than solar flares.

events are mainly due to protons. Since they were detected following solar flares, it was assumed that the protons were created at or near the Sun, and therefore dubbed solar cosmic rays. We prefer to call them solar protons to avoid confusion with the cosmic-ray protons that come from outside the solar system, which are sometimes called Galactic cosmic rays.

Only exceptionally rare and energetic solar protons reach the ground. They need energies of about 1,000 million electron volts, or 1 GeV, to overcome resistance by the Earth's magnetic field and pass beyond this barrier. Lower-energy protons are deflected by this magnetism, and channeled into the polar regions where they enhance the ionosphere. The protons that cause such polar cap absorptions, or PCAs, have a maximum effect with an energy near 10 MeV, and such events are about ten times more frequent than the proton events that reach the ground.

Flare-associated neutrons were detected in the 1980s with the Gamma Ray Spectrometer aboard the Solar Maximum Mission (SMM) satellite and in the 1990s using the Compton Gamma Ray Observatory. The neutrons are produced by flare protons with energies of up to 1,000 million electron volts, or 1 GeV. Neutrons with very high energy of about 1 GeV have been detected by ground-based neutron monitors following exceptionally intense solar flares.

Energetic, flare-associated, interplanetary electrons have also been watched during solar radio bursts for several decades. The impulsive radio bursts are generated by high-speed electrons with energies of 10 to 100 keV, or 10,000 to 100,000 electron volts. Radio astronomers spotted these electrons leaving the Sun, stimulating oscillations of lower and lower frequency as they passed through and jostled the progressively more rarefied coronal atmosphere (Sect. 6.2). Spacecraft observations of these type III radio bursts have been used to track the electrons as they move away from the Sun, confirming that open magnetic field lines connect flares directly into the interplanetary medium.

Once the high-speed electrons leave the Sun, they follow a curved path rather than a straight line. The flaring electrons that trigger the radio emission must follow the interplanetary magnetic field lines and usually do not have enough energy to cross them. As the solar wind streams radially outward, pulling the Sun's magnetism with it, the rotating star twists the magnetic fields into a spiral shape within the plane of the Sun's equator (Sect. 2.4, Fig. 2.5). The spiral trajectories were first inferred from type III radio bursts in the 1970s using spacecraft in the ecliptic, or orbital plane of the planets, and then mapped by tracking the type III bursts from above when the Ulysses spacecraft passed over the poles of the Sun (Fig. 6.20).

Energetic charged particles generated during solar flares will only threaten our planet if they occur at just the right place on the Sun, at one end of a spiral magnetic field line that connects the flaring region to the Earth. Given the right circumstances, with a flare near the west limb and the solar equator, the magnetic spiral acts like an interplanetary highway that connects the flaring electrons to the Earth. Moving along this magnetic conduit at about half the speed of light, a 100-keV electron generated during a type III radio burst travels from the Sun to the Earth in about 20 minutes.

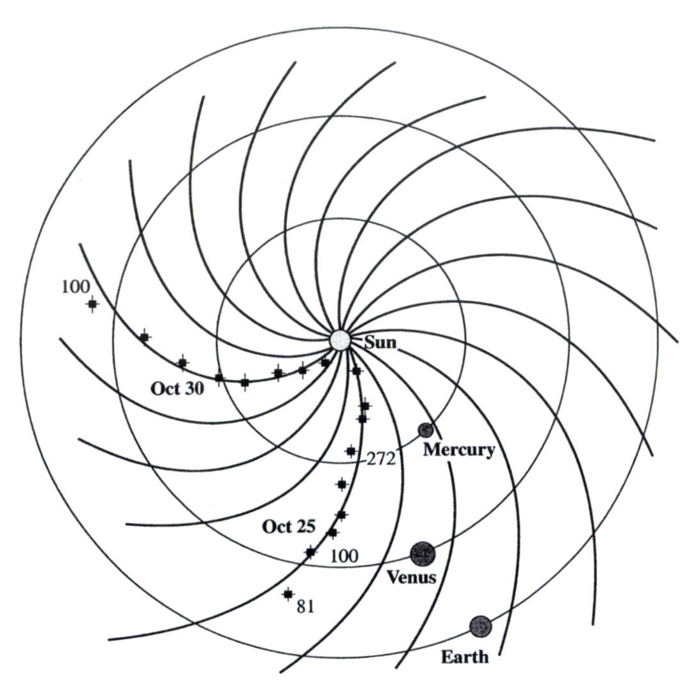

Fig. 6.20. Interplanetary Electrons. The trajectory of flare electrons in interplanetary space as viewed from above the Polar Regions using the Ulysses spacecraft. The squares and crosses show Ulysses radio measurements of type III radio bursts on 25 October 1994 and 30 October 1994. As the high-speed electrons move out from the Sun, they excite radiation at successively lower plasma frequencies. The numbers denote the observed frequency in kiloHertz, or kHz. Since the flaring electrons are forced to follow the interplanetary magnetic field, they do not move in a straight line from the Sun to the Earth, but instead move along the spiral pattern of the interplanetary magnetic field, shown by the solid curved lines. The magnetic fields are drawn out into space by the radial solar wind, and attached at one end to the rotating Sun. The approximate locations of the orbits of Mercury, Venus and the Earth are shown as circles. (Courtesy of Michael J. Reiner, and adapted from M. J. Reiner, J. Fainberg and R. G. Stone, 1995. Ulysses is a project of international collaboration between ESA and NASA)

These interplanetary electrons have been directly sampled by *in situ* measurements using interplanetary probes and Earth-orbiting satellites. Impulsive bursts of high-speed electrons, with energies greater than 40 keV, were, for example, found in deep space in 1965 by James A. Van Allen and Stramatios M. Krimigis using a detector on NASA's Mars-bound spacecraft Mariner 4. The energetic particles were presumably hurled into space by solar flares that occurred at approximately the same time. Subsequent observations indicate that solar flares toss out interplanetary electrons, with energies of 10 to 100 keV, about 100 times a month during the maximum of the solar activity cycle.

Much more energetic, interplanetary electrons, from relatively rare solar-flare events, have been recorded with energies up to 100 million electron volts, or 100 MeV. They were first observed by Peter Meyer and Rochus Vogt in 1962 with a balloon-borne detector and in more detail throughout the 1960s with NASA's Interplanetary Monitoring Platform (IMP) series of spacecraft. Such non-thermal particles were previously invoked to explain the synchrotron radiation of type IV radio bursts (Sect. 6.2). Interplanetary proton and electron events with particle energies greater than 10 MeV occur just a few times a month near solar maximum.

Self-contained clouds of electrons, protons and magnetic fields can also be ejected from the Sun into surrounding space. Such interplanetary plasma clouds, propagating from the Sun to the Earth, were proposed in 1919 by Frederick A. Lindemann as the cause of pronounced fluctuations in the Earth's magnetic field, called geomagnetic storms. As suggested by Philip Morrison in 1954, ionized magnetic clouds emitted by the active Sun will also deflect cosmic rays from an Earth-bound path, thereby accounting for world-wide decreases in cosmic-ray intensity, lasting for days, and correlated roughly with geomagnetic storms. Today we recognize that these plasma clouds originate at the Sun as coronal mass ejections.

A shock wave will develop between gas ejected from the Sun and the interplanetary material already present. Such shock waves were inferred from type II radio bursts in the 1950s (Sect. 6.2), proposed by Thomas Gold in 1959, and first detected in interplanetary space by Charles P. Sonett and colleagues from Mariner 2 measurements in 1962.

Direct *in situ* magnetic field and plasma measurements of the solar wind have now routinely demonstrated the existence of interplanetary shocks and magnetized plasma clouds for over thirty years. As an example, in 1981 Leonard F. Burlaga and his colleagues used data from five spacecraft (Voyager 1 and 2, Helios 1 and 2 and IMP 8) to describe interplanetary magnetic clouds that expand as they move away from the Sun. Such clouds become about 50 times larger than the Sun by the time they reach the Earth. They can twist and spiral out, like a giant corkscrew, carrying intense, helical magnetic fields with them. The magnetic clouds can remain rooted in the Sun at both ends, producing a sort of magnetic bottle as it moves into space, much as Gold had supposed in 1959.

When coronal mass ejections were discovered in the 1970s, it was realized that they generate many of the interplanetary shocks. A subsequent comparison of space-borne coronagraph data with type II radio bursts, detected from space at low frequencies, indicted that all of the energetic interplanetary shocks were indeed associated with fast coronal mass ejections. Strong ejections plow into the slower-moving solar wind, like a car out of control, serving as pistons to drive huge shock waves billions of meters ahead of them (Fig. 6.21).

Many of the largest solar energetic particle events are linked to prolonged shocks, driven by coronal mass ejections, that propagate far from the Sun, rather than to rapid, impulsive solar flares that accelerate particles close to the Sun. The coronal mass ejections are associated with gradual soft X-ray flares, lasting hours to days, and with prominence eruptions. Shock waves propelled by coronal mass ejections carry along electrons and ions in the interplanetary

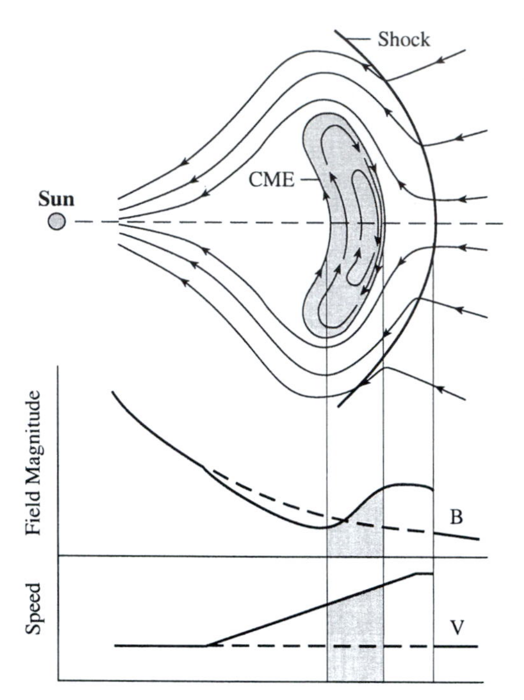

Fig. 6.21. Interplanetary Shock Wave. A cross-sectional cut through an interplanetary shock wave disturbance driven by a very fast coronal mass ejection (CME) (*top*) and the corresponding radial variation of solar wind speed, V, and magnetic field strength, B, along the centerline of the disturbance (*bottom*). As it moves away from the Sun (*top left*) a fast coronal mass ejection (CME, *top right*) pushes an interplanetary shock wave before it. The CME expands to become a magnetic cloud that is much larger than the Sun. The fast CME produces a speed increase all the way to the shock front, where the wind's motion then slows down precipitously to its steady, unperturbed speed. Compression, resulting from the relative motion between the fast CME and its surroundings, produces strong magnetic fields in a broad region extending sunward from the shock. The strong magnetic fields and high flow speeds commonly associated with interplanetary disturbances driven by fast CMEs are what make such events effective in stimulating geomagnetic activity. As suggested by Giuseppe Cocconi, Thomas Gold and their colleagues in the late 1950s, the magnetic field lines of a CME can form an extended loop or magnetic bottle anchored on the Sun. Such a closed field region is shown here by lines that drape about the CME and with arrows that denote the directions and positions of the magnetic fields. (Adapted from J. T. Gosling, et al., 1991)

medium, crossing magnetic field lines, and accelerating particles as they go, much as ocean waves propel surfers. When slow coronal mass ejections move outward into interplanetary space, with speeds below that of the ambient solar wind, no shock waves are generated and energetic solar particle events are not observed.

To sum up, solar energetic particles found in the interplanetary medium are of two basic types, the impulsive and gradual events, with the properties described in Table 6.9. The impulsive events have their origin during rapid flares, lasting from a few minutes to an hour; the gradual flares last for hours. The impulsive particle events are rich in electrons, helium-3 and iron, and they are not associated with coronal mass ejections. The gradual particle events are proton rich and are well associated with coronal mass ejections. Impulsive events are about 100 times more frequent than the gradual ones. Solar energetic particles associated with both impulsive and gradual flares can endanger astronauts and destroy satellite electronics (Sect. 7.2). The shock waves, ejected mass and magnetic fields associated with coronal mass ejections can produce intense geomagnetic storms, create power surges in terrestrial transmission lines and trigger brilliant auroras in the polar skies (Sect. 7.2).

Table 6.9. Properties of impulsive and gradual solar energetic particle events

	Impulsive Events	Gradual Events
Particles:	Electron-rich	Proton-rich
$^3He/^4He$	≈ 1	≈ 0.0005
Fe/O	≈ 1	≈ 0.1
H/He	≈ 10	≈ 100
Duration of X-ray Flare	Impulsive (minutes, hard X-rays)	Gradual (hours, soft X-rays)
Duration of Particle Event	Hours	Days
Radio Bursts	Types III and V**	Types II and IV
Coronagraph	Nothing Detected	Coronal Mass Ejections, 96%
Solar Wind	Energetic Particles	Very Energetic Particles
Longitudinal Extent	< 30 degrees	≈ 180 degrees
Events Per Year	$\approx 1{,}000$ at activity maximum	≈ 100 at activity maximum

* Adapted from Gosling (1993) and Reames (1997).
** Impulsive type III and V bursts can be followed by type II and IV.

Table 6.10. Key events in understanding explosive solar activity*

Date	Event
1852	Edward Sabine demonstrates that global magnetic disturbances of the Earth, now called geomagnetic storms, vary in tandem with the 11-year sunspot cycle.
1859–60	In 1859 Richard C. Carrington and Richard Hodgson independently observe a solar flare in the white light of the photosphere and in 1860 publish the first account of such a flare. There was no perceptible change in the sunspots after the "sudden conflagration", leading Carrington to conclude that it occurred above and over the sunspots. Seventeen hours after the flare a large magnetic storm begins on the Earth.
1908	George Ellery Hale uses the Zeeman splitting of spectral lines to measure intense magnetic fields in sunspots. They are thousands of times stronger than the Earth's magnetism.
1919	Frederick Alexander Lindemann (later Lord Cherwell) suggests that an electrically neutral plasma ejection from the Sun is responsible for non-recurrent geomagnetic storms.

1919 George Ellery Hale and his colleagues show that sunspots occur in bipolar pairs with an orientation that varies with a 22-year period.

1924–31 George Ellery Hale develops the spectroheliscope that enables the entire Sun to be scanned visually at selected wavelengths, most notably in the light of the hydrogen alpha, or Hα, spectral line at 656.3 nanometers.

1930–60 Hydrogen alpha, or Hα, observations of chromosphere brightenings, by V. Bumba, Helen W. Dobson, Mervyn Archdall Ellison, Ronald G. Giovanelli, Harold W. Newton, Robert S. Richardson, A. B. Severny and Max Waldmeier, show that chromosphere flares occur close to sunspots, usually between the two main spots of a bi-polar group, that magnetically complex sunspot groups are most likely to emit flares, and that hydrogen alpha flare ribbons lie adjacent and parallel to the magnetic neutral line.

1935–37 J. Howard Dellinger suggests that the sudden ionosphere disturbances that interfere with short-wave radio signals have a solar origin.

1944 Robert S. Richardson proposes the term solar flare for sudden, bright, rapid, and localized variations detected in the hydrogen alpha, or Hα, light of the chromosphere.

1946 Edward V. Appleton and J. Stanley Hey demonstrate that meter-wavelength solar radio noise originates in sunspot-associated active regions, and that sudden large increases in the Sun's radio output are associated with chromosphere brightenings, also known as solar flares.

1946–48 Ronald G. Giovanelli develops a theory of solar flares involving the magnetic fields in the solar atmosphere above sunspots, including electric currents at magnetic neutral points.

1946, 50 Scott E. Forbush and his colleagues describe flare-associated transient increases in the cosmic ray intensity at the Earth's surface, and attribute them to very energetic charged particles from the Sun.

1947 Ruby Payne-Scott, D. E. Yabsley and John G. Bolton discover that meter-wavelength solar radio bursts often arrive later at lower frequencies and longer wavelengths. They attributed the delays to disturbances moving outward at velocities of 0.50 to 0.75 million meters per second, exciting radio emission at the local plasma frequency.

1948–49 Soft X-rays from the Sun were first detected on 6 August 1948 with a V-2 rocket experiment performed by the U. S. Naval Research Laboratory, reported by T. R. Burnight in 1949.

1949 Alfred H. Joy and Milton L. Humason show that main sequence (dwarf M) stars other than the Sun emit flares.

1950–59 John Paul Wild and his colleagues use a swept frequency receiver to delineate Type II radio bursts, attributed to shock waves moving out during a solar flare at about a million meters per second, and Type III radio bursts, due to outward streams of high-energy electrons. The electrons are accelerated at the onset of a solar flare and move at nearly the velocity of light.

1950 Hannes Alfvén and Nicolai Herlofson argue that synchrotron radiation of high-speed electrons spiraling about magnetic fields might generate the observed radio emission from discrete cosmic sources, and Karl Otto Kiepenheuer reasons that the synchrotron radiation mechanism can account for the radio emission of our Galaxy.

1951–63 Herbert Friedman and his colleagues at the U. S. Naval Research Laboratory use rocket and satellite observations to show that intense X-rays are emitted from the Sun, that the X-ray emission is related to solar activity, and that X-rays emitted during solar flares are the cause of sudden ionosphere disturbances.

1953 James W. Dungey proposes a magnetic neutral point discharge theory for solar flares.

1954 Philip Morrison proposes that magnetized clouds of gas, emitted by the active Sun, account for world-wide decreases in the cosmic-ray intensity observed at Earth, lasting for days and correlated roughly with geomagnetic storms.

1954–58 Scott E. Forbush demonstrates the inverse correlation between the intensity of cosmic rays arriving at Earth and the number of sunspots over two 11-year solar activity cycles.

1957 André Boischot discovers moving type IV radio bursts, and Boischot and Jean-Francoise Denisse explain them in terms of magnetic clouds of high-energy electrons propelled into interplanetary space.

1958–59 During a balloon flight on 20 March 1958, Laurence E. Peterson and John Randolph Winckler observed a burst of high energy, gamma-ray radiation (200 to 500 keV) coincident with a solar flare, suggesting non-thermal particle acceleration during such explosions on the Sun.

1958 Alan Maxwell and Govind Swarup call attention to U-type radio bursts, a spectral variation of type III fast-drift bursts that first decrease and then increase in frequency, suggesting motions away from and into the Sun along closed magnetic field lines.

1959 Thomas Gold argues that solar flares will eject material within magnetic clouds that remain magnetically connected to the Sun, and suggests that an associated shock front can produce sudden geomagnetic storms. He also coined the term magnetosphere for the region in the vicinity of the Earth in which the Earth's magnetic field dominates all dynamical processes involving charged particles.

1960 Thomas Gold and Fred Hoyle show that magnetic energy must power solar flares, and argue that flares are triggered when two magnetic loops of opposite sense or direction interact, merge and suddenly dissipate the stored magnetic energy.

1960–61 Gail E. Moreton uses rapid, time-lapse hydrogen alpha, or Hα, photography to discover wave-like chromosphere disturbances initiated by solar flares. These Moreton waves move away from impulsive flares for distances of about a billion meters, with velocities of around a million meters per second.

1961–68 Mukul R. Kundu demonstrates the similar time profiles of centimeter wavelength, impulsive radio bursts and hard X-ray radiation from solar flares. In 1963, Kees de Jager and Kundu explained the similarity in the profiles in terms of the same energetic electrons producing the radio and hard X-ray emission. In 1967–68, Roger L. Arnoldy and colleagues demonstrated a correlation between the flux of centimeter-wavelength and hard X-ray radiation from energetic solar flares.

1962–64 Interplanetary shocks associated with solar activity are detected using instruments aboard the Mariner 2 spacecraft in 1962, reported by Charles P. Sonett and colleagues in 1964.

1964 T. R. Hartz obtains the first spacecraft observations of solar type III bursts using a swept frequency receiver from 1.5 to 10 MHz.

1964 Harry E. Petschek clarifies the process of magnetic field reconnection and shows that it could occur rapidly even in a highly conducting plasma.

1966–68 Peter A. Sturrock develops a magnetic reconnection model for solar flares, accounting for particle acceleration, ejected plasma and the formation of two bright ribbons in the chromosphere.

1967–68 Roger L. Arnoldy, Sharad R. Kane and John Randolph Winckler demonstrate a flux correlation for centimeter-wavelength impulsive radio bursts and hard X-ray solar flares.

1968 Giuseppe Vaiana and his colleagues show that the soft X-ray emission of a solar flare corresponds spatially with the hydrogen alpha emission, with roughly the same size, indicating a close link between the two phenomena.

1968 Werner M. Neupert uses soft X-ray flare data, obtained with the third Orbiting Solar Observatory (OSO 3), to confirm that soft X-rays slowly build up in strength, and to show that the rise to maximum intensity resembles the time integral of the rapid, impulsive microwave burst.

1970–73 K. C. Hsieh, John A. Simpson, Joan Hirshberg, William F. Dietrich and their colleagues demonstrate that impulsive solar flares can produce a large relative abundance of helium in the solar wind with an enrichment of the helium-3 isotope.

1971–73 The first good, space-based observation of a coronal disturbance or transient, now called a coronal mass ejection (CME), was obtained on 14 December 1971 using the coronagraph aboard NASA's seventh Orbiting Solar Observatory (OSO 7), reported by Richard Tousey in 1973.

1972–73 Edward L. Chupp and his colleagues detect solar gamma ray lines for the first time using a monitor aboard NASA's seventh Orbiting Solar Observatory (OSO 7). They observed the neutron capture (2.223 MeV) and positron annihilation (0.511 MeV) lines associated with solar flares. The 2.223 MeV line had been anticipated theoretically by Philip Morrison.

1973–74 Yutaka Uchida explains chromosphere Moreton disturbances in terms of magnetohydrodynamic waves responsible for type II bursts.

1973–74 The manned, orbiting solar observatory, Skylab, is launched on 14 May 1973, and manned by three person crews until 8 February 1974. Skylab's Apollo Telescope Mount contained twelve tons of solar observing instruments that spatially resolved solar flares at soft X-ray and ultraviolet wavelengths.

1974 John Thomas Gosling and colleagues report observations of coronal mass ejections (CMEs), then called coronal disturbances or coronal transients, with the coronagraph aboard Skylab, noting that some CMEs have the high outward speed of up to a million meters per second needed to produce interplanetary shocks. They also found that many CMEs are not associated with solar flares, but instead with the eruption of quiescent filaments or prominences; but over-the-limb events would appear as prominences and not flares as such.

1974 Richard T. Hansen and colleagues observe sudden depletions of background regions of the inner solar corona using the Mark I coronagraph at Mauna Loa.

1974 Tadashi Hirayama develops a theoretical model of two-ribbon solar flares in which magnetic reconnection begins under a prominence and expands upwards, setting up the configuration for the main phase of the flare and incorporating the three-dimensional appearance of a prominence eruption.

1976 Roger A. Kopp and Gerald W. Pneuman develop a theory for two-ribbon flares in which coronal magnetic fields above active regions come together after the outward expulsion of a prominence.

1977 Roberto Pallavicini, Salvatore Serio, and Giuseppe S. Vaiana use Skylab observations to define two classes of soft X-ray flares: compact, brief (minutes) events, and extensive, long-enduring (hours) ones associated with soft X-ray arcades, filament eruptions, and coronal mass ejections.

1977–78 Franz Dröge and Cornelius Slottje independently discover microwave spikes that are a few milliseconds in duration, estimating sizes of less than a million meters across and brightness temperatures of a million billion (10^{15}) degrees Kelvin.

1980, 1984–89 The Solar Maximum Mission (SMM) satellite is launched on 14 February 1980 to study the physics of solar flares during a period of maximum solar activity. It excelled in X-ray and gamma ray spectroscopy of solar flares, as well as observing the white light emission of coronal mass ejections in 1980 and from 1984 to 1989.

1981–82 The Japanese spacecraft Hinotori, meaning fire-bird, was launched on 21 February 1981 and operated until 11 October 1982. It created images of solar flare X-rays with an energy of around 20 keV, and measured solar flare temperatures

	of between 10 and 40 million (1 to 4×10^7) degrees Kelvin using soft X-ray spectroscopy.
1981	Robert P. Lin and colleagues find that solar flares can produce thermal sources with high enough temperatures to be detectable as hard X-rays.
1980–89	George Doschek, Ester Antonucci and their colleagues use P78-1 and Solar Maximum Mission (SMM) observations of soft X-ray spectral lines to show that the impulsive phase of solar flares is associated with upward flows of heated chromosphere material with velocities of several hundred thousand meters per second. Such an upflow is called chromospheric evaporation. It was independently confirmed in 1982 by Katsuo Tanaka and colleagues using a spectrometer aboard the Hinotori spacecraft.
1980, 1984–89	Arthur J. Hundhausen and colleagues use the coronagraph aboard the Solar Maximum Mission (SMM) satellite to specify the mass, velocity, energy, shape and form of a large number of coronal mass ejections, fully reported in the literature in the 1990s.
1980–82	Edward L. Chupp and his colleagues use the Gamma Ray Spectrometer (GRS) on the Solar Maximum Mission (SMM) satellite to detect energetic solar neutrons near the Earth following a solar flare which occurred on 21 June 1980.
1980–84	Kenneth A. Marsh and Gordon J. Hurford use the Very Large Array, or VLA, in 1980 to resolve a two-centimeter burst source and locate it at the top of a flaring coronal loop. This implies that the initial flare energy release occurs near the loop apex, and may indicate trapping of the energetic electrons there. In 1984 Robert F. Willson and Kenneth R. Lang used the VLA to show that flaring emission at 20 centimeters wavelength also originates near the apex of coronal loops, marking the site of flare energy release near the loop tops in the low corona.
1981–82	Peter Hoyng, André Duijveman and their colleagues use instruments aboard the Solar Maximum Mission (SMM) satellite to resolve hard X-ray solar flares into double sources found at the two footpoints of coronal loops. Similar structures were found by Tatsuo Takakura and his colleagues using the Hinotori spacecraft, but most of the detected flares were apparently single rather than double.
1981-83 1997	In 1981–83 Arnold O. Benz and colleagues discover decimeteric type III bursts that drift rapidly from low to high frequencies, indicating downward directed electron beams. This suggests flare energy release and electron acceleration and injection in the low corona. The acceleration site is subsequently pinpointed at the demarcation between the downward moving and upward moving electron beams. In 1997, Markus J. Aschwanden and Benz show that the electron densities inferred for the acceleration sites are smaller than that in the underlying soft X-ray coronal loops, implying energy release above the loops.
1982	Russell A. Howard and colleagues detect an Earth-directed, halo coronal mass ejection using the Solwind coronagraph aboard the P78-1 satellite.
1982	Zdenek Svestka and colleagues use instruments aboard the Solar Maximum Mission (SMM) satellite to discover giant X-ray post-flare arches above eruptive flares.
1982–90	Edward L. Chupp, Hermann Debrunner and colleagues report the observation of time extended neutron emission at the Earth from the 3 June 1982 flare giving signals in both the SMM detector and the neutron monitor on Jungfraujoch.
1983	David J. Forrest and Edward L. Chupp use Solar Maximum Mission (SMM) satellite observations to demonstrate the simultaneous acceleration of relativistic electrons (hard X-rays) and energetic ions (gamma rays) to within a few seconds. This was confirmed by Masato Yoshimori and colleagues using observations from the Hinotori spacecraft.

1985 David J. Forrest and colleagues report the observation of gamma rays from the decay of neutral mesons during the 3 June 1982 solar flare.

1985, 95 Raghunath K. Shevgaonkar and Mukul R. Kundu resolve a double, loop-footpoint source at two-centimeters wavelength using the Very Large Array during a impulsive solar flare in 1985. A decade later, Kundu and his colleagues use the Nobeyama radioheliograph and the Yohkoh spacecraft to show that the two-centimeter and hard X-ray sources coincide spatially. The two non-thermal radio sources were circularly polarized with opposite polarities, indicating oppositely directed magnetic fields, consistent with the footpoints of a single coronal loop.

1987 Hilary V. Cane, Neil R. Sheeley Jr., and Russell A. Howard show that strong interplanetary shocks are associated with fast coronal mass ejections moving at speeds greater than 0.5 million meters per second. They used the third International Sun-Earth Explorer (ISEE 3) low-frequency (< 1 MHz) radio data and Solwind coronagraph observations.

1988 Marcos E. Machado and colleagues use X-ray images, taken with the Solar Maximum Mission (SMM) spacecraft, to show that the interaction of magnetic loops in the corona is an essential ingredient in triggering flare energy release.

1990 Eric R. Priest and Terry G. Forbes give a model for prominence eruption from twisted magnetic fields, including magnetic reconnection and the creation of currents below the erupting prominence.

1990–95 Donald V. Reames presents a two-class picture of solar energetic particle events associated with either solar flares or coronal mass ejections.

1991–99 The Soft X-ray Telescope (SXT) on the Yohkoh satellite, launched on 31 August 1991, shows the magnetically structured, dynamic nature of the inner corona more clearly than ever before.

1991–93 Gottfried Kanbach and colleagues report observations with the Compton Gamma Ray Observatory of a meson decay, gamma-ray flare on 11 June 1991 lasting 8 hours.

1992 Saku Tsuneta and his colleagues discover a cusp geometry in Yohkoh Soft X-ray Telescope (SXT) images of a long-lasting (hours) soft X-ray flare at the solar limb on 21 February 1992, suggesting magnetic reconnection in a neutral sheet near the loop top.

1994 Satoshi Masuda and his colleagues report that a solar flare included an unexpected, hard X-ray source in the corona, above the closed soft X-ray flaring loop and the two expected hard X-ray sources at the footpoints of that loop. This event was observed with the Yohkoh Hard X-ray Telescope (HXT) and Soft X-ray Telescope (SXT) on 13 January 1992.

1994 Taro Sakao uses Yohkoh Hard X-ray Telescope (HXT) data to show that hard X-ray bursts occur simultaneously at the footpoints of coronal loops, to within 0.2 seconds. This confirms that high-speed electrons are accelerated in closed magnetic loops, with the emission of hard X-rays as the electrons stream downwards along the magnetic field lines and enter the dense lower layers of the solar atmosphere.

1994–97 Yoichiro Hanaoka and Kazunari Shibata use data from Yohkoh's Soft X-ray Telescope (SXT) to show that solar flares are associated with the magnetic interaction of coronal loops, confirming Solar Maximum Mission (SMM) observations by Marcos E. Machado and colleagues in 1988.

1996 Markus J. Aschwanden and colleagues use electron time-of-flight measurements with the Compton Gamma Ray Observatory (CGRO) to confirm the existence of a coronal acceleration site for the 13 January 1992 flare event and for all other flares observed with both CGRO and Yohkoh.

---

* See the References for complete references to these seminal papers.

7. The Sun–Earth Connection

Overview

We are protected from the full blast of the Sun's violent activity by a dipolar magnetic field that diverts charged particles around the Earth and forms a cavity, called the magnetosphere, within the Sun's relentless winds. The Earth's magnetic barrier is nevertheless imperfect, and some of the tempestuous solar wind can buffet and even penetrate the magnetosphere. Intense, non-recurrent geomagnetic storms, accompanied by exceptionally bright auroras, occur at times of high solar activity when fast coronal mass ejections hit the Earth's magnetosphere and have the right magnetic alignment with it. When the Sun is near a lull in its 11-year activity cycle, magnetic waves formed by the Sun's interacting winds strike the Earth's magnetic field and connect with it, producing moderate geomagnetic storms and less bright auroras.

High-speed protons and electrons, generated during explosive solar activity, can cripple satellites and endanger space-walking astronauts. Intense radiation due to enhanced solar activity heats the Earth's upper atmosphere, altering satellite orbits and disrupting communications. When encountering the Earth, coronal mass ejections can compress the magnetosphere below the orbits of geosynchronous satellites, that hover above one place on Earth, and also generate power surges on transmission lines that could cause electrical power blackouts of entire cities. With adequate warning, we can defend ourselves from violent space weather that is driven by the Sun and its winds. National centers and defense agencies therefore continuously monitor the Sun from ground and space to forecast threatening solar activity. As an example, when large, twisted sigmoid (S or inverted S) shapes appear in soft X-ray images of the Sun, they can give advance notice, by a few days, of mass ejections that might collide with the Earth.

Solar X-rays and extreme ultraviolet radiation both produce, and significantly alter, the Earth's ionosphere. The X-rays fluctuate in intensity by two orders of magnitude during the Sun's 11-year magnetic activity cycle, producing increased ionization, greater heat, and expansion of our upper atmosphere near activity maximum. This affects the ability of the ionosphere to transmit and reflect radio communication waves. The ozone layer in our stratosphere is both created and modulated by solar ultraviolet radiation, while also protecting us from dangerous ultraviolet sunlight.

The total solar irradiance of the Earth, the so-called solar constant, rises and falls in step with the 11-year solar cycle, but with a total recent change of only about 0.1 percent. When sunspots cross the visible solar disk, they produce, in themselves, a brief dimming of the irradiance, amounting to a few tenths of one percent for just a few days; a brightness increase caused by faculae and plage exceeds the overall sunspot decrease at times of high solar activity. Variations in the Sun's activity and brightness may be linked to changes in terrestrial climate on time scales of decades and centuries. During the past 130 years, the global sea temperature has varied in tandem with the 11-year cycle of solar magnetic activity, and the land air temperature has been correlated with the length of the solar cycle. These temperature variations might be attributed to solar-driven changes in cloud cover, caused by the Sun's 11-year modulation of the amount of cosmic rays reaching Earth. Further back in time, over the past millennium, tree ring and ice core studies show rings high in radiocarbon and snow high in beryllium, precipitated over century-long periods when the Sun was inactive and the climate was cold.

During the past one million years, the world's climate has been dominated by the deep glacial cold of major ice ages, each lasting about 100 thousand years, punctuated by a warm interglacial of 10 to 30 thousand years in duration. The ice ages seem to be caused by three astronomical cycles, which combine to alter the angles and distances from which sunlight strikes the Earth. When the temperature increases it is most likely amplified by increased atmospheric concentrations of carbon dioxide and other greenhouse gases. Future global warming of a few degrees Celsius is forecast for the Earth's surface temperature in the twenty-first century if current human increases in concentrations of such heat-trapping gases are not curtailed. The Sun's luminosity has been steadily increasing, by some 30 percent over its lifetime of 4.6 billion years, but some global thermostat has conspired to keep most of the Earth's water from freezing solid or boiling away during this time. Nevertheless, the Sun will keep on getting brighter, boiling away our oceans in 3 billion years and ballooning into a giant star in another four billion years.

7.1 Earth's Magnetic Storms and the Aurora Lights

We are protected from cosmic forces by an invisible magnetic barrier with two poles, generated inside our planet and extending out into the space that surrounds it. As demonstrated by William Gilbert in 1600, the Earth is itself a great magnet. The magnetic fields emerge from the south geographic pole, loop through nearby space, and re-enter at the north geographic pole (Fig. 7.1); the north geographic pole corresponds to the south magnetic pole and *vice versa*. The Earth's rotation axis is inclined only 12 degrees with respect to its magnetic axis, so compass needles that point toward the magnetic south pole also tilt in the direction of the geographic North Pole.

A mathematical description of this dipolar magnetic field was devised by Carl Friedrich Gauss in 1838, and compared with observations to show that the field originates deep inside the Earth's core. More than a century later, Walter M. Elsasser proposed that currents within the molten, electrically conducting, rotating core produce the Earth's magnetic field by dynamo action, much as the internal solar dynamo creates the Sun's magnetism.

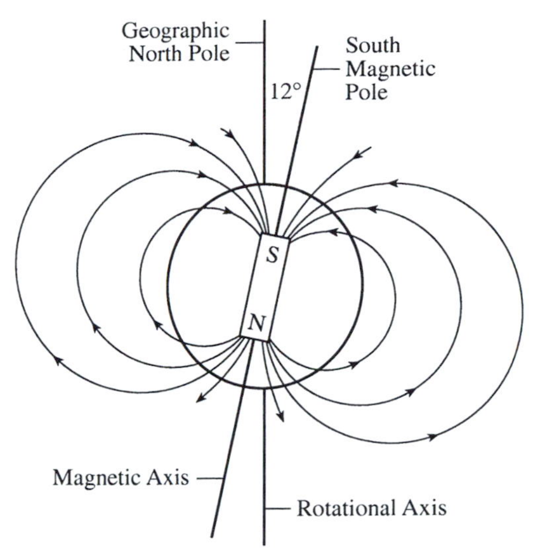

Fig. 7.1. Dipole Field of Earth. The Earth's magnetic field looks like that which would be produced by a bar magnet at the center of the Earth, with the North Magnetic Pole corresponding to the South Geographic Pole and *vice versa*. Magnetic field lines loop out of the South Geographic Pole and into the North Geographic Pole. A compass needle will always point along a field line. The lines are close together where the magnetic force is strong, and spread out where it is weak. The magnetic axis is tilted at an angle of 12 degrees with respect to the Earth's rotational axis. Notice that the poles of the magnet are inverted with respect to the geographic poles, following the custom of defining positive, north magnetic polarity as the one in which magnetic fields point out, and negative, south magnetic polarity as the place where magnetic fields point in. This dipolar (two poles) configuration applies near the surface of the Earth, but further out the magnetic field is distorted by the solar wind

The space that is threaded by the Earth's magnetism is not a vast and tranquil world of complete emptiness. Our planet is instead immersed within a hot, gusty, electrically charged solar wind that blows out from the Sun in all directions and never stops, carrying with it a magnetic field rooted in the Sun. The direct measurement of this perpetual gale of electrons, protons and heavier ions, and magnetic fields was one of the main early accomplishments of the space age (Sect. 2.2). Fortunately, the Earth's magnetic field protects us from the full force of the solar wind, diverting most of it around our planet at a distance far above the atmosphere, like a rock in a stream or a windshield that deflects air around a car.

Although the solar wind is exceedingly rarefied, far less substantial than a terrestrial breeze or even a whisper, it possesses the power to bend and move things in its path. Measurements from NASA's first Interplanetary Monitoring Platform (IMP-1) in the mid-1960s showed that the never-ending flow from the Sun transforms the outer edges of the Earth's magnetic field into the shape of a comet or a teardrop (Fig. 7.2). The Sun's variable wind produces a shock wave when it first encounters the Earth's magnetism, forming a bow shock.

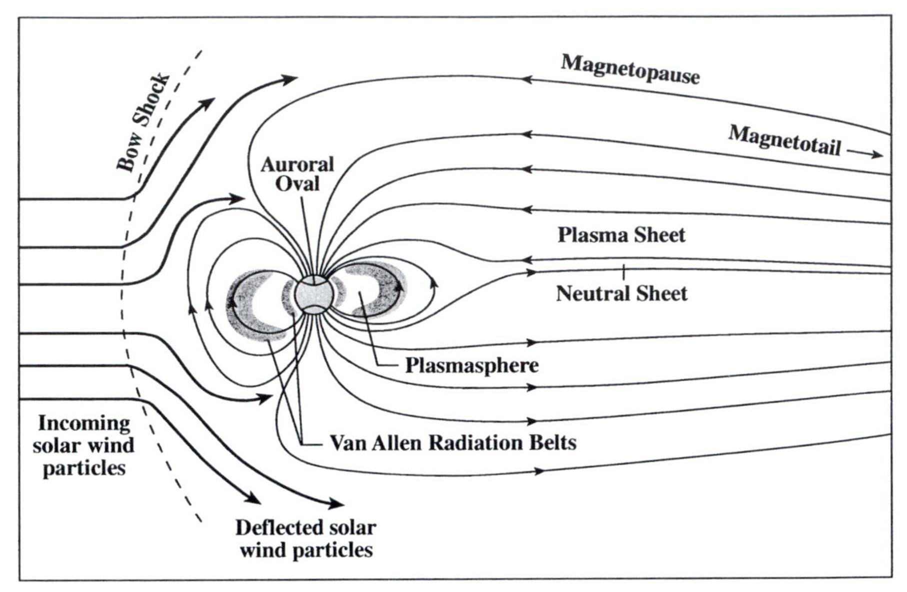

Fig. 7.2. Earth's Magnetosphere. The Earth's magnetic field carves out a hollow in the solar wind, creating a protective cavity, called the magnetosphere. A bow shock forms some 14 to 16 Earth radii on the sunlit side of our planet. Its location is highly variable since it is pushed in and out by the gusty solar wind. The magnetopause marks the outer boundary of the magnetosphere, at the place where the solar wind takes control of the motions of charged particles. The solar wind is deflected around the Earth, pulling the terrestrial magnetic field into a long magnetotail on the night side. The magnetic field points roughly toward the Earth in the northern half of the tail and away in the southern. Because of these opposed orientations, the field strength drops to nearly zero in the center of the tail, at the neutral current sheet. One consequence of the weakened field is that the plasma pressure increases to maintain a balance, becoming greatest in the plasma sheet. The Earth, its auroras, atmosphere and ionosphere, and the two Van Allen radiation belts all lie within this magnetic cocoon

The solar wind pushes the magnetic field toward the Earth on the side facing the Sun, and drags and stretches the terrestrial magnetic field out into a long magnetotail on the night side of Earth. The magnetic field points roughly toward the Earth in the northern half of the tail and away in the southern. The field strength drops to nearly zero at the center of the tail, where the opposite magnetic orientations meet.

Fortunately for life on Earth, the terrestrial magnetic field shields us from the full blast of the Sun's relentless wind, deflecting most of the solar wind around the Earth and hollowing a cavity in it. Nowadays this magnetic cocoon is called the magnetosphere, a name coined by Thomas Gold in 1959. As previously noted, the Earth's magnetosphere is not precisely spherical. It has a bow shock facing the Sun and a magnetotail in the opposite direction. So the term magnetosphere does not refer to form or shape, but instead implies a sphere of influence. The magnetosphere of the Earth, or any other planet, is that region surrounding the planet in which its magnetic field dominates the motions of energetic charged particles such as electrons, protons and other ions.

Although the solar wind never actually reaches the Earth's surface, it can cause dramatic changes in the Earth's magnetic field. Under constant buffeting by the tempestuous solar wind, the terrestrial magnetic fields are buckled, distorted and reshaped, producing invisible storms that rage within the magnetic fields far above our atmosphere. The large and rapid variations in the Earth's magnetic field were indicated long ago by wide, irregular movements in the direction that compass needles point. These oscillations were independently discovered by George Graham and Anders Celsius in 1724, and precisely measured late in the eighteenth century by Baron Alexander von Humboldt, who called them magnetic storms – a term we use today.

A solar-terrestrial connection was next suggested in the mid-nineteenth century when the sunspot cycle was discovered and related to magnetic storms. The decade-long variation in the number of sunspots, reported by Samuel Heinrich Schwabe in 1843, became widely known when Humboldt included it in his third, 1851 volume of *Kosmos*, an encyclopedic compilation of information about the physical world. Edward Sabine was then analyzing magnetic measurements obtained at army stations in Britain and its colonies, and his wife was translating the relevant edition of *Kosmos*. In 1852, the astronomer John Herschel wrote to Sabine about the 11-year sunspot period discussed in Humboldt's book, and in reply Colonel Sabine reported that the frequency of global magnetic storms rose and fell with the number of sunspots.

The identification of the period and phase of the sunspot variation with those of magnetic disturbances over the whole globe was a fantastic discovery. It seemed to imply that geomagnetic storms are caused by the Sun, and unrelated to the Earth's internal core where the magnetic fields originate. However, there was no known physical mechanism that could explain the solar-terrestrial interaction, and it took nearly a century to understand the connection. It turned out that sunspots do not produce geomagnetic storms, and particles produced during solar flares were eventually discarded as the responsible agent. Coronal mass ejections are now thought to produce the most intense geomagnetic storms, but the weaker repetitive variety has a more complex origin.

The search for a *modus operandi* connecting the Sun to the Earth's magnetic field can be dated back to at least 1892, when Sir William Thomson (Baron Kelvin) argued that the "supposed connection between magnetic storms and sunspots is unreal". He showed that the geomagnetic storms could not be due to the direct magnetic action of the Sun, and concluded "that the seeming agreement between the two periods had been mere coincidence". By 1908 George Ellery Hale had measured the strong magnetic fields in sunspots, that are thousands of times stronger than the Earth's magnetic field. Yet, as Hale noticed, the intense sunspot magnetic fields are still inadequate to account for geomagnetic storms by direct magnetic action.

Huge shock waves are pushed ahead of the fast ejections. When directed toward the Earth, these shocks ram into the terrestrial magnetic field and trigger the initial phase, or sudden commencement, of a large geomagnetic storm. The main storm phase is due to the magnetic fields and high speed of the coronal mass ejection.

Energy is transferred from the relentless solar wind to the magnetosphere at a rate proportional to the magnetic reconnection rate, which increases with the strength and speed of the interplanetary magnetic field. The energy gained drives currents that make the intense magnetic storm and accelerates both the infiltrating solar wind particles and local particles to make polar auroras. The Earth intercepts about 70 coronal mass ejections per year when solar activity is at its peak, and several of them will produce great magnetic storms and exceptionally intense auroras.

The magnetic reconnection in the tail effectively breaks a hole in the Earth's magnetic barrier and allows solar wind particles and energy to pour into the magnetosphere. At such times, the accelerated electrons hurtle along magnetic conduits connected to the upper atmosphere in both Polar Regions, generating spectacular northern and southern lights, or auroras, in an oval centered at each magnetic pole (Fig. 7.4 and Focus 7.1).

The clear-cut association between geomagnetic storms and enhanced solar activity falls apart when one examines the details of low-level storms. That is, there are two basic types of geomagnetic storms, the great sporadic ones, attributed to coronal mass ejections that occur most often near solar activity maximum, and moderate recurrent events, discovered by E. Walter Maunder in 1904, whose elusive source remained a mystery for more than half a century. The weaker variety of geomagnetic storms recurs with a 27-day period corresponding to the apparent rotation period of the Sun at low solar latitudes, so they must be related to the spinning Sun. Yet, unlike the great sporadic storms, the small, recurrent ones do not exhibit a well-defined correlation with sunspots. As emphasized by Julius Bartels in the 1930s, these modest events sometimes occur when there are no visible sunspots.

It wasn't until the 1960s and 1970s that the recurrent geomagnetic storms were identified with long-lived, high-speed streams in the solar wind. Near the minimum in the 11-year solar activity cycle, the fast wind streams that emanate from coronal holes can then extend to the plane of the solar equator, periodically sweeping past the Earth and producing moderate geomagnetic storms every 27 days.

Since the Sun's slow-speed winds are localized in the plane of the solar equator, they also play a role in recurrent storms. When the fast streams overtake the slower moving gas, shock waves and intense magnetic fields are produced that rotate with the Sun (Fig. 7.5). It reminds one of the turmoil created at the confluence of two rivers. Such Co-rotating Interacting Regions, or CIRs, were first detected during studies of the geomagnetic field by Charles P. Sonett and colleagues in the early 1970s, and investigated by Edward J. Smith and John H. Wolfe in 1976, using Pioneer 10 and 11 data. As emphasized by Nancy U. Crooker and Edward W. Cliver in 1994, the interaction of high-speed

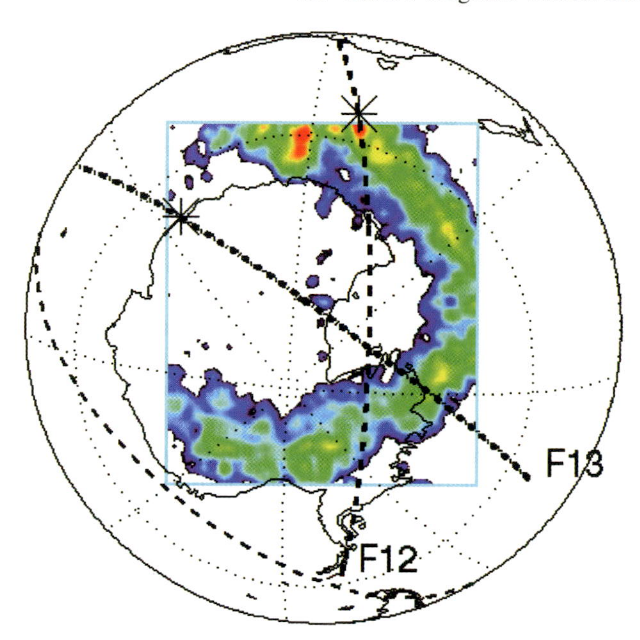

Fig. 7.4. Aurora Oval in X-rays. The POLAR spacecraft looks down on an aurora from high above the South Pole, obtaining an X-ray image on 27 May 1996 with its Polar Ionospheric X-ray Imaging Experiment (PIXIE). This image shows the X-ray emission of electrons precipitating down along the polar magnetic fields into the upper atmosphere. The huge, luminous X-ray ring is projected against a map of Antarctica, and centered on the south geomagnetic pole. The X-ray bremsstrahlung (braking radiation) of the electrons has photon energies in the range of 2 keV to 10 keV; it is emitted when the electrons pass near ions in the ionosphere. The two dashed lines denote the orbits of the Defense Meteorological Satellite Program (DMSP) F12 and F13 spacecraft that were passing through the PIXIE field of view at 840 thousand meters altitude, obtaining direct measurements of the precipitating electrons. These measurements are in good agreement with the intensity of the X-ray emission observed with PIXIE, which can also be used to infer the density of the precipitating electrons. (Courtesy of David L. Chenette, Lockheed Martin Advanced Technology Center, the PIXIE team, and NASA)

Focus 7.1

The Auroras – Cosmic Neon Signs

Rivers of green and red light transform the cold Arctic and Antarctic sky, shimmering far above the highest clouds. These northern and southern lights, named the *aurora borealis* and *aurora australis* in Latin, have been noticed for centuries. Low-level auroras can be seen by residents in far northern locations every clear and dark winter night. Rare, brilliant auroras, associated with great magnetic storms, extend down toward the Earth's equator.

The multi-colored auroral light show is principally caused by energetic electrons bombarding the upper atmosphere with energies of about 6 keV, or 6,000 electron volts, and speeds of about 50 million meters per second. When the electrons slam into the rarefied upper atmosphere, they excite the oxy-

endanger spacecraft or astronauts that venture into space. The blustery solar gale also constantly buffets and reshapes the Earth's magnetic cocoon, or magnetosphere, sending high-energy particles into it. This Sun-driven space weather can destroy satellite electronics, disrupt radio navigation and communication systems, endanger humans in high-altitude aircraft and in space, and induce damaging currents in electric power grids on Earth.

Time-varying storms in space can disrupt or wipe out Earth-orbiting spacecraft that are becoming more and more important to governments, corporations, and the lives of ordinary citizens of our planet. We now depend on satellites for weather information, communications, navigation, exploration, search and rescue missions, research, and national defense. Geostationary satellites, that hover above the same place on the Earth, relay and beam down signals that are used for aviation and marine navigation, money and commodity exchanges, and world-wide telephone communications. Other satellites move in lower orbits and whip around the planet, scanning air, land and sea for environmental change, weather forecasting and military reconnaissance. All of these spacecraft can be temporarily or permanently disabled by our tempestuous Sun. Some of these failures are not just inconveniences, but can have major economic impacts and potentially result in the loss of lives.

With their miniaturized circuits and reduced overall size, modern satellites have become vulnerable to direct hits by energetic charged particles. They have already knocked several communication and weather satellites out of commission. When a very energetic proton, above 10 MeV in energy, strikes a spacecraft, it can destroy its electronic components. Metal shielding and radiation-hardened computer chips are used to guard against this persistent, ever-present threat to satellites, but nothing can be done to shield solar cells. Since they use sunlight to power spacecraft, solar cells must be exposed to space.

High-speed electrons can move right through the metallic skin of spacecraft, sending phantom signals inside. The spurious commands can produce erroneous instrumental data or even send the satellite spinning out of control. The danger was fully realized long ago, in 1962, when a 1.4-megaton thermonuclear bomb, code named Starfish, was exploded 400,000 meters up in space, producing high-energy electrons (1 MeV and greater) that decommissioned satellites in low-altitude orbits at the time.

Once they arrive at the Earth's magnetosphere, charged particles are deflected at right angles to both the magnetic field lines and the direction of the particle's motion. This normally shields all but the polar caps from their direct effect. So, you might wonder how the threatening protons and electrons penetrate the Earth's magnetic defense and reach satellites that are orbiting beneath it.

When a coronal mass ejection encounters the Earth, the force of impact can compress the day side magnetosphere down to as much as half its normal size. High-flying geosynchronous spacecraft, that orbit the Earth at the same rate as the planet spins, can then find themselves outside the magnetosphere's protective embrace, exposed to the full brunt of the gusty solar wind and its charged, energized ingredients. Enhanced fluxes of energetic particles can also enter through the back door, at the magnetotail (Sect. 7.1), damaging spacecraft within the magnetosphere.

In fact, the inner magnetosphere is always filled with a veritable shooting gallery of electrons and protons, trapped within the Van Allen radiation belts that girdle the Earth's equator (Focus 7.2, Fig. 7.6). Curiously, these radiation belts can present the greatest hazard to satellites when the Sun is least active. High-speed particles are then accelerated in the magnetosphere in response to the high-speed solar wind streams, and these particles may inflate the Van Allen belts.

The greatest electronic danger zone exits in the inner radiation belt above the East Coast of South America. This region, known as the South Atlantic Anomaly, exists because the Earth's magnetic dipole is offset from its center by about 500,000 meters. Consequently, one side of the inner Van Allen radiation belt comes closer to our planet's surface than the other side does. Energetic particles therefore move closer to Earth's surface above the South Atlantic than elsewhere, so this region is anomalous because of its proximity.

The Van Allen radiation belts are not static, fixed entities, and their threat to Earth-orbiting satellites can change with time. As an example, data from

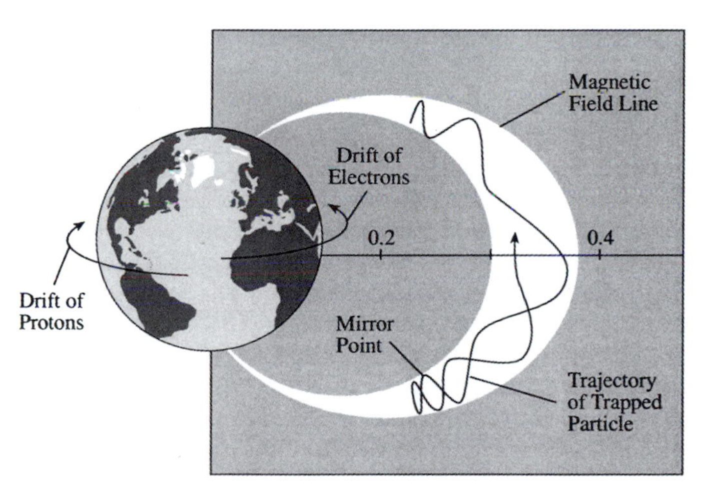

Fig. 7.6. Trapped Particles. Charged particles can be trapped by Earth's magnetic field. They bounce back and fourth between mirror points in either hemisphere at intervals of seconds to minutes, and they also drift around the planet on time scales of hours. As shown by Carl Størmer in 1907, with the trajectories shown here, the motion is turned around by the stronger magnetic fields near the Earth's magnetic poles. Because of their positive and negative charge, the protons and electrons drift in opposite directions. Two such belts of trapped electrons and protons encircle the Earth's equator at distances of about 1.5 and 6 Earth radii from the planet's center (also see Fig. 7.2). They are named after James A. Van Allen whose discovered them in 1958 using an instrument aboard the first American satellite, Explorer 1. The inner Van Allen belt contains a mixture of trapped charged particles: protons with energies greater than 10 million electron volts, or 10 MeV, and electrons exceeding 500 thousand electron volts, or 0.5 MeV. The outer belt also contains protons and electrons, most of which have energies under 1.5 million electron volts

the Combined Release and Radiation Effects Satellite, or CRRES for short, published by Edward G. Mullen and colleagues in 1991, showed that solar-initiated disturbances of the terrestrial magnetic field can inject high-energy (> 20 MeV) protons into the magnetosphere. The protons became trapped within the Earth's magnetic environment, producing an enlarged, double-peaked inner radiation belt that could impair or degrade spacecraft over long periods of time.

Solar flares and coronal mass ejections also threaten satellites by altering the Earth's atmosphere. The enhanced ultraviolet radiation from a solar flare heats the atmosphere and causes it to expand outward from the Earth. This brings higher densities to a given altitude, increasing the drag exerted on a satellite and pulling it into a lower orbit. Similar or greater effects are caused by coronal mass ejections that produce major geomagnetic storms. They create unpredictable, erratic heating and expansion of the terrestrial atmosphere by powerful currents, resulting in significant perturbations of low-altitude satellites.

Rising solar activity, for example, sent both the Skylab Space Station and the Solar Maximum Mission (SMM) satellite into a premature and fatal spiral toward the Earth. The Space Shuttle is also vulnerable to changes in atmospheric drag which affects re-entry calculations and can threaten the safety of the vehicle and crew. Navigation and global positioning systems also depend on accurate knowledge of atmospheric changes caused by the intense radiation of solar flares and by coronal mass ejections via geomagnetic storms.

These events alter the ionization structure of the atmosphere, interfering with high frequency, or short wave, radio communications (Sect. 7.3). The U. S. Air Force takes an active interest in the ability of solar flares or coronal mass ejections to disrupt contact with airplanes flying over oceans or remote countries.

Space weather, that is driven by the Sun and its winds, can also endanger humans in space. They risk damaging exposure to both solar energetic particles and galactic cosmic rays coming from interstellar space. At times of high solar activity, the frequency and intensity of solar explosions is greater, but the enhanced solar winds cut off the flow of cosmic rays into the solar system. The opposite conditions apply near the minimum in solar activity. So, there is some risk no matter when an astronaut takes a trip into space.

On Earth we are protected from these particles by our magnetic field and atmosphere, which deflect or absorb all but the most energetic cosmic rays. Astronauts in near-Earth orbit are shielded by the terrestrial magnetic field which deflects solar energetic particles and cosmic rays, but they are exposed to electrified particles trapped in the Van Allen radiation belts. Despite the fact that they are within the magnetosphere, the eyes of astronauts sometimes light up at night. Heavy ions in the radiation belts apparently rip through the satellite walls and enter their eyeballs, making them glow inside.

High-altitude aircraft crews and passengers on polar routes are also susceptible to the hazards of space weather. The Earth's magnetic field guides energetic particles, hurled out from solar explosions, into the magnetic poles.

Astronauts will not enjoy the protection of Earth's magnetic field during trips to the Moon or Mars. They could suffer serious consequences from solar energetic particles, even within their spacecraft, resulting in cataracts, skin cancer or even lethal radiation poisoning. The high-speed solar protons could kill an unprotected astronaut that ventures into space for repairs or to take a walk on the Moon or Mars. During a 1972 Apollo mission, astronomers walking on the Moon just missed a solar flare that could have been lethal and would certainly have prevented the mission from continuing.

During future space missions, solar astronomers will keep careful watch over the Sun, providing timely warnings of solar flares and coronal mass ejections. The astronauts can then move inside storm shelters. Current recommendations advise that spacecraft contain aluminum cylinders with walls of aluminum at least 9 centimeters thick. Future settlers on the Moon or Mars might want to build underground shelters for protection from Sun-driven space weather.

There are also indirect consequences of explosive solar activity down here on the ground. When the blast of a coronal mass ejection slams into the Earth's magnetosphere, it produces electric currents that can cause electrical power outages capable of paralyzing entire cities (Focus 7.3).

So, there is no getting around it, space weather is here to stay. The danger blowing in the Sun's wind is threatening all kinds of things in our technological age, including potential damage to satellites, communications, navigation, surveillance, humans in space and electric power grids. All of these effects are of such vital importance that national centers employ space weather forecasters to warn of threatening solar activity. The Space Environment Center of the National Oceanic and Atmospheric Administration (NOAA) collects and distributes the relevant data, using satellites and ground-based telescopes to monitor the Sun and relay information about conditions in interplanetary space. The U. S. Air Force operates a global system of ground-based solar telescopes and taps into the output of national, space-borne ones, to continuously monitor solar activity, hoping to forecast events that might severely disrupt military communications and to give operators time to find alternative means of contact. The National Science Foundation has organized a National Space Weather Program to coordinate and disseminate research.

Focus 7.2

The Van Allen Radiation Belts

Using simple radiation detectors aboard the Explorer 1 and 3 satellites in 1958, James A. Van Allen and his colleagues discovered that untold trillions of energetic charged particles surround our planet, moving within two torus-shaped belts that encircle the Earth's equator but do not touch it. These regions are sometimes called the inner and outer Van Allen radiation belts. Van Allen used the term "radiation belt" because the charged particles were then known as corpuscular radiation; the nomenclature does not imply either electromagnetic radiation or radioactivity.

In 1907 Carl Størmer published tedious numerical calculations, performed before the advent of computers, that showed how electrons and protons can be trapped and suspended in space by the Earth's dipolar magnetic field. The energetic charged particles spiral around the magnetic field lines and bounce back and forth between the north and south magnetic poles (Fig. 7.6). It takes about one minute for an energetic electron to make one trip between the two polar mirror points. The spiraling electrons also drift eastward, completing one trip around the Earth in about half an hour. There is a similar drift for protons, but in the westward direction.

The inner Van Allen belt contains protons with energies greater than 10 MeV and electrons exceeding 0.5 MeV. The outer belt also contains protons and electrons, most of which have energies below 1.5 MeV. By way of comparison, a proton in the solar wind usually has an energy of about 1 keV, or 0.001 MeV.

Although the particles in the solar wind are not usually as energetic as the electrons and protons found in the inner radiation belt, solar flares or coronal mass ejections can accelerate charged particles to energies of GeV, or 1,000 MeV. These particles have energies that are much greater than the most energetic particles found in the radiation belts. Many of the electrons and protons in the Van Allen belts might therefore originate in solar activity that produces violent gusts in the solar wind. The blustery solar wind might also help explain how electrified particles get into the radiation belts in the first place.

When a coronal mass ejection strikes the Earth's magnetosphere, its shock front can compress the magnetic field, permitting the solar wind to flow closer to the Earth on its sunlit side. The gusty wind's particles also enter the magnetosphere through the night side, at the magnetotail (Sect. 7.1). Once inside, the charged particles are trapped and cannot easily get out. Thus, solar-initiated disturbances can inject very energetic particles into the inner and outer Van Allen radiation belts.

There is another explanation that involves neutrons. They are produced when very energetic solar particles and galactic cosmic rays bombard the Earth's atmosphere, that lies beneath the radiation belts. The collisions tear neutrons out of the nuclear centers of atoms in our air. These neutrons are hurled out in all directions unimpeded by magnetic fields.

Once they are liberated, the neutrons cannot stand being left alone. When outside the atomic nucleus, a free neutron lasts only 10.25 minutes on average, before it falls apart and decays into an electron and proton. A small fraction of the neutrons produced in our atmosphere move out into the inner belt before they disintegrate, producing electrons and protons when they are in it. These electrically charged particles are immediately snared by the magnetic fields and remain stored within them, accumulating in substantial numbers over time. A 10 MeV proton will, for example, remain stored within the strong magnetic fields near Earth for a decade.

Bulletins, alerts, and warnings about solar activity are now routinely disseminated to a broad range of customers including satellite operators, power companies, telecommunications companies, navigational systems companies and research institutions. In a broader context, the customers include everyone that uses these services: the general public, industries, and governments, particularly space and defense agencies.

What everyone wants to know is how strong the storm is and when it is going to hit us. Somewhat like hurricanes on Earth, general space weather can now be predicted up to three days in advance. The exact warning time will depend on the type of solar hazard, since they travel with different velocities and on various trajectories in space (Fig. 7.7). Intense radiation from powerful solar flares moves from the Sun to the Earth in just 8 minutes, traveling at the speed of light. Energetic particles, accelerated during the flare process or by the shock waves of coronal mass ejections, can reach the Earth within an hour or less (for energies above 10 MeV). A coronal mass ejection arrives at

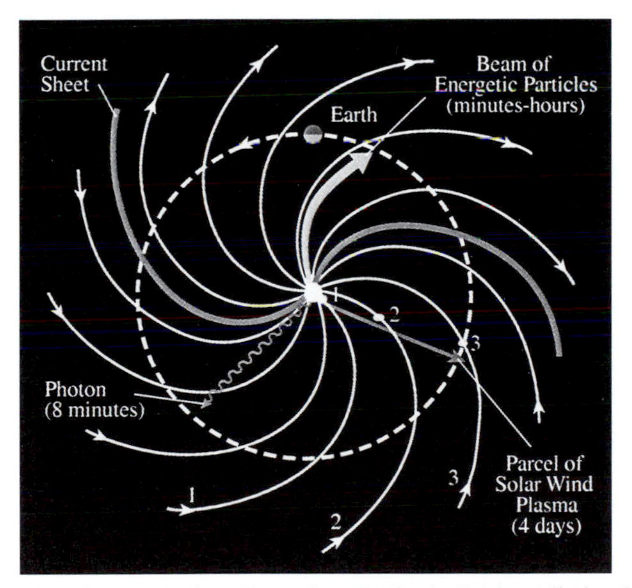

Fig. 7.7. Travel Time From Sun To Earth. In the ecliptic plane, shown here, the radial flow of the solar wind and the rotation of the Sun combine to wind the solar magnetic field into a spiral. Intense radiation, or photons, generated during solar flares passes right through the interplanetary magnetic field and arrives at the Earth just 8 minutes after being emitted from the Sun. In contrast, solar flares beam energetic charged particles across a narrow trajectory that follows the interplanetary magnetic spiral, and the time for the particles to reach Earth depends on their energy and velocity, taking roughly an hour for a particle energy of about 10 MeV. A coronal mass ejection, or CME, with an average speed of 450 thousand meters per second takes about 4 days to travel from the Sun to Earth's orbit; the CME can energize particles across a wide swath in interplanetary space. The heliospheric current sheet separates magnetic fields of opposite polarities, or directions, denoted by the arrows on the spiral lines. (Courtesy of Frances Bagenal)

the Earth as a dense cloud of magnetic fields, electrons and protons one to four days after leaving the Sun.

If we had early warning that a solar storm was on the way, just several hours or a few days in advance, we could defend ourselves and be protected from their effects. System operators could then power down sensitive electronics on satellites, putting them to sleep until the danger passed. Airplane pilots and cellular telephone customers could be warned of potential communication failures. Sensitive navigation and positional systems might be temporarily shut down. The launch of manned space flight missions could be postponed, and walks outside spacecraft or on the Moon or Mars might be delayed. Utility companies could reduce load in anticipation of geomagnetically induced currents, in that way trading a temporary "brown out" for a potentially disastrous "black out".

We are just beginning to understand the vital link and dynamic interplay between the Sun and the Earth by viewing them as an interconnected whole. Previous individual, localized observations have resulted in a blurred, incomplete picture of solar-terrestrial interactions, but now a variety of spacecraft are making coordinated, simultaneous measurements of the Sun, the solar wind, and the Earth's magnetosphere. They are providing a new global perspective of the intricate coupling between the Sun and Earth under the auspices of an International Solar Terrestrial Physics (ISTP) program, that draws on the data and resources of the worldwide scientific community.

Focus 7.3

Turning off the Lights

When our planet takes a direct hit by a coronal mass ejection, the Earth's reverberating, fluctuating magnetic fields can induce powerful electric currents in the high atmosphere and along the ground. When such an induced current surges through long-distance power transmission lines, it can plunge major urban centers, like New York City or Montreal, into complete darkness, causing social chaos and threatening safety. The direct-current surge is capable of causing massive failures in electrical power grids, permanently damaging multi-million dollar equipment in power generation plants, and producing hundreds of millions of dollars in losses from unserved power demand or disruption of factories. The blackouts could happen almost anywhere, occurring globally and simultaneously with little or no warning.

As electric utility companies rely more and more on enormous power networks that connect widely separated geographical areas, with complex electronic controls and technology, they become increasingly susceptible to Earth-directed coronal mass ejections and the resulting geomagnetically induced currents. They therefore have a strong vested interest in ongoing efforts to forecast space weather and provide warning of impending storms, thereby permitting operators to shut down the vulnerable electrical systems and protect the power grids from collapsing.

For the first time ever, we can now track every move of possibly destructive events from their beginning on the Sun, to their passage through space, and their ending impact on Earth (Fig. 7.8). The SOHO and Yohkoh satellites have been regularly watching the Sun, monitoring the solar corona for signs of an impending explosion. Solar flares can outshine the entire Sun at the ul-

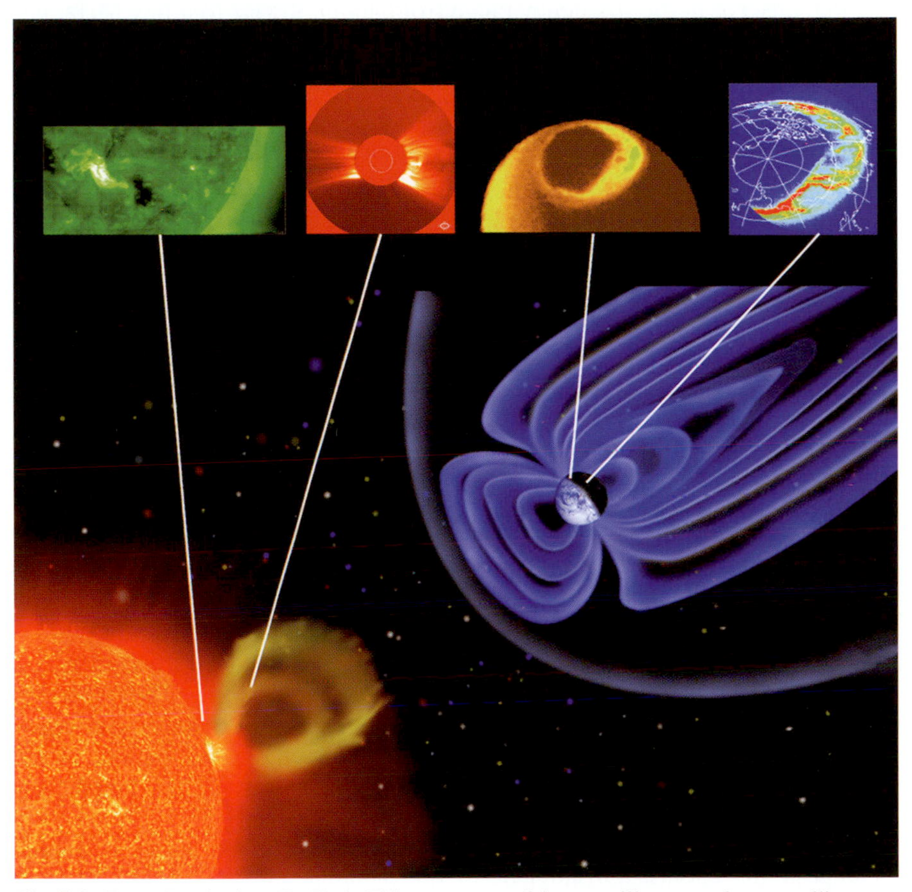

Fig. 7.8. From Beginning To End. This montage of images illustrates how satellites can follow the solar-terrestrial connection, monitoring a solar storm from its birth on the Sun to its interaction with the Earth. The Yohkoh satellite might observe the X-ray emission from tightly coiled magnetic loops, that release their pent-up energy as a coronal mass ejection. SOHO's LASCO coronagraph can detect the actual mass ejection. Other satellites can track the bubble of magnetized gas on its way to Earth, and then record the collision with our magnetosphere. As an example, a coronal mass ejection that occurred on 6 January 1997 was recorded by about 20 satellites in the International Solar-Terrestrial Physics (ISTP) program during its four-day journey to Earth and when it hit the Earth's magnetosphere. A radio experiment on board the WIND spacecraft tracked the shocks driven by the mass ejection through the interplanetary medium. The GEOTAIL satellite observed the magnetic connections when the solar ejection collided with the terrestrial field, and the resultant auroras were captured in POLAR satellite images. (Adapted from the cover of Geophysical Research Letters **25**, No. 14, 1998)

traviolet and X-ray wavelengths detected by these spacecraft. Coronagraphs aboard SOHO routinely detect mass ejections directed toward the Earth or elsewhere in space. The progression of mass ejections through space can be seen by the very long-wavelength radio emissions that they produce, measuring the apparent speed and trajectory of the associated shock waves. This must be done using radio instruments on other spacecraft, such as the WIND satellite, that observe in the frequency range from about 30 kHz to a few MHz, not detectable from the ground.

Key areas of the Earth's magnetosphere and the solar wind are monitored from satellites placed within them. Recent examples include:

- SOHO, stationed about 1 percent of the distance to the Sun, or at about 200 Earth radii from the Earth in the direction of the Sun; the edge of the magnetosphere is located at about 10 Earth radii from the Earth in the direction of the Sun.
- WIND, also located in the solar wind.
- POLAR, that looks down at the Earth's magnetic poles where auroras form.
- GEOTAIL, that skims along the outer edge of the magnetosphere and dives deep into the magnetotail.

The combined set of measurements, taken at a multitude of points in space, is specifying how energy is generated on the Sun and transferred to the space near Earth, with its complex magnetic and atmospheric environment. They are also telling us how energy is coupled back and forth between the gusty, incident solar wind and the magnetosphere on the one hand and between the magnetosphere and the Earth's upper atmosphere on the other.

The ultimate goal is to learn enough about solar activity to predict when the Sun is about to unleash its pent-up energy, and direct energetic particles, intense radiation, and magnetic fields toward us. The forecasts will probably involve magnetic changes that precede solar flares and coronal mass ejections, especially in the low corona where the energy is released or down below where internal motions pull and twist the magnetism about.

The Sun's sudden and unexpected eruptions used to be about as unpredictable as the passionate outbursts of my daughter and wife, but scientists now think they know a lot more about at least some of the solar eruptions. For instance, in 1997 Alphonse C. Sterling and Hugh S. Hudson used Yohkoh observations to show that coronal mass ejections can rise out of severely twisted soft X-ray structures. In 1999, Richard C. Canfield, Hugh S. Hudson and David E. Mc Kenzie used Yohkoh soft X-ray images obtained in 1993 and 1997 to show a correlation between the appearance of a large twisted feature and the probable occurrence of an eruption a few days later. When the bright, X-ray emitting coronal loops are distorted into an S or inverse S shape, collectively called sigmoids, an explosion from that region becomes more likely, particularly for the larger ones (Figs. 7.9 and 7.10). They also tend to occur when the areas of the underlying sunspots are large. In other words, when the magnetism gets stirred up into a complex, stressed situation, it might be the

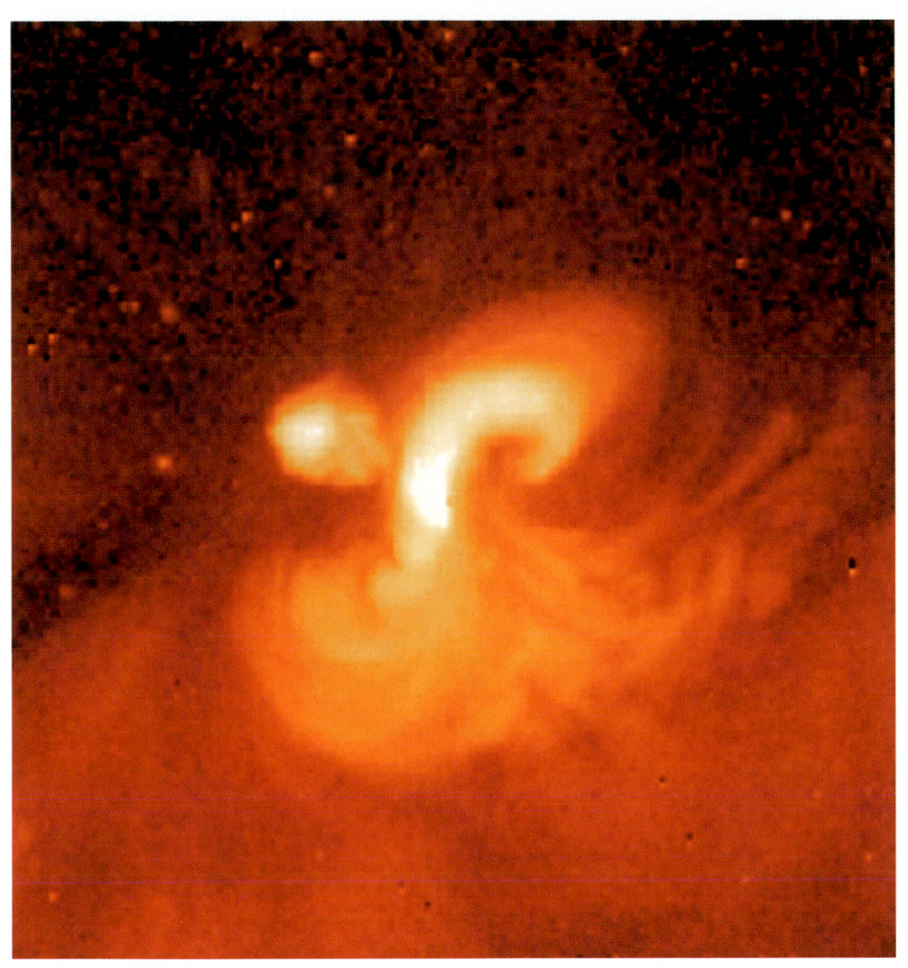

Fig. 7.9. The Sun Getting Ready to Strike. The strong magnetic fields that constrain hot, X-ray emitting gas in this active region have been contorted into an S, or sigmoid, shape. When the coronal magnetic fields get twisted like this, they become dangerous, like a coiled rattlesnake waiting to strike. Statistical studies indicate that the appearance of such a large S or inverted S shape in soft X-rays is likely to be followed by an eruption in just a few days. The twisted coronal loops subsequently release their pent-up magnetic energy in a solar explosion also detected in X-rays. This image was taken on 8 June 1998 at 15 hours 19 minutes UT with the Soft X-ray Telescope (SXT) aboard the Yohkoh satellite (Courtesy of Richard C. Canfield, NASA, ISAS, the Lockheed-Martin Solar and Astrophysics Laboratory, the National Astronomical Observatory of Japan, and the University of Tokyo)

prelude to an explosion on the Sun, so there might be some parallels to human behavior after all.

It is also critically important to know if the material sent out from the solar explosions is headed toward Earth. The high-energy charged particles that accompany solar flares follow the spiral pattern of the interplanetary magnetic

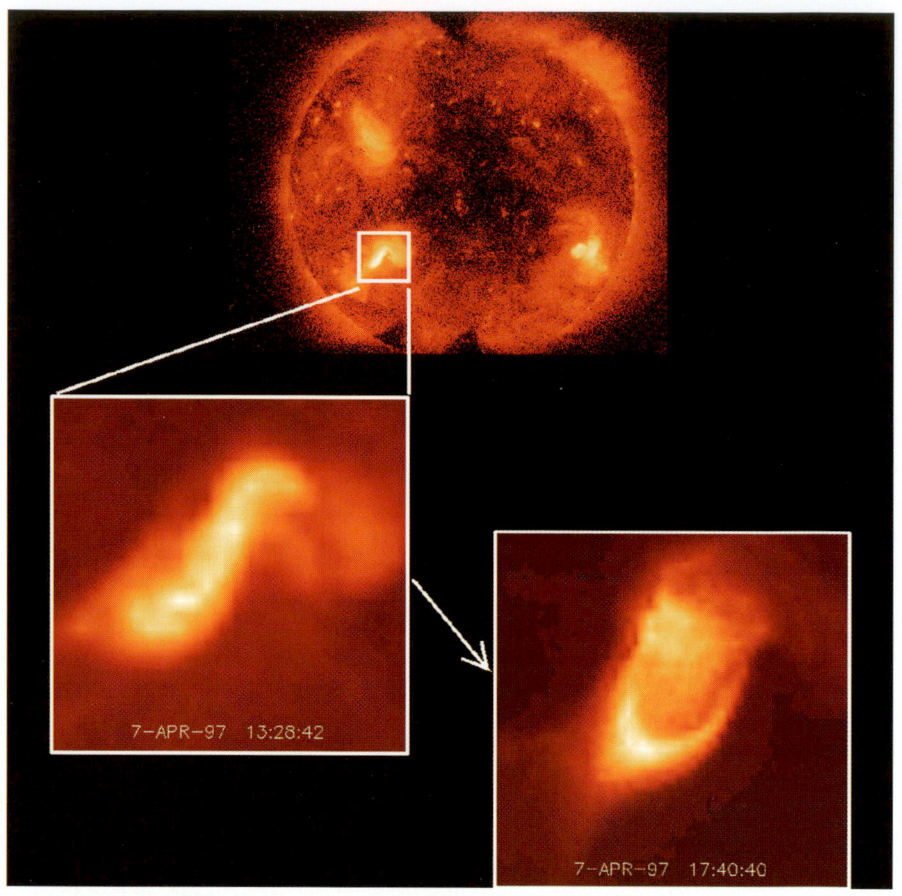

Fig. 7.10. Pre-Eruption Twist and Post-Eruption Cusp. The full-disk X-ray image, taken with Yohkoh's Soft X-ray Telescope (SXT), shows the Sun with a sigmoid present on 7 April 1997. It produced a halo Coronal Mass Ejection (CME) detected by SOHO's LASCO coronagraph the following day. The inset (*left*) shows the soft X-ray sigmoid at 13 hours 29 minutes UT on 7 April, just before eruption of the CME. The other inset (*right*), taken at 17 hours 41 minutes UT on the same day, shows the soft X-ray cusp and arcade formed just after the CME took place. (Courtesy of Richard C. Canfield, Alphonse C. Sterling, NASA, ISAS, the Lockheed-Martin Solar and Astrophysics Laboratory, the National Astronomical Observatory of Japan and the University of Tokyo)

field, so they must be emitted from active regions near the west limb and the solar equator to be magnetically connected with the Earth. Solar flares emitted from other places on the Sun are not likely to hit Earth, but they could be headed toward interplanetary spacecraft, the Moon, Mars or other planets. Coronal mass ejections that are shot out from near the visible edge of the Sun are not likely to impact Earth, but threaten other parts of space.

Mass ejections are most likely to hit the Earth if they originate near the center of the solar disk, as viewed from the Earth, and are sent directly toward the planet. The outward rush of such a mass ejection appears in coronagraph images as a ring or halo around the occulting disk, but the coronagraph data are unable to determine if the halo-like ejection is travelling toward or away from the observer. The Earth-directed ones may nevertheless be preceded by coronal activity at ultraviolet and X-ray wavelengths near the center of the solar disk as viewed from the Earth. The expulsion can additionally be associated with waves running across the visible disk (Fig. 7.11), like tidal waves or tsunami going across the ocean. Soft X-ray signatures include a scar-like dim-

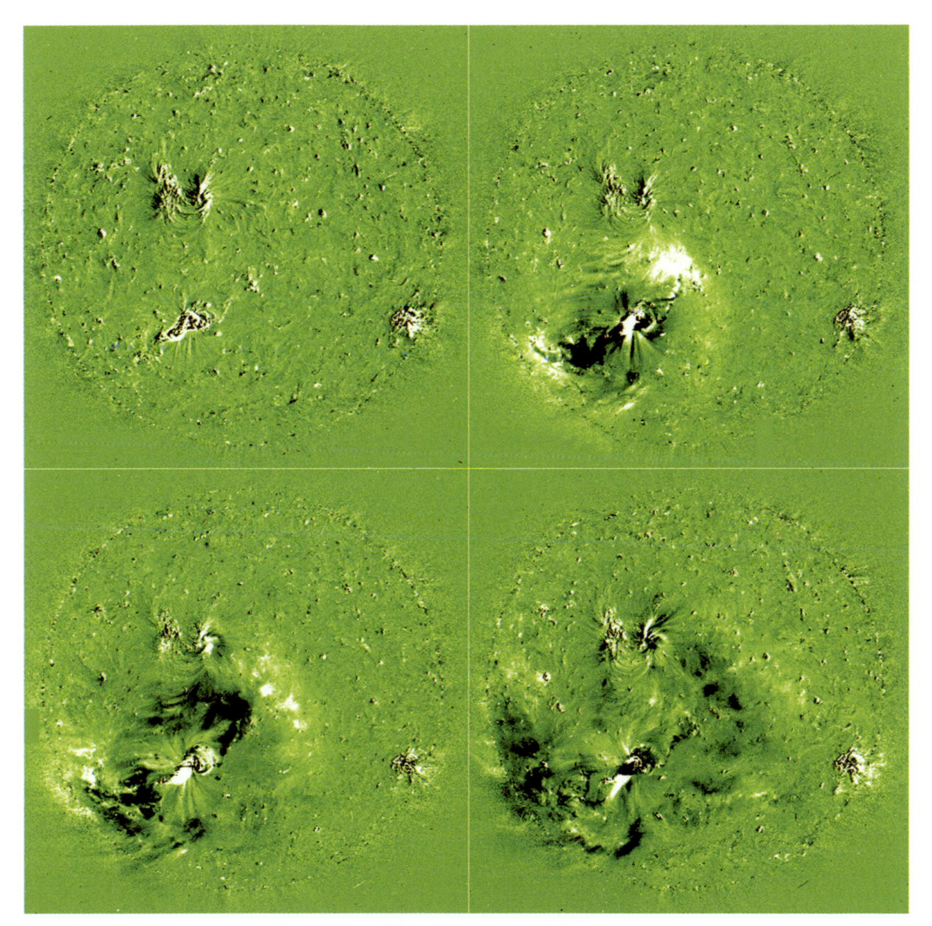

Fig. 7.11. Waves Across the Sun. This sequence of images, taken with the EIT instrument on SOHO, shows a wave running across the solar disk in the corona at supersonic velocities of about 1.5 billion meters per hour. It originated in the vicinity of a solar explosion, or flare, on 7 April 1997. The images were taken at 19.5 nanometers in the emission lines of iron ions (Fe XII) formed at 1.5 million degrees. Each image is the difference between the one taken at the time shown and the previous one. X-ray images of the Sun on the same day are shown in Fig. 7.10. (Courtesy of the SOHO EIT consortium. SOHO is a project of international collaboration between ESA and NASA)

ming left behind where the mass has been extracted, and arcades of magnetic loops that stitch the wound back together again (Sect. 6.5).

By closely observing the Sun, scientists may someday be able to anticipate changes in the Earth's upper atmosphere. It can be radically transformed by ultraviolet and X-ray radiation during active times on the Sun, and we now turn to this interesting aspect of the Sun-Earth connection.

7.3 Solar Ultraviolet and X-rays Transform Our Atmosphere

Unlike the charged particles in the solar wind, the Sun's electromagnetic radiation passes right through the Earth's magnetic fields without noticing them. Short-wavelength radiation, that contributes only a tiny fraction of the Sun's total luminosity, is mainly absorbed high in the Earth's atmosphere, transforming its physical constitution. In contrast, the most intense radiation from the Sun is emitted at visible wavelengths that penetrate our transparent air, to warm the ground and light our days, and it does not noticeably affect the upper atmosphere.

The Sun is a magnetic variable star whose radiation varies across the electromagnetic spectrum. This variability is controlled by the Sun's magnetic field, which varies not only over the 11-year solar cycle but on shorter and longer time scales as well. The variations are greatest at the short ultraviolet and X-ray wavelengths that we cannot see. The ultraviolet emission doubles from activity minimum to maximum, while the X-ray brightness of the corona increases by a factor of 100 (Fig. 7.12). Explosions in the corona, called solar flares, can outshine the entire Sun at X-ray wavelengths by a factor of 1,000, for a few minutes and sometimes for many hours (Sect. 6.3). Virtually all of this activity goes unseen at visible wavelengths.

Our thin atmosphere is pulled close to the Earth by its gravity, and suspended above the ground by molecular motion. Like all gases, air is highly compressible, and the atmosphere near the ground is compacted to its greatest density and pressure by the weight of the overlying air. At greater heights there is less air pushing down from above, so the compression is less and the density and pressure of the air falls off into the near vacuum of space.

The temperature of our atmosphere tends to decrease at higher altitudes where the air expands in the lower pressure and becomes cooler. The average air temperature drops below the freezing point of water (273 degrees Kelvin) only 1,000 meters above the Earth's surface, and bottoms out at about 10 times this height. Although common experience does not suggest it, the temperature of our atmosphere then increases at higher altitudes where it becomes hotter than the ground (Fig. 7.13).

The Sun's ultraviolet and X-ray radiation is absorbed in the upper atmosphere to heat and expand it. Moreover, the increased solar output at these wavelengths during the maximum in the Sun's 11-year activity cycle can cause the temperature of the upper atmosphere to soar to more than twice the values encountered at activity minimum (Fig. 7.14). The density of the upper atmosphere can increase by a factor of ten near the Sun's activity peak. Smaller changes in the Earth's global surface temperature, observed during the past century, correlate with the length of the sunspot cycle (Sect. 7.4).

Earth's atmosphere consists of neutral, unionized gases, mostly nitrogen and oxygen molecules, extending up to about 100,000 meters. Above this altitude, the air becomes highly ionized by the Sun's ultraviolet and X-rays. The energetic, short-wavelength radiation tears electrons off the atoms and molecules in the upper atmosphere, and creates the ionosphere, a permanent, spherical shell of electrons and ions.

The ionosphere was postulated in 1902 to explain Guglielmo Marconi's transatlantic radio communications. Since radio waves travel in straight lines, and cannot pass through the solid Earth, they get around the curved Earth

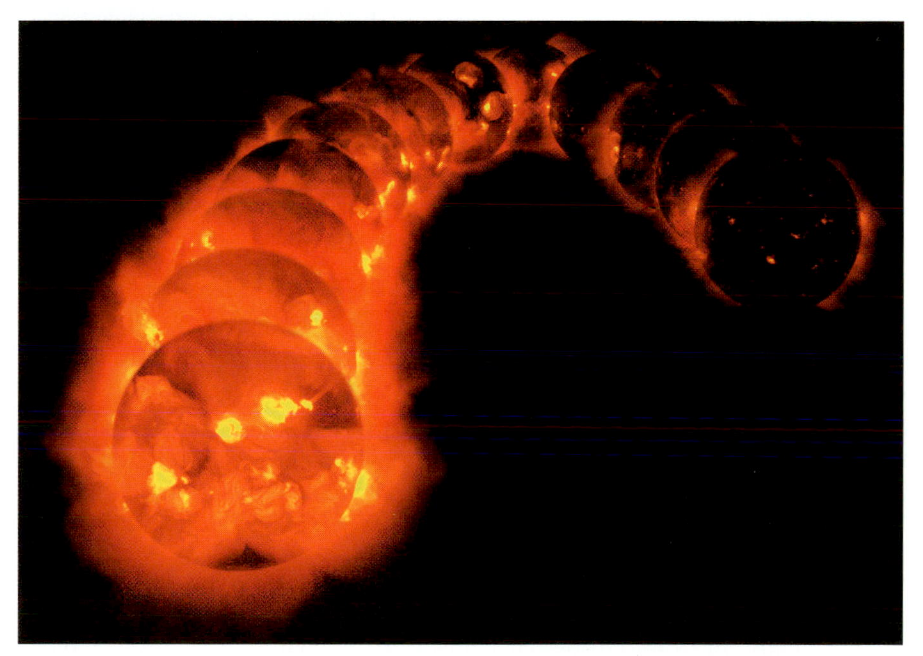

Fig. 7.12. X-ray View of the Solar Cycle. A montage of images from the Soft X-ray Telescope (SXT) aboard Yokhoh showing dramatic changes in the corona as the Sun's 11-year magnetic activity cycle changed from maximum to minimum. The 12 images are spaced at 120-day intervals from the time of the satellite's launch in August 1991, at the maximum phase of the 11-year sunspot cycle (*left*), to late 1995 near the minimum phase (*right*). The bright glow of X-rays near activity maximum comes from very hot, million-degree coronal gases that are confined within powerful magnetic fields anchored in sunspots. Near the cycle minimum, the active regions associated with sunspots have almost disappeared, and the Sun's magnetic field changes from a complex structure to a simpler configuration. In response, the coronal gases became less agitated and cooled down, resulting in an overall decrease in X-ray brightness by 100 times. (Courtesy of Gregory L. Slater and Gary A. Linford, NASA, ISAS, the Lockheed-Martin Solar and Astrophysics Laboratory, the National Astronomical Observatory of Japan, and the University of Tokyo)

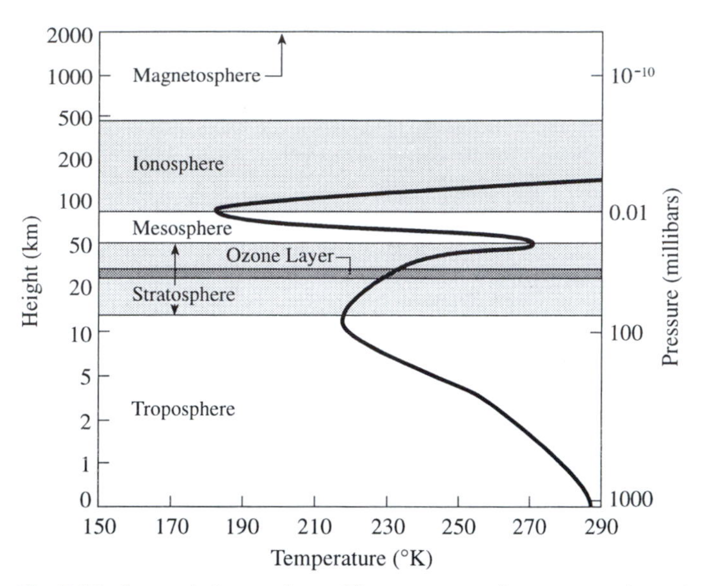

Fig. 7.13. Layered Atmosphere. The pressure of our atmosphere (*right scale*) decreases with altitude (*left scale*). This is because fewer particles are able overcome the Earth's gravitational pull and reach higher altitudes. The temperature (*bottom scale*) also decreases steadily with height in the ground-hugging troposphere, but the temperature increases in two higher regions that are heated by the Sun. They are the stratosphere, with its critical ozone layer, that is mainly heated by ultraviolet radiation from the Sun, and the ionosphere that is largely due to solar X-ray radiation

by reflection from electrons in the ionosphere. Verification of the existence of the ionosphere did not come until 1925 when radio transmissions were used to detect and measure it (Focus 7.4). Marconi was awarded the Nobel Prize in physics in 1909 for his "wireless telegraphy", and Edward V. Appleton received that prize in 1947 for his discovery and investigation of the radio-reflecting layer. It wasn't until the early 1950s that instruments on sounding rockets were used to measure the amount of solar ultraviolet and X-rays coming in from outside our atmosphere, showing that there is enough of the short-wavelength radiation to create the ionosphere (Sect. 2.4).

The ionosphere's ability to mirror radio waves depends on the level of solar activity. Near the maximum of the 11-year cycle of magnetic activity, there is more ultraviolet and X-ray radiation produced by the Sun. The ionosphere becomes more highly ionized as the result of this increased solar radiation, and it can reflect radio waves with higher frequencies then (Fig. 7.14).

Solar flares and coronal mass ejections also occur more frequently during the maximum of the cycle, producing changes in the Earth's ionosphere. The flares emit blasts of energetic radiation that can transform the ionosphere and temporarily disrupt radio communications. Energetic protons accelerated during solar flares may be magnetically directed on a collision course with Earth, where they can be sent into the polar regions and create very large amounts

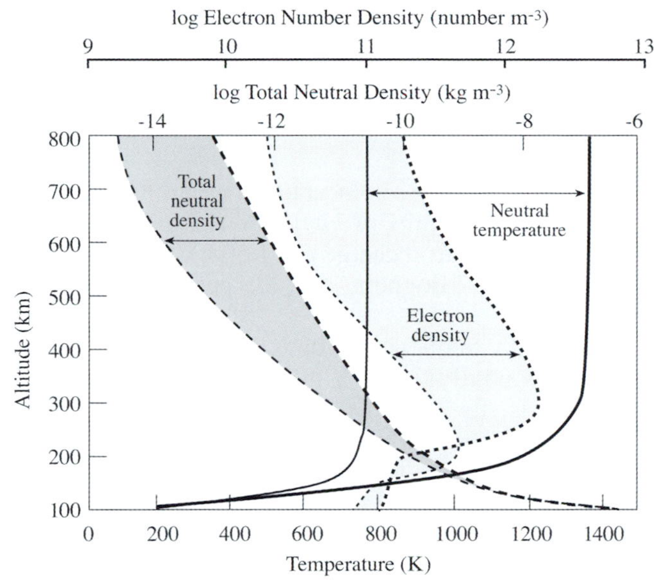

Fig. 7.14. Changes in Solar Heating. The Sun's ultraviolet and X-ray radiation heats the Earth's upper atmosphere to temperatures that exceed those on the terrestrial surface. Enhanced magnetic activity on the Sun produces increased short-wavelength radiation from the Sun, and this results in yet higher temperatures and greater densities at a given altitude in our atmosphere. During the Sun's 11-year activity cycle, the upper-atmosphere temperatures fluctuate by factors of two, and neutral and electron densities by factors of ten. (Courtesy of Judith Lean)

of ionization there. When coronal mass ejections hit the Earth, they can compress the magnetosphere and cause intense geomagnetic storms. Powerful electric currents can then couple the high latitude ionosphere with the magnetosphere, causing considerable heating of the ionized and neutral (unionized) gases.

The varying ultraviolet radiation from the Sun has other important consequences for life on Earth, for it produces and modulates the ozone layer within the middle atmosphere, or stratosphere. When ultraviolet rays with wavelengths of about 200 nanometers strike a molecule of ordinary diatomic oxygen that we breathe (O_2), they split it into its two component oxygen atoms (two O). Some of the freed oxygen atoms then bump into, and become

Focus 7.4

Radio Reflection by the Ionosphere

The ionosphere acts like a mirror that reflects radio waves up to a maximum frequency that depends on the density of electrons in the ionosphere. These are free electrons that have been liberated from atoms and molecules in our upper atmosphere by the action of solar ultraviolet and X-rays. The crucial upper frequency is equal to the plasma frequency, the natural frequency of oscillation for the ionosphere. The exact expression for the square of the plasma frequency, ν_p^2, is:

$$\nu_p^2 = e^2 N_e/(4\pi^2 \varepsilon_0 m_e) = 81 N_e \,\text{Hz}^2 ,$$

where the electron density, denoted by N_e, is in units of electrons per cubic meter, and the plasma frequency is in units of Hertz, or Hz, and one Hz is equal to one cycle per second. The electron charge $e = 1.60 \times 10^{-19}$ Coulomb, the electron mass $m_e = 9.11 \times 10^{-31}$ kilograms, and the permittivity of free space $\varepsilon_0 = 10^{-9}/(36\pi)$ farad m^{-1}.

Radio scientists usually express frequency in units of MegaHertz, abbreviated MHz, where one MHz is equivalent to a million Hertz, or 1 MHz = 10^6 Hz. Your FM radio station is, for example, identified by its call letters and the frequency of its broadcasts in MHz. The product of the frequency and the wavelength of the radio waves is equal to the velocity of light, 299,792,454 meters per second.

Radio waves with frequencies below the plasma frequency are reflected from the ionosphere and cannot pass through it. This means that cosmic and solar radiation at low frequencies and very long wavelengths cannot be observed from the ground. The ionosphere is transparent to radio radiation above the plasma frequency, so these waves can be used for communication with high-flying satellites or to observe the radio Universe.

The maximum transmission frequency depends on the solar activity cycle. At activity maximum, the Sun produces more ultraviolet radiation and X-rays, the ionosphere is more highly ionized by them, and the free electron density is enhanced (Fig. 7.14). The plasma frequency then increases, so higher radio frequencies are reflected from the ionosphere near the maximum in the 11-year cycle of solar activity. The height of the reflecting layer in the ionosphere can be inferred from the time between the transmission of a radio pulse and the reception of its reflected echo, and the electron density can be determined from the highest frequency at which a return signal is still received. The peak electron density in the ionosphere can reach 100 billion (10^{11}) to a million million (10^{12}) electrons per cubic meter, depending on the level of solar activity (Fig. 7.14).

The ionosphere has been divided into three layers, designated D, E and F, that reflect waves of different frequencies or wavelengths. The uppermost ionosphere layer, at an altitude of roughly 250,000 meters, reflects frequencies between 3 and 30 MHz, or wavelengths from 10 to 100 meters, that are used for radio broadcasting and over-the-horizon radar surveillance. Electrons at lower altitudes, from 95,000 to 140,000 meters in the E region, reflect medium frequencies from 0.3 to 3 MHz with wavelengths of 100 to 1,000 meters. The lowest D layer at heights of less that 90,000 meters reflects the lowest frequencies between 0.003 and 0.3 MHz, or wavelengths of 1,000 to 100,000 meters, that are used in some navigation systems such as LORAN and OMEGA.

attached to, an oxygen molecule, creating an ozone molecule (O_3) that has three oxygen atoms instead of two. The Sun's ultraviolet rays thereby produce the globe-circling layer of ozone at an altitude near 25,000 meters.

The ozone layer protects us by absorbing most of the ultraviolet and keeping its destructive rays from reaching the ground. However, the Sun's ultraviolet output varies with solar activity, producing cyclic variations in the global ozone abundance. The total global amount of ozone becomes enhanced, depleted and enhanced again from 1 to 2 percent during the 11-year solar activity cycle, modulating the protective ozone layer at a level comparable to human-induced ozone depletion by chemicals wafting up from the ground.

The ozone-destroying chemicals, called chlorofluorocarbons or CFCs for short, were banned by international agreement in the late 1980s, but the ozone layer is not expected to regain full strength until well into the latter half of the twenty-first century. Future monitoring of the expected recovery will require measurements of the natural fluctuations in global ozone caused by the varying solar ultraviolet output.

Variations in the intensity of solar ultraviolet radiation produce temperature fluctuations in the stratosphere where the ultraviolet is absorbed. In the late 1980s, Karin Labitzke and Harold Van Loon showed that these temperature changes drive high altitude winds and winter storms that beat to the 11-year solar rhythm. These winds might circulate down to lower levels in the Earth's atmosphere, where they could interact more directly with our weather.

The ultraviolet and X-ray variations are more pronounced then are the changes in visible sunlight. The tiny fluctuations are far too small to be directly detected with your eye. Yet, solar radiation achieves its maximum flux levels at these visible wavelengths that are essential for photosynthesis and warming the globe. The spectral region from 300 to 10,000 nanometers contains 99 percent of the solar radiation. Roughly 30 percent of the incoming radiation is reflected or scattered back to space, but the remaining 70 percent is absorbed by the ocean and land surface.

It is visible sunlight that passes through to the lower atmosphere where all our weather and climate occur. Global circulation of the air, driven by differential solar heating of the equatorial and polar surface, create complex, wheeling patterns of weather in this region, leading to the designation troposphere from the Greek *tropo* for turning. As we shall next see, the variable Sun may have affected the climate in this part of our air during the past millennium.

7.4 Varying Solar Activity and Climate Change

Day after day the Sun rises and sets in an endless cycle, an apparently unchanging ball of fire that illuminates our days and warms our world. Our lives depend on the continued presence and steady output of its light and heat. A measurement of this life-sustaining energy is called the "solar constant". It is the average amount of radiant solar energy per second per unit area reaching the top of Earth's atmosphere at a mean distance of one astronomical unit. The solar constant describes the total solar radiation at all wavelengths, and it can be used with the known distance and radius of the Sun to infer its luminosity (Focus 3.2).

Until the early 1980s, it was not known if the Sun was anything but rock-steady because no variations could be reliably detected from the ground. The required measurement precision could not be attained here on Earth because one could not accurately specify the variable amount of sunlight absorbed and scattered by our atmosphere. Nevertheless, reliable as the Sun appears, it fades and brightens, producing a slightly varying luminosity.

Stable detectors placed aboard satellites above the Earth's atmosphere precisely measure the solar constant and provide conclusive evidence for small variations in it. These instruments have been monitoring the Sun's total irradiance of the Earth since 1978 (Fig. 7.15). The mean value of the solar constant, from 1978 to 1998, was 1,366.2 Joule per second per square meter, with an uncertainty of about ± 1.0 in the same units. Moreover, this irradiance is almost always changing, in amounts of up to a few tenths of a percent and on time scales from 1 second to 10 years and possibly centuries. For instance, the mean value of the solar constant was 1,365.6 during the two measured minimums in the 11-year solar magnetic activity cycle. More than 95 percent of this fluctuating radiation is at visible and infrared wavelengths that penetrate to the troposphere and the Earth's surface.

The magnetic fields in sunspots produce a dimming of the light coming from that part of the Sun. When solar rotation carries sunspots across the visible solar disk, they therefore create a decrease in the solar irradiance detected at Earth of a few tenths of a percent that lasts a few days. The concentrated magnetism in the dark sunspots acts as a valve that blocks the outflow of energy from inside the Sun, making the sunspots cooler than their surroundings and producing the observed reductions in the solar constant.

Yet, the satellite observations show that the total solar irradiance at the Earth varies in step with the 11-year cycle of magnetic activity, increasing by about 0.1 percent between activity minimum and maximum. That is, the Sun becomes brighter overall when there are more spots, and *vice versa*. This seems counterintuitive, since the dark sunspots produce a decrease in the solar output, and sunspots are more numerous near the peak of the activity cycle.

The increase in luminous output is attributed to bright, localized magnetic structures, called *faculae* from the Latin for little torches or *plage* after the French word for beach. Both faculae and plage mark small, bright regions of concentrated magnetic flux that appear next to sunspots, and also within the magnetic network across the entire solar disk. The faculae are bright regions in the photosphere that are barely detectable in visible solar light. Excess heating of the solar chromosphere above the facular areas creates emission plages, which are easily detected by their enhanced radiation in two violet emission lines of ionized calcium. Designated the H and K lines, they have respective wavelengths of 396.85 and 393.37 nanometers. So, faculae and plage are two words for the same magnetic brightening that is most easily monitored in calcium H and K images of the Sun's chromosphere.

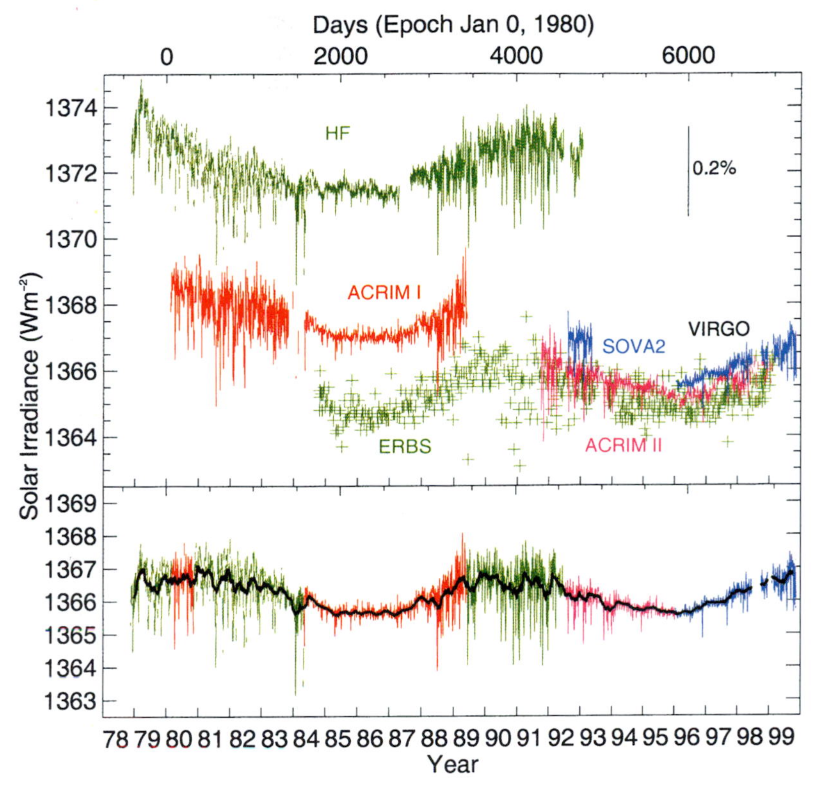

Fig. 7.15. The Sun Is A Variable Star. Observations with very stable and precise detectors on several Earth-orbiting satellites show that the Sun's total radiative input to the Earth, termed the solar irradiance, is not a constant, but instead varies over time scales of days and years. Measurements from five independent space-based radiometers since 1978 (*top*) have been combined to produce the composite solar irradiance (*bottom*) over two decades. They show that the Sun's output fluctuates during each 11-year sunspot cycle, changing by about 0.1 percent between maximums (1980 and 1990) and minimums (1987 and 1997) in magnetic activity. Temporary dips of up to 0.3 percent and a few days' duration are due to the presence of large sunspots on the visible hemisphere. The larger number of sunspots near the peak in the 11-year cycle is accompanied by a rise in magnetic activity that creates an increase in luminous output that exceeds the cooling effects of sunspots. Here the total irradiance just outside our atmosphere, called the solar constant, is given in units of watts per square meter, where one watt is equivalent to one Joule per second. The data are from the Hickey–Frieden (HF) radiometer of the Earth Radiation Budget (ERB) experiment on the Nimbus-7 spacecraft, the two Active Cavity Radiometer Irradiance Monitors (ACRIM I and II) placed aboard the Solar Maximum Mission satellite and the Upper Atmosphere Research Satellite (UARS), respectively, the radiometer on the Earth Radiation Budget Satellite (ERBS), and the Variability of solar IRradiance and Gravity Oscillations (VIRGO) radiometers flying on the SOlar and Heliospheric Observatory (SOHO). The SOVA2 is part of the SOlar VAriability experiment (SOVA) on the European Retrievable Carrier (EURECA). Offsets among the various data sets are the direct result of uncertainties in the absolute radiometer scale of the radiometers. Despite these offsets, each data set clearly shows varying radiation levels that track the overall 11-year solar activity cycle. (Courtesy of Claus Fröhlich, also see T. J. Quinn and C. Fröhlich, 1999, and C. Fröhlich and J. Lean, 1998)

As solar activity increases, there is an increase in sunspots, faculae and plage. The irradiance increases during the 11-year cycle are attributed to the changing emission of faculae and plage. Dimming on short time scales is related to the appearance of sunspots and the evolution of solar active regions. The irradiance decrease caused by sunspots is roughly balanced by the increase due to faculae or plage in the active regions near them, and the excess brightness from the magnetic network outside these regions more than compensates for the sunspot deficit. In other words, the total long-term irradiance increase by the bright faculae or plage overwhelms the short-term decrease by dark sunspots as solar activity increases.

Fingerprints of the Sun's varying activity are showing up all over the Earth's climate record, apparently warming and cooling our globe every decade for the past century. As an example, George C. Reid demonstrated in 1987 that the global sea-surface temperature varied with the 11-year sunspot cycle over the previous 130 years. Warren White and his colleagues refined the measurements a decade later, showing that the global sea-surface temperatures are swinging up and down by about 0.05 degrees Centigrade in time with the 11-year activity cycle. The temperature fluctuations are confined to the upper Sun-warmed layer of the oceans, disappearing at depths below about 100 meters. The sea-surface temperatures increase in step with the total solar irradiance of Earth, at least for the last two cycles that it has been observed, but by amounts that may be two or three times warmer than direct solar heating can account for.

Decade-long fluctuations in the global land temperature exhibit a strong correlation with the Sun's activity as measured by the length of the solar cycle. The yearly mean air temperature over land in the Northern Hemisphere has moved higher or lower, by about 0.2 degrees Centigrade, in close synchronism with the solar-cycle length during the past 130 years (Fig. 7.16). Short cycles are characteristic of greater solar activity that seems to produce global warming, while longer cycles signify decreased activity on the Sun and cooler times at the Earth's surface. This striking correspondence is a relatively recent discovery, published by Eigil Friis-Christensen and Knud Lassen in 1991.

Unless some unknown amplifying mechanism is magnifying the Sun's direct heating effect, it is difficult to understand how the feeble changes in the Sun's brightness, that have been observed so far, can have a noticeable effect on the Earth's climate. Some scientists have therefore reasoned that the Sun may be influencing the Earth by powerful indirect routes geared to the solar activity cycle. One such idea, suggested by Edward P. Ney in 1957, involves changes in the solar wind, Galactic cosmic rays and terrestrial clouds.

At times of enhanced activity on the Sun, the solar wind is pumped up with intense magnetic fields that extend far out into interplanetary space, blocking more cosmic rays that would otherwise arrive at Earth (Sect. 2.3). The resulting decrease in cosmic rays means that fewer energetic charged particles penetrate to the lower atmosphere where they may help produce clouds, particularly at higher latitudes where the shielding by Earth's magnetic field is less. The reduction in clouds, that reflect sunlight, would explain why the Earth's surface temperature gets hotter when the Sun is more active. In 1997

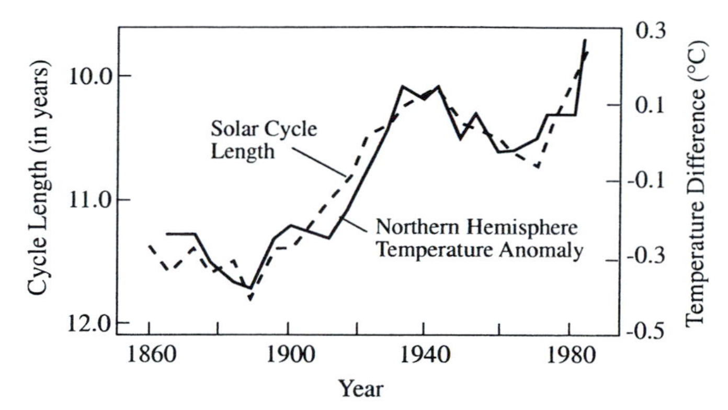

Fig. 7.16. Sunspot Cycle and Global Surface Temperatures. Variations in the length of the sunspot cycle (*dashed line*) closely fit changes in the air temperature over land in the Northern Hemisphere (*solid line*). Shorter sunspot cycles seem to be associated with increased temperatures and more intense solar activity. This suggests that solar activity is at least partly responsible for the rise in global temperatures over the last century, and that the Sun can substantially moderate or enhance global warming brought about by human increases of carbon dioxide and other greenhouse gases in the atmosphere. [Adapted from Eigil Friis-Christensen and Knud Lassen, Science **254**, 698–702 (1991)]

Henrik Svensmark and Eigil Friis-Christensen indeed found that during the previous 11-year activity cycle of the Sun, the cloudiness on a global scale decreased with increasing solar activity and a lower intensity of cosmic rays entering the atmosphere.

Evidence that the varying solar wind may affect the Earth's climate on decadal and century time-scales was additionally provided by Edward W. Cliver, Valentin Boriakoff and Joan Feynman in 1998. They reported a correlation between geomagnetic activity, driven by the Sun's gusty winds, and time variations in the global surface temperature on the Earth over the past 120 years.

Persistent, substantial changes in the Sun's total radiative output may nevertheless be directly warming and cooling the Earth. The mean global land temperatures have, for example, been on the rise by total of about 0.6 degrees Centigrade over the past 100 years, but with large fluctuations on the way (Fig. 7.17). Solar variability provides a reasonable match to many of the detailed ups and downs of the temperature record, and probably made a significant contribution to temperature changes in the first half of the century. So, many of the temperature fluctuations on Earth during the early part of the 20th century could be directly related to the brightening and dimming of the Sun after all. Nevertheless, global warming by man-made greenhouse gases now seems to be taking over and controlling the Earth's global thermostat.

The exponential increase in the atmospheric concentration of man-made carbon dioxide, which is produced by burning coal, oil and natural gas, is at least partly responsible for the unusual temperature increase during the last two decades of the twentieth century. Carbon dioxide and other greenhouse

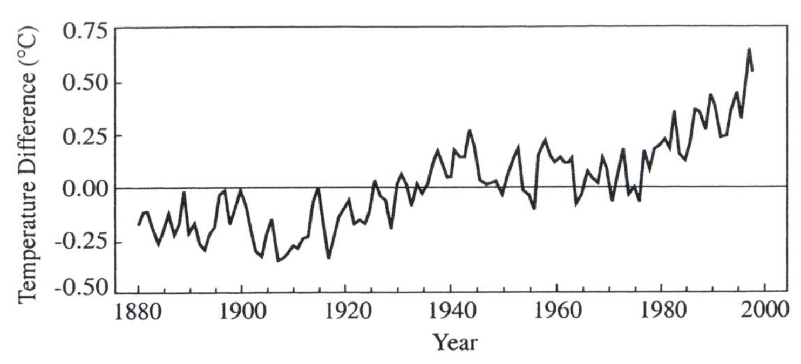

Fig. 7.17. Global Temperature Variation. The annual global-mean temperature near the Earth's surface has increased approximately 0.6 degrees Celsius, or 0.6 °C, since the beginning of the twentieth century, occurring from 1910 to 1940 and from 1980 to present. Here the data are plotted as a temperature difference from the long-term mean value. At more than 0.5 degrees Celsius above the mean, recent global temperatures maintain a warming trend for the past two decades. The global mean temperature for the period 1880 to 1997 is 13.8 degrees Celsius (56.9 degrees Fahrenheit). Natural temperature fluctuations prohibit a clear detection of human-induced warming in the early part of the twentieth century, but human increases of greenhouse gases in the atmosphere may be noticeably contributing to the recent rise in temperatures. [Courtesy of Michael Changery, the National Climatic Data Center (NCDC) of the National Oceanic and Atmospheric Administration (NOAO)]

gases trap heat radiation near the planet's surface and elevate its temperature by the greenhouse effect (Focus 7.5). To fully understand the complex implications of these heat-trapping gases for future global warming, scientists have examined historical records of the variable brightness of the Sun and other stars.

Since the Sun's total radiative output has only been precisely measured from space since 1978, earlier brightness changes must be inferred from historical records of solar activity and observations of Sun-like stars. This two-part reconstruction of the past involves the use of ancient sunspot observations to estimate the variable radiation from solar active regions, and extrapolations from the brightness variations of other stars to describe the changing luminosity of the quiet, inactive Sun.

Reconstruction of the varying solar irradiance of Earth, published by Judith Lean, Juerg Beer and Raymond Bradley in 1995 indicate that the Sun's changing brightness dominated the climate between 1600 and 1800. The Sun also played a significant role in the global warming observed from 1874 to 1970, but heat-trapping gases in the Earth's atmosphere began to have a dominant effect after 1970 (Fig. 7.18).

These results were followed up in 1999 by Michael E. Mann, Raymond E. Bradley and Malcolm K. Hughes with a more detailed comparison of the surface temperature estimates and the various forces that might have affected our climate during the past 1,000 years (Fig. 7.19). Variations in the Sun's radiative output, and therefore in the amount of that radiation reaching the Earth, produced significant global temperature variations during the 17th to

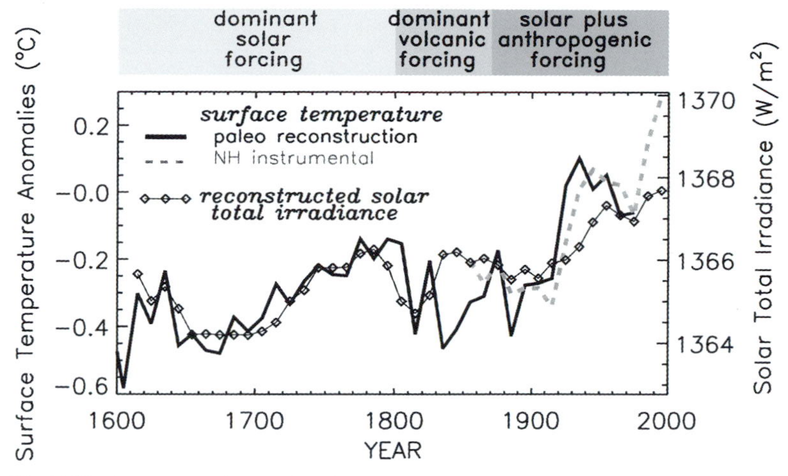

Fig. 7.18. Reconstructing Global Temperature. The global temperature records closely trace a reconstruction of the Sun's brightness over the past 400 years. From 1600 to about 1800, the temperature fluctuations are mainly attributed to variations in solar activity, while cooling by haze from volcanoes plays a role during the next 100 years. After the mid-1970's, when about half the century's global temperature increase took place, the warming resulted largely from heat-trapping greenhouse gases emitted by industrial economies. Here the decadally averaged values of Northern Hemisphere summer surface temperatures (*scale at left*) are compared with the reconstructed solar total irradiance (*symbols, scale at right*). The dark solid line is paleoclimate temperature data, primarily tree rings, and the gray dashed line is instrumental temperature data. [Courtesy of Judith Lean and adapted from J. Lean. J. Beer, and R. S. Bradley, Geophysical Research Letters **22**, 3,195–3,198 (1995); also see J. Lean (1997)].

19th centuries; cooling associated with stratospheric volcanic aerosols reached peak influence in the early and mid 19th centuries. The Sun has apparently had only a modest influence on climate variability in the 20th century, with global warming by greenhouse gases dominating any other climate-changing force in the late 20th century. These findings are substantiated by several independent studies in 1999, including those by Thomas J. Crowley and Kwang-Yul Kim, Paul E. Damon and Alexei N. Peristykh and Simon F. B. Tett and colleagues. So, greenhouse warming may eventually dominate the climate if we do not do something about it.

The Sun's activity can nevertheless substantially enhance or moderate this warming, and there isn't very much we can do about the Sun's changing output except monitor it. Solar radiation may, in fact, have dominated the Earth's climate over the past centuries. The last millennium was, for example, characterized by a succession of cool and warm periods, each of roughly a century in duration, that may have been associated with changes in solar activity. The most conspicuous long cold spells, with summer temperatures between 0.25 and 0.5 degrees Centigrade below the long-term average, were between 1100 and 1350, between 1400 and 1800 (the Little Ice Age), and between 1800 and 1890. These transformations in climate may have been driven by the Sun,

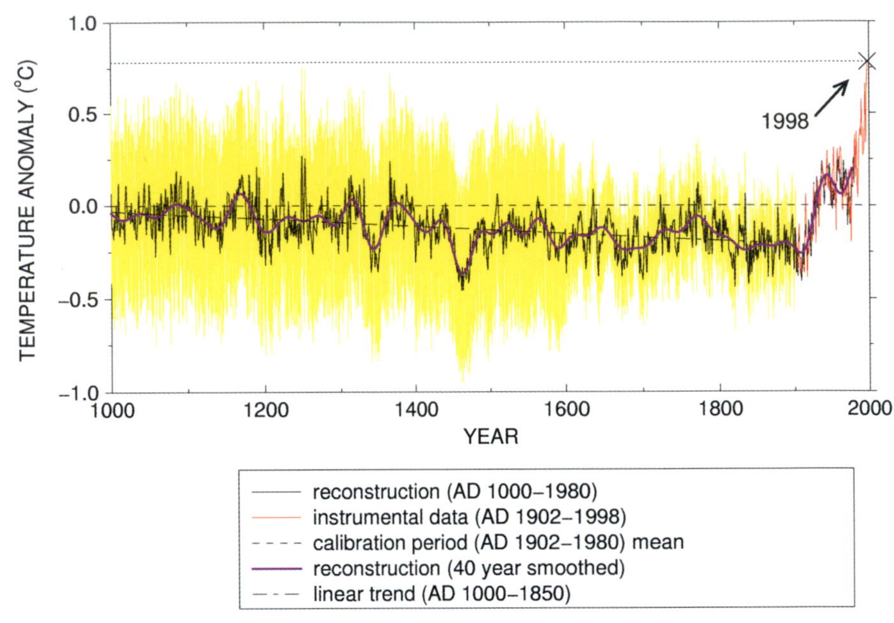

Fig. 7.19. A Thousand Years of Hot and Cold. The Northern Hemisphere has apparently been warmer in the 20th century than in any other century of the last thousand years, according to this reconstruction of the hemisphere temperature record. The sharp upward jump in the temperature during the last 100 years was recorded by thermometers at and near the Earth's surface. Earlier fluctuations were reconstructed from "proxy" evidence of climatic change contained in tree rings, lake and ocean sediments, and ancient ice and coral reefs. The farther back in time the reconstruction is carried, the larger the range of possible error, denoted by the light shaded regions. [Courtesy of Michael E. Mann and adapted from M. E. Mann, R. S. Bradley and M. K. Hughes, Geophysical Research Letters **26**, 759–762 (1999)]

whose past activity level is recorded in sunspot observations, tree rings and ice cores. The observed brightness variations of Sun-like stars confirm that pronounced changes in solar activity might have occurred in the past, leading to global warming and cooling, and that the Sun could again undergo similar drastic changes in the future (Focus 7.6).

The period from 1645 to 1715, now known as the Maunder Minimum, was particularly outstanding, in that sunspot activity dropped to unusually low levels and the world experienced simultaneously one of the coldest periods of the Little Ice Age in Europe. The newly invented telescope was used by Galileo Galilei to systematically investigate sunspots early in the 17th century, but very few were seen with telescopes just half a century later. The dark spots had virtually disappeared from the face of the Sun, and scarcely any could be found for seventy years. During the 1880s and 1890s, first Gustav F. W. Spörer and then Edward Walter Maunder drew attention to this dearth of sunspots, and Maunder than extended the discussion in a paper entitled the "The Prolonged Sunspot Minimum, 1645–1715", published in 1922.

Several authors have proposed that a ponderous reduction in the Sun's radiative output accompanied the long drop in sunspot activity, thereby lowering the temperature across the globe. Although no one was monitoring the solar irradiance of Earth back then, the total range of the Sun's variable brightness over long periods of time can be estimated from observations of solar-type stars (Focus 7.6). Such stars go through bright, active phases and quiet, steady ones of reduced output. The Sun is linked to these stars by using plage emission, in the calcium H and K lines, as an index of stellar brightness. By

Focus 7.5

The Greenhouse Effect

Our planet's surface is now comfortably warm because the atmosphere traps some of the Sun's heat and keeps it near the surface. The thin blanket of gas acts like a one-way filter, allowing sunlight through to warm the surface, but preventing the escape of some of the heat into the cold, unfillable sink of space. Much of the ground's heat is re-radiated out toward space in the form of longer infrared waves that are less energetic than visible ones and thus do not pass through the atmosphere's gas as easily as sunlight.

This is why cloudy nights tend to be warmer than clear nights; escaping infrared heat radiation is blocked by the water vapor in the clouds, keeping the ground warm at night. The glass windows of a greenhouse were thought to retain heat and humidity inside of a greenhouse, so the atmospheric warming has also come to be called the greenhouse effect.

As first shown by John Tyndall in 1861, it is the minor ingredients in air, like water vapor and carbon dioxide, that absorb heat at infrared wavelengths, retaining it and raising the temperature of the planet. Methane and nitrous oxide also act as greenhouse gases, but with even lower abundance than water or carbon dioxide. The methane is released from swamps, coal mines and even cows, and nitrous oxide, or laughing gas, originates from fertilizers. The main ingredients of the atmosphere, nitrogen (78 percent) and oxygen (21 percent), play no part in the warming of the Earth's surface, since these diatomic molecules do not absorb noticeable amounts of infrared radiation.

The greenhouse effect is literally a matter of life and death. Without its atmosphere, the Earth is heated by the Sun to only −18 degrees Celsius, which is well below the freezing point of water at 0 degrees Celsius. Fortunately for life on Earth, the greenhouse gases in the air warm the planet by as much as 33 degrees Celsius, and this extra heat keeps the oceans from becoming frozen over. Most of this "natural" greenhouse warming comes from water molecules (60 to 70 percent) and carbon dioxide provides only a few degrees.

Since the industrial revolution, humans have released heat-trapping gases into the atmosphere at an ever-increasing rate, creating an "unnatural" greenhouse effect. The amount of carbon dioxide in our air has, for example, grown by more than 13 percent since 1958, when direct monitoring started, and it is still steadily accumulating as the result of rapid growth of the world population and increased burning of fossil fuels like coal and oil. There is therefore growing concern that it could produce significant global warming.

The smooth, inexorable increase in atmospheric carbon dioxide cannot, however, explain all the ups and downs in the global temperature record for the past century. Although the Earth's average surface temperature has risen by about 0.6 degrees Celsius over this period, astronomers think that the Sun, our ultimate source of energy, might be varying enough to account for much of the observed climate change. Reconstruction of the Sun's brightness over the last four hundred years exhibits an intriguing match to the global temperature record (Fig. 7.18), indicating that the Sun could have been responsible for as much as half of the warming of the past century. However, such investigations also demonstrate that global warming by man-made greenhouse gases is largely responsible for the unprecedented rise in temperatures in recent decades.

A number of influences, both natural and man-made, complicate speculations about our climate's future. The natural ones include solar radiation, whose fluctuations can cause the Earth's surface temperature to rise and fall, and an atmospheric haze cast aloft by erupting volcanoes, which lowers the surface temperature by reflecting sunlight. Clouds provide another uncertainty. They reflect back to space part of the solar radiation, and thus have a cooling effect, while the water molecules in clouds trap infrared radiation emitted from the Earth's surface, causing warming. The cooling effects of sulfate particles emitted from industrial smokestacks can offset the warming of man-made, heat-trapping gases. Some researchers have also speculated that the carbon dioxide in the lower atmosphere may already be absorbing as much heat as is possible, so future increases in this greenhouse gas could mainly affect the upper atmosphere that does not play as important a role in determining the surface temperature.

So, the big questions are not yet answered, and we do not have the precise information required to know exactly how much of the recent warming of the Earth is due to human activity and how much to natural causes. That should nevertheless not be an invitation to complacency. If current emissions of carbon dioxide and other greenhouse gases go unchecked, their concentrations in the atmosphere are likely to be double pre-industrial levels sometime in the twenty-first century. If that happens, many experts forecast, the average surface temperature of the globe will rise by about 2 degrees Celsius, making the Earth hotter than it has been in millions of years. Sea levels could rise enough to turn parts of Florida into Atlantis, inundate Venice and submerge island nations, but the true threat remains difficult to gauge. Still, it is likely that the increased heat and violent weather will drastically change the climate we are used to, and some of us will think that the world is melting down in a pool of sweat. The continued accelerated burning of fossil fuels will someday cause great damage to the environment, so both the developing and industrial nations should now do more to stop it.

matching the Sun's current plage brightness to those of stars in their active phase, one can then use the lowest brightness of the other stars to estimate the solar irradiance during the sunspot-free Maunder Minimum, when the Sun was presumably in its quiet phase. Such procedures have been used to show that the Sun was approximately 0.25 percent dimmer at the time of the Maunder Minimum then currently, and was therefore capable of explaining the estimated drop of about 0.5 degrees Celsius in global mean temperature. Moreover, the increase in solar brightness since the Maunder Minimum can account for most of the global warming during the subsequent 255 years, from 1715 to 1970.

Radioactive isotopes record prolonged solar magnetic activity and related climate change further back in time. During periods of increased activity on the Sun, when the Earth was presumably warmer, the magnetic fields in the solar wind had a larger shielding effect on Galactic cosmic rays. This prevented the energetic charged particles from entering the Earth's atmosphere and producing radioactive isotopes. In contrast, high amounts of the radioactive elements were produced when the Sun was inactive and the climate was cold.

Carbon 14, dubbed radiocarbon and designated ^{14}C, is the first radioactive isotope to be used to reconstruct past solar activity. It appears in annual tree rings dating back to 8,000 years ago. Radiocarbon is produced in the Earth's atmosphere by a nuclear reaction in which energetic neutrons interact with nitrogen, the most abundant substance in our air. The neutrons are themselves the products of interactions between cosmic rays and the nuclei of air molecules.

Each radiocarbon atom, ^{14}C, joins with an oxygen molecule, O_2, in the air to produce a form of carbon dioxide, designated by $^{14}CO_2$, which is chemically indistinguishable from the ordinary carbon dioxide, $^{12}CO_2$, found in larger amounts in the air. Through photosynthesis, live trees assimilate both types of carbon dioxide and deposit them into their outer rings. The ratio of the amounts of the two kinds of carbon, $^{14}C/^{12}C$, provides an inventory of atmospheric radiocarbon, and therefore of solar activity, at the time of take up. That time can be determined at any later date from the age of the annual tree ring. Just count the number of tree rings that have been subsequently formed at the rate of one ring per year.

Focus 7.6

Brightness Variations of Sun-Like Stars

Spacecraft observations of the varying solar brightness have only been possible since 1978, and they indicate it has varied by about 0.1 percent over the past two 11-year cycles of solar magnetic activity. The full range of possible fluctuations in the Sun's brightness over much longer periods of time can be estimated only by comparing the Sun to other solar-type stars which are close in mass, composition and age.

In a pioneering study, begun in 1966 and published a decade later, Olin C. Wilson observed that Sun-like stars do undergo rhythmic changes in magnetic activity on time scales of years, much like our Sun. The magnetic variations of roughly 100 such stars were detected by monitoring their chromospheric emission in the H and K lines of ionized calcium, at the visible wavelengths of 396.8 and 393.4 nanometers respectively. This work has been continued for decades, and extended to include measurements of stellar brightness variability.

In 1990, Richard R. Radick, G. W. Lockwood and Sallie L. Baliunas used nearly a quarter century of brightness and magnetic activity observations to conclude that stars that are much younger than the Sun vary in a much different way than older ones. Younger stars, like children, are generally more active than older ones. The fast rotation of youth presumably generates stronger magnetic fields and enhances magnetic activity; but perhaps because of enhanced starspot dimming, the younger stars tend to become fainter as their magnetic activity increases. In contrast, the older stars, including the Sun, become brighter at times of increased magnetic activity, a conclusion that was reaffirmed by Qizhou Zhang and colleagues in 1994.

The comparisons with Sun-like stars prompted Sallie L. Baliunas and Robert L. Jastrow to speculate in 1990 that the Sun might have undergone substantial luminosity variations on time scales of centuries, associated with dramatic changes in the Earth's climate. Detailed investigations of stars with masses and ages very close to those of the Sun have subsequently substantiated this conclusion in greater detail.

Solar-type stars exhibit two fundamental types of behavior: a bright, active cyclic one and a faint, inactive, non-cyclic one. About one third of the monitored stars are in a dead calm, with no detectable magnetic activity cycle similar to the Sun's 11-year sunspot cycle. This finding implies that the Sun and similar stars can undergo prolonged intervals of exceptional inactivity, suggesting a phase similar to the Sun's Maunder Minimum from 1645 to 1715 in which sunspots virtually disappeared.

The comparisons with solar-type stars also permit estimates for the possible total range in brightness variation of the Sun in past and future centuries, from the inactive, non-cycling lulls to the active, cycling luminosity highs. Such calculations can then be used to infer extreme consequences for the climate on Earth. In 1995, for example, Judith Lean, Juerg Beer and Raymond Bradley used such stellar comparisons in combination with proxies of solar activity to estimate the Sun's effect on the Earth's climate since 1610 (Fig. 7.18). They concluded that the Sun's low luminosity during the Maunder Minimum could account for a simultaneous long, cold spell on Earth, and that the Sun's subsequent brightness increase might account much of the global warming since then. Rising temperatures in the ending decades of the millennium are nevertheless attributed mainly to the greenhouse effect of gases dumped into the atmosphere by humans.

Wood from the world's longest-lived trees, the bristlecone pines *Pinus aristata*, has been used to derive the amounts of radiocarbon present in the air for 8,000 years into the past. In the 1970s and early 1980s, Hans E. Suess and his colleagues showed that "wiggles" with an approximate 200-year period were superimposed on the general trends in the tree-ring radiocarbon data, attributing the fluctuations to variations in solar activity. The prolonged dips and rises in the radiocarbon record, respectively corresponding to long increases and decreases in solar magnetic activity, respectively coincide with intervals of warmer and colder climate.

Subsequent examination of the radiocarbon data in the 1990s, by Paul E. Damon, Charles P. Sonett and others, confirm that periods of 11, 88 and 208 years are in large part due to solar wind modulation of cosmic-ray production of ^{14}C. The 11-year periodicity corresponds to the well-known cycle of solar magnetic activity. A periodicity of around 80 years is named the Gleissberg cycle, after Wolfgang Gleissberg's 1958 detection of it in records of the number of sunspots, and subsequently in the frequency of auroras. The 208-year fluctuation corresponds to the Suess "wiggles". Damon and Alexei N. Peristykh reported in 1999 that the Gleissberg cycle may be correlated with ancient climate records and the 20th century measurements of the Earth's global surface temperature.

In an important paper entitled "The Maunder Minimum", published in 1976, John A. Eddy used the tree-ring data to show that this 70-year period of very few sunspots was also a time of increased radiocarbon production. This reinforced the connection between the paucity of sunspots and a dramatic reduction in the level of solar activity, which he further documented by ancient eclipse observations of coronas without extensive streamers and historical records of relatively weak aurora activity. Moreover, the Maunder Minimum was not an isolated incident. Eddy showed that episodes of abnormally low solar activity lasting many decades are a fairly common aspect of the Sun's behavior, concluding that the Sun has spent nearly a third of the past two thousand years in a relatively inactive state (Fig. 7.20). Extended periods of solar variability must therefore be considered to be a permanent feature of the Sun, and can be expected to occur again in the future.

The changing Sun has apparently been drastically altering the climate for thousands of years. The Little Ice Age (1400–1800), for example, overlaps the Spörer Minimum (1400–1530) and Maunder Minimum (1645–1715). During this long period of unusual cold weather, alpine glaciers expanded, the Thames River and the canals of Venice regularly froze over, and painters depicted unusually harsh winters in Europe (Fig. 7.21). Moreover, an extended period of high solar activity, between about 1100 and 1250, coincided with the relatively warm weather that seems to have made Viking migrations to Greenland and the New World possible. The temperate climate during this Medieval Maximum enabled Icelandic settlers to flourish on Greenland, but both the green plants and the colony expired in the frozen climate of the Little Ice Age.

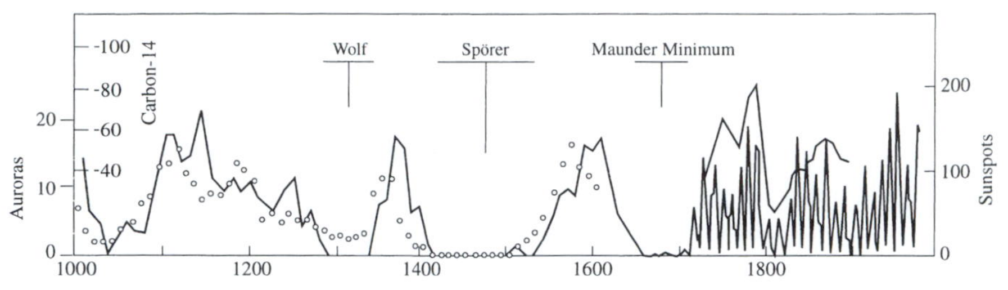

Fig. 7.20. Lapse in Solar Activity. Three independent indices confirm the existence of long-period changes in the level of solar activity. The observed annual mean sunspot numbers (*scale at right*) are shown from AD 1650 to AD 1990. The dearth of sunspots between 1645 and 1715 is now known as the Maunder Minimum. The 11-year solar activity cycle can be followed as a modulation in the record of sunspot numbers after about AD 1700. The curve extending from AD 1000 to AD 1900 is a proxy sunspot number index derived from measurements of carbon-14 in tree rings. Increased carbon-14 is plotted downward (*scale at left-inside*), so increased solar activity and larger proxy sunspot numbers correspond to reduced amounts of radiocarbon in the Earth's atmosphere. Open circles are an index of the occurrence of auroras in the Northern Hemisphere (*scale at left-outside*). (Courtesy of John A. Eddy)

The Earth's global environmental history has been well preserved in the polar ice caps, whose record complements and extends the tree-ring evidence for past Sun-climate connections. The ice contains the radioactive isotope of beryllium, ^{10}Be, that has been deposited there by snows, the later snows compressing earlier ones into ice. Like radiocarbon, the ^{10}Be is produced by nuclear reactions between energetic neutrons and air molecules, a consequence of cosmic rays entering the atmosphere. When there are fewer cosmic rays arriving at Earth, smaller quantities of ^{10}Be are produced in the air and are precipitated by snows at the polar regions. The amount of ^{10}Be found in the ice therefore measures the flux of incoming cosmic rays, which in turn mirrors the degree of solar magnetic activity. Moreover, the beryllium content is immune to the industrial contamination that may affect the carbon-14.

Deep cores have been drilled and extracted from the glacial ice in Greenland and Antarctica. The beryllium content of these tubes of ice generally confirms the radiocarbon results. The amounts of the two radioactive isotopes move up and down together over the past 5,000 years, and they exhibit a common 200-year periodicity. This strongly supports the explanation that these century-long fluctuations are caused by magnetic field changes in the solar wind that modulate the cosmic-ray flux. If these changes in solar magnetic activity are indeed linked to substantive variations in the Sun's brightness, they might explain the changes in climate.

Half of any newly-produced radioactive carbon, ^{14}C, will decay into normal, stable carbon, ^{12}C, in 5,730 years, so you cannot use radiocarbon to look into the past for much longer than this time. There just will not be any measurable radiocarbon left from more than 50,000 years ago. In contrast, cores of deep-sea sediments and glacial ice archive the local environment for hundreds of thousands of years, unlocking the secrets of the major ice ages.

Fig. 7.21. Hunters in the Snow. This section of a painting by Pieter Bruegel the Elder depicts a time when the average temperatures in Northern Europe were much colder than they are today. This chilly spell, known as the Little Ice Age, extended from 1400 to 1800. Severe cold occurred during the Maunder minimum, from 1645 to 1715, when there was a conspicuous absence of sunspots and other signs of solar activity. This picture was painted in 1565, near the end of another dearth of sunspots, called the Spörer minimum. (Courtesy of the Kunsthistorisches Museum, Vienna)

7.5 Climate Change over Millions and Billions of Years

During the past one million years, called the late Quaternary period, the world's climate has been dominated by the deep glacial cold of ice ages. They are punctuated every 100 thousand years or so by a relatively short interval of unusual warmth, called an interglacial, lasting 10, 20 and even 30 thousand years. At the height of each long glaciation, the great polar ice sheets advance down to lower latitudes, retreating during the next interglacial stage. We now live in such a warm interglacial interval, called the Holocene period, in which human civilization has flourished. Still, the die is cast for the next glaciation, and the ice will come again.

The Holocene, which has already lasted 11,000 years, is, by far, the longest stable warm period recorded in Antarctica during the past 420,000 years, and it is not expected to last more than a few thousand years more. When we enter the next ice age, continents will become larger than they are today, because of the steep drop in sea level – about 100 meters, and people might then walk from England to France, from Siberia to Alaska, and from New Guinea to Australia. The next time it happens, the advancing glaciers will bury Copenhagen, Detroit and Montreal under mountains of ice. Thus, studies of past ice ages may have profound implications for future civilization.

The rhythmic alteration of glacial and interglacial intervals is set in motion by the way the Earth moves about the Sun, periodically altering the amount and distribution of sunlight received by Earth over tens of thousands of years. When less sunlight is received in far northern latitudes, the winter temperatures are milder there, but so too are summer temperatures. So, less polar ice then melts in the summer, and over time the winter snows are compressed into more ice to make the glaciers grow.

To be exact, three astronomical cycles combine to alter the angles and distance at which sunlight strikes the far northern latitudes of Earth, triggering the ice ages. This explanation was fully developed by Milutin Milankovitch from 1920 to 1941, so the astronomical cycles are now sometimes called the Milankovitch cycles. They involve periodic wobbles in the Earth's rotation and changes in the tilt of its axis and the shape of its orbit, occurring over tens of thousands of years (Fig. 7.22). The wobble, or precession, of the Earth's spin axis changes over periods near 23,000 years, changing the Earth's orientation to the Sun. The planet also nods back and forth, with a 41,000-year periodic variation of its axial tilt, or obliquity. The greater the tilt, the more intense the seasons in both hemispheres, with hotter summers and colder winters. In the longest cycle, the shape of the Earth's orbit stretches slightly and its eccentricity changes, from more circular to more elliptical and back again, over a period of 100,000 years. As its path becomes more elongated, the Earth's distance from the Sun varies more during each year, intensifying the seasons in one hemisphere and moderating them in the other. The interplay of the cycles means that each glacial and interglacial period follow an uneven course.

The elegant astronomical theory of the ice ages can be documented by sediments that have accumulated at the bottom of the oceans. Researchers investigate the relative amounts of the heavier and lighter forms, or isotopes, of oxygen locked up in the fossilized shells of tiny marine creatures that are found in the deep-sea sediments. The changing ratios of the two forms tell how much ice was present when the sea animals lived, and therefore record the advance and retreat of the glacial ice.

There are eight protons in the nucleus of every oxygen atom, but two isotopes contain eight and ten neutrons, giving atomic mass numbers of 16 and 18, respectively. Molecules of water that contain the rare, heavier sort of oxygen, denoted ^{18}O, move more slowly than those of water with the more abundant, lighter type, symbolized by ^{16}O, so their relative amounts depend upon how cold it is. During the freezing glaciations, more of the water containing the common, light form of oxygen becomes trapped in the ice, leading to an increase in the relative amount of the rare, heavier sort of oxygen in the ocean waters and the shells of creatures living in it. So, the ratio of the amounts of the two oxygen isotopes found in their skeletal remains can be

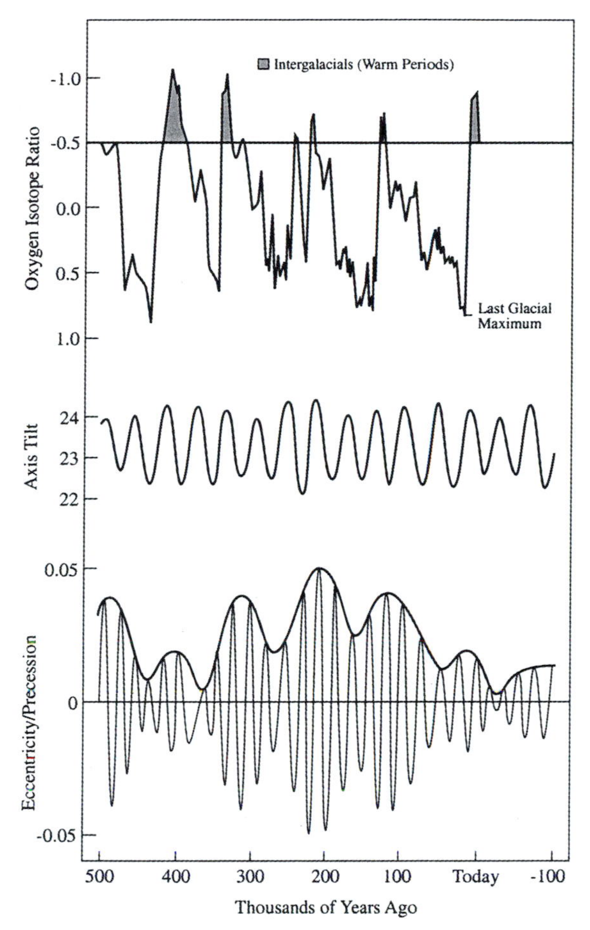

Fig. 7.22. Pacemakers of the Ice Ages. Changes in the Earth's orbital shape or eccentricity, and variations in its axial tilt and wobble, control the timing of global cooling and warming, and the advance and retreat of glaciers. They alter the angles and distances from which solar radiation reaches Earth. The global ebb and flow of ice is inferred from the presence of lighter and heavier forms, or isotopes, of oxygen in the fossilized shells of tiny marine animals found in deep-sea sediments. During glaciations, the shells are enriched with oxygen-18 because oxygen-16, a lighter form, is trapped in glacial ice. The changing ratios of the two forms allow researchers to infer the change in global ice volume over the past 500 thousand years. Cold, glacial periods are interspersed with relatively brief, warm interglacials every 100,000 years or so. The relative abundance of oxygen 18 and oxygen 16 (*top*) is compared with periodic 41,000-year variations in the tilt of the Earth's axis (*middle*) and in the shape, or eccentricity – the longer 100,000-year variation, and wobble, or precession – the shorter 23,000-year variation, of the Earth's orbit (*bottom*). The three overlapping astronomical cycles control the timing of the ice ages by altering the amount and distribution of sunlight at the Earth. Such a comparison was first performed by J. E. Hays, J. Imbrie and N. J. Shackleton in 1976. Similar conclusions are obtained in analyzing the oxygen isotope ratios in air bubbles trapped in Antarctica ice cores (J. R. Petit, et al. 1999)

used to infer the proportion of the world's water frozen within the glacial ice sheets.

In a key paper, published in 1976 and entitled "Variations in the Earth's Orbit: Pacemaker of the Ice Ages", James D. Hays, John Imbrie and Nicholas J. Shackleton used the oxygen abundance ratio in cores of deep-sea sediments to work out a timetable for the glacial ice flows. They showed that the amount of ice rose and fell with all three astronomical cycles during the past 500,000 years. Major climatic changes repeated every 41,000 and 100,000 years, just as Milankovitch had predicted. The greatest switch from cold to warm periods, and back to cold again, occurs roughly every 100,000 years. The deep-sea sediment record also included two periods of 19,000 and 23,000 years instead of the single predicted wobble cycle, but in the following year André Berger used refined wobble, or precession, calculations to confirm the astronomical cause for the twin cycle that had been observed.

Cores extracted from the glacial ice in Greenland and Antarctica provide the longest natural archive of the Earth's past climate. Representative examples are the ice cores drilled at Camp Century, Greenland as early as 1966, and those extracted from the Vostok station in Antarctica since 1982, reaching a record depth of 3,623 meters in 1998. The deepest ice core records have taken us back 420,000 years covering four glacial-interglacial cycles. These ancient dates are obtained by counting annual layers of ice, like growth rings of a tree, and on ice flow modeling. More precise, recent dates are determined from seasonal ice variations, supplemented by well-dated volcanic eruptions whose debris is found in the ice.

The record of changing deuterium content along an ice core is used to describe local temperature variations in the atmosphere. Deuterium, denoted by both D and ^{2}H, is a rare heavy form of hydrogen. The changing ratio of heavy hydrogen and normal light hydrogen, symbolized by ^{1}H, acts like a fossil thermometer, measuring the temperature in the air where the snow formed as well as at the surface where it fell. A high ratio corresponds to warm temperatures, and a low value signifies a colder local environment.

The fluctuating temperatures recorded in the deepest Vostok ice core have confirmed the deep-sea evidence for the cause of the ice ages during the past 420,000 years (Fig. 7.23). The greatest temperature change at Antarctica is the periodic 100,000-year, orbital-eccentricity switch between a cold glacial and a warm interglacial, with a total temperature change of about 5 degrees Celsius for the last change between them. A smaller temperature change occurs at 41,000-year intervals, corresponding to the axial-tilt periodicity. Although the shorter wobble periodicity of about 23,000 years does not account for much in the Antarctica temperature record, it is a prominent feature in the ice core's oxygen isotope record that reflects changes in global ice volume (Fig. 7.22). The ice-core data therefore strongly supports the idea that changes in the Earth's orbit and spin axis cause variations in the intensity and distribution of sunlight arriving at Earth, which in turn initiate natural climate changes and trigger the ebb and flow of glacial ice.

Air trapped in the polar ice cores constitutes an archive for reconstructing the relation between climate and greenhouse gases in the past (Fig. 7.23).

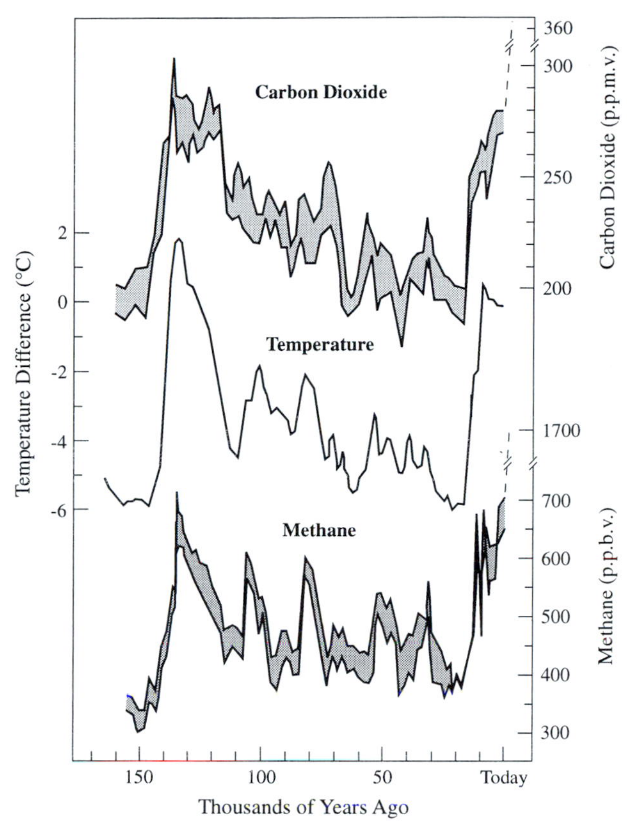

Fig. 7.23. Ice Age Temperatures and Greenhouse Gases. Measurements along the 2,083-meter Vostok ice core indicate that changes in the atmospheric temperature over Antarctica closely parallel variations in the atmospheric concentrations of two greenhouse gases, carbon dioxide and methane, for the past 160,000 years. When the temperature rises, so does the atmospheric concentration of these two greenhouse gases, and *vice versa*. This strong correlation has been extended by a deeper Vostok ice core, to 3,623 meters in depth and the past 420,000 years (J. R. Petit et al., 1999). Major transitions between the coldest and hottest air temperatures, of roughly 5 degrees Celsius during each of the last four glacial-interglacial cycles, are accompanied by increases in atmospheric concentrations of carbon dioxide, rising from 180 to 280–300 p.p.m.v. (parts per million by volume) and in atmospheric methane, increasing from 220–350 to 650–770 p.p.b.v. (parts per billion by volume). This high concentration indicates that carbon dioxide and methane may have contributed to the glacial-interglacial changes by amplifying orbital forcing of climate change. The ice core data does not include the past 200 years, shown as dashed lines at the right. The present-day levels of carbon dioxide and methane (about 360 p.p.m.v. and 1,700 p.p.b.v.) are unprecedented during the past four 100,000-year glacial-interglacial cycles. [Adapted from C. Lorius et al., 1990 – also see J. M. Barnola et al. (1987) and J. Jouzel et al. (1987)]

Microscopic air bubbles have been captured in the light and fluffy snow before it turned to ice at the poles, preserving pristine examples of the ancient atmospheric gas. The entrapped air contains greenhouse gases, carbon dioxide, written CO_2, and methane, CH_4, as well as a record of the heavy sort of oxygen, the ^{18}O, that reflects changes in global ice volume.

The deepest Vostok ice core indicates that the Antarctica air-temperature changes are associated with varying concentrations of atmospheric carbon dioxide and methane throughout the past 420,000 years. It demonstrates a remarkable similarity between the temporal variations in the local climate, or temperature, and in the important greenhouse gases, carbon dioxide and methane. The temperatures go up whenever the levels of carbon dioxide and methane do, and they decrease together as well.

Scientists cannot however, yet agree whether the increase in greenhouse gases preceded or followed the rising temperatures. Some investigators conclude that the elevated levels of carbon dioxide follow the warming in Antarctica by about 600 years during the onset of the last three warm interglacials. The rising temperatures might cause carbon dioxide to be released from the ocean water, and it could be absorbed back into the ocean when the climate cools. Methane is locked up in frozen wetlands during the glacial cold, but in warm times it is liberated into the atmosphere as the wetlands thaw.

Other researchers conclude, from the same Vostok ice-core data, that one cannot accurately date the onsets, and that increased levels of carbon dioxide and methane might be causing the temperatures to go up by the natural greenhouse effect. The interpretation is clouded by uncertainties in timing the changes, which often cannot be specified with an accuracy better than 1,000 years. Within this uncertainty, we can probably only conclude that the temperature and greenhouse gases increase at about the same time.

The carbon dioxide level does increase when the methane does, and with synchronism between the Northern and Southern Hemisphere, indicating a pervasive, global influence. Moreover, the ice-core data definitely show that both the Antarctica air temperature and the atmospheric greenhouse-gas concentrations increase several thousands of years before the ice volume is reduced.

The natural increase in carbon dioxide apparently resolves a difficulty in explaining the ice ages by the astronomical cycles. Although the largest climate variations occur every 100,000 years, the corresponding rhythmic stretching of the Earth's orbit is far too small to directly create the observed temperature changes. Calculations indicate that the associated changes in the distribution and amount of sunlight received at Earth, cannot, by themselves, alter the climate enough. In fact, the shorter astronomical cycles have a greater direct effect on the change in sunlight. So, the 100,000-year orbital change has to be leveraged by some other factor and a favored explanation is that the build up of greenhouse gas amplifies effects triggered and timed by the astronomical rhythm.

Other suggested explanations that might enhance the weak orbital effects include non-linear mechanisms associated with ice-sheet growth and decay or with changes in the flow of air and sea. They might be able to create a large climate change from a relatively small change in received sunlight. After all, the glaciers and recurrent ice ages have only been around for a million years or so, and the Earth has been free of ice over most of the rest of its 4.5-billion-year history. The cyclic variations in incident sunlight must have been present during all that time, but the land and oceans were being reshaped by drifting, colliding and broken continents.

For now, the overall scenario deduced from the Antarctica ice-core data is that variations in the intensity and distribution of solar radiation arriving at Earth initiate the changes, permitting our world to snap out of long glacial epochs. An increase in greenhouse gases probably amplifies this effect, producing more warming than would otherwise be possible. Melting of the large ice sheets in the Northern Hemisphere would have reduced the amount of sunlight reflected back into space, and further enhanced the warming.

Perhaps the most startling outcome of the ice-core studies is the realization that human civilization has grown so large and powerful that it can dominate just about anything in the natural world, including the Earth's atmosphere. The concentrations of carbon dioxide and methane have now risen to unprecedented levels in our air, vastly exceeding those at any time during the past 420,000 years. Although the levels of these greenhouse gases have fluctuated up and down every 100,000 years or so, the changes have always been within stable bounds, at least during the last 420,000 years. Now mankind has pushed the world well beyond these limits, primarily by burning coal, oil and natural gas.

The amounts of carbon dioxide in the deep ice cores (Fig. 7.23) have never exceeded 300 p.p.m.v. (parts per million by volume), and those for methane have never surpassed 780 p.p.b.v. (parts per billion by volume). The pre-industrial levels of about 280 p.p.m.v. and 650 p.p.b.v. are therefore within the bounds of previous observed interglacial periods. In contrast, the concentrations of these two gases were 360 p.p.m.v and 1,700 p.p.b.v., respectively, at the onset of the twenty-first century. These are enormous increases, comparable to any detected natural change, and elevated way beyond anything observed before. The uniquely elevated concentrations of these gases are of relevance to the continued debate about the Earth's future climate (Focus 7.5), especially when we consider that these greenhouse gases seem to have contributed significantly to recent climate changes.

Still, when stepping back and viewing the Earth on a cosmic time scale of billions of years, we realize that the Sun may continue to play a dominant role in terrestrial change, for the solar constant has been steadily increasing as the result of the Sun's long, slow evolution. Well-accepted models of stellar evolution, begun in the 1950s, indicate that the Sun has slowly grown in luminous intensity with age, rising steadily in brightness by about 30 percent from the time of formation of the Sun to the present. In other words, the Sun began its life about 4.5 billion years ago shining with about 70 percent of the brightness it has today, and it has been slowly increasing in brightness ever since.

The Sun's steady inexorable brightening is a consequence of the conversion of hydrogen into helium in its core. As the amount of helium increases, so does the density, producing a higher core temperature to accommodate the composition change and still support the weight of overlying material. The extra heat increases the rates of the nuclear reactions that make the Sun shine, producing the gradual increase in its luminosity.

To put this evolutionary brightening in perspective, it amounts to a 1 percent increase in the solar radiative energy output in about 150 million years, or only 0.0000023 (2.3×10^{-6}) percent during the past 350 years. There is no way that this small change will ever be directly measured. In contrast, the observed solar constant moved up and down by 0.1 percent during the past two 11-year activity cycles, and estimates suggest that the Sun has grown brighter by 0.25 percent over the past 350 years due to increased solar magnetic activity. Changes in Earth's orbital parameters produce even larger variations in the amount of solar radiation received at some places on Earth over thousands of years, up to 30 percent at some latitudes.

Nevertheless, the slow, gradual evolutionary brightness increase ought to have important terrestrial consequences over cosmic periods of time. For every 1 percent increase in the Sun's irradiance, there should be a global increase in the Earth's surface temperature of approximately 2 degrees Centigrade, so the Sun should have turned the heat up by about 60 degrees over the past 4.5 billion years.

In 1972, Carl Sagan and George Mullen showed that a dim young Sun is in conflict with the temperature history of the Earth. Assuming an unchanging atmosphere, with the same composition and reflecting properties as today, the decreased solar luminosity would have caused the Earth's global surface temperature to drop below the freezing point of water about 2 billion years ago. This contradicts geological evidence that indicates the Earth never was this cold. Sedimentary rocks, which must have been deposited in liquid water, date from 3.8 billion years ago, and there is fossil evidence in these rocks for the emergence of life at least 3.5 billion years ago. Moreover, if the oceans ever did freeze completely over, they would be unlikely to melt. The high reflecting power of ice would require a solar constant higher than the present value to thaw out a completely glaciated Earth.

The discrepancy between the Earth's warm climatic record and an initially dimmer Sun has come to be known as the faint-young-Sun paradox. It can be resolved if the Earth has a long-lasting climate control system that maintained relatively constant surface temperatures throughout the four billion years of recorded geological history. Although we aren't certain about just how it happened, some global thermostat must have kept the planet from freezing over early in its history, and counteracted ever since the tendency of the surface temperature to rise as the Sun grew brighter and hotter.

A possible solution of the faint-young-Sun paradox is a substantially stronger atmospheric greenhouse in the past, that has slowly weakened over time. If the Earth's primitive atmosphere contained a thousand times more carbon dioxide than it does now, the greater heating of the enhanced greenhouse effect could have kept the oceans from freezing. The Earth could only maintain a temperate climate by turning down its greenhouse effect as the Sun grew warmer and turned up the Earth's heat. So, if this explanation is correct, our planet's atmosphere, rocks, and ocean combined to decrease the amount of carbon dioxide over time. The recent increase in atmospheric carbon dioxide caused by the burning of fossil fuels is currently negligible compared to this hypothetical long-term depletion.

An alternative explanation of the paradox involves the ancient oceans, during the first 1.5 billion years on Earth, and the regulatory effects of plants and animals thereafter. The very young Earth contained little dry land, and the greater ocean surface would have absorbed more of the incoming solar radiation than it does today. Then, for the past three billion years, plants and animals could have developed the capability to control the environment, transforming the atmosphere and regulating the surface temperature. According to this *Gaia* hypothesis, developed by James E. Lovelock and Lynn Margulis in the 1970s, it is life that continues to control the environment, making it comfortable for living things in spite of adverse physical and chemical change. For example, there was little or no oxygen in our atmosphere billions of years ago, but oxygen now makes up about a fifth of our air. If plants did not con-

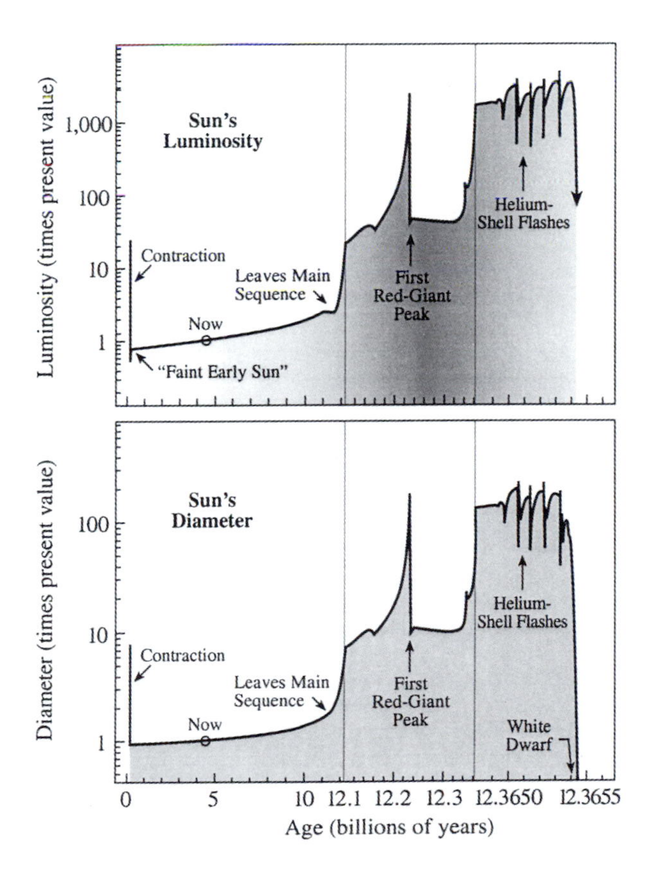

Fig. 7.24. Fate of the Sun. In about 8 billion years the Sun will become much brighter (*top*) and larger (*bottom*). Note the different time scales, expanded near the end of the Sun's life to show relatively rapid changes. [Courtesy of I-Juliana Sackmann and Arnold I. Boothroyd, adapted from K. R. Lang, 1999; also see Sackmann, Boothroyd and Kraemer (1993)]

tinuously replenish the oxygen, animals that breathe oxygen would use it all up.

In the end, there may be no escape from the Sun's control over life on Earth. From both astrophysical theory and observations of other stars, we know that the Sun will grow enormously in size and luminosity billions of years from now (Fig. 7.24). Astronomers calculate that the Sun will be hot enough in 3 billion years to boil the Earth's oceans away, leaving the planet a burned-out cinder, a dead and sterile place. Four billion years thereafter, our star will balloon into a giant star, engulfing the planet Mercury and becoming 2,000 times brighter than it is now. Its light will be intense enough to melt the Earth's surface, and to turn the icy moons of the giant planets into globes of liquid water. The only imaginable escape would then be interplanetary migration to distant moons or planets with a warm, pleasant climate.

Table 7.1. Key events in the discovery of solar-terrestrial interactions*

Date	Event
1600	William Gilbert, physician to Queen Elizabeth I of England, publishes a small treatise demonstrating that "the terrestrial globe is itself a great magnet".
1610–13	Galileo Galilei first systematically studies sunspots through a telescope.
1716	Edmund Halley suggests that aurora rays delineate terrestrial magnetic field lines, and that the auroras are due to a magnetized fluid that circulates poleward along the Earth's dipole magnetic field lines.
1724	George Graham discovers large, irregular fluctuations in compass needles, later called magnetic storms; they were seen at about the same time by Anders Celsius – see Celsius (1747).
1733	Jean Jacques d'Ortous de Mairan argues for a connection between the occurrence of auroras and sunspots.
1799–1804	Alexander von Humboldt makes regular, precise measurements of the strength, dip and inclination of the Earth's magnetic field during his voyage to South America. The term "magnetic storm" came into common usage as the result of Humboldt's scientific analyses and reports.
1838	Carl Friedrich Gauss publishes a mathematical description of the Earth's dipolar magnetic field, using it with observations to show that the magnetism must originate deep down inside the Earth's core.
1842	Joseph Alphose Adhémar suggests that the ice ages might be caused by variations in the way the Earth moves around the Sun, basing his theory on changes in the tilt of the Earth's rotation axis over long periods of time.
1844	Samuel Heinrich Schwabe demonstrates that the number of sunspots varies from a maximum to a minimum and back to a maximum again in a period of about 11 years.
1852	Edward Sabine demonstrates that global magnetic disturbances of the Earth, now called geomagnetic storms, vary in tandem with the 11-year sunspot cycle.
1859–60	In 1859 Richard C. Carrington and Richard Hodgson independently observe a solar flare in the white light of the photosphere and in 1860 publish the first account of such a flare. Seventeen hours after the flare a large magnetic storm begins on the Earth.

1875	James Croll publishes his theory of climate change based on long-term changes in the shape of the Earth's orbit and the wobble of its spin axis, arguing that ice ages develop when winters are colder than average.
1887–89	Gustav Friedrich Wilhelm Spörer draws attention to a long, continued "damping-down" in solar activity between 1645 and 1715, when almost no sunspots were seen.
1892	William Thomson (Baron Kelvin) shows that geomagnetic storms cannot be due to the direct magnetic action of the Sun, arguing that the "supposed connection between magnetic storms and sunspots is unreal".
1892, 1900	George Francis Fitzgerald and Oliver Lodge independently suggest that terrestrial magnetic disturbances might be due to electrified particles emitted by the Sun.
1896–1913	Kristian Birkeland argues that polar auroras and magnetic activity are due to beams of electrons from the Sun.
1899–1902	Guglielmo Marconi successfully sends radio signals across the Atlantic Ocean. Arthur E. Kennelly and Oliver Heaviside independently postulate the existence of an electrically conducting atmospheric layer, now called the ionosphere, to explain Marconi's radio transmission. The radio waves found their way around the curved Earth because they were reflected from the ionosphere.
1905	Edward Walter Maunder shows that geomagnetic storms tend to recur at 27-day intervals, the rotation period of the Sun (relative to Earth) at low latitudes, and argues that the recurrent storms are due to narrow streams emanating from active areas on the Sun.
1908	George Ellery Hale measures intense magnetic fields in sunspots, thousands of times stronger than the Earth's magnetism, but points out that the sunspot magnetic fields are still inadequate to account for geomagnetic storms by direct magnetic action.
1911	Arthur Schuster shows that a beam of electrons from the Sun cannot hold itself together against the mutual electrostatic repulsion of the electrons.
1919	Frederick Alexander Lindemann (later Lord Cherwell) suggests that an electrically neutral plasma ejection from the Sun is responsible for non-recurrent geomagnetic storms.
1920–41	Milutin Milankovitch describes how the major ice ages might be produced by rhythmic fluctuations in the wobble and tilt of the Earth's rotational axis and the shape of the Earth's orbit with periods of 23, 41 and 100 thousand years. They cause cold northern summers that prevent winter snow from melting and produce the ice ages.
1922	Edward Walter Maunder provides a full account of the 70-year dearth of sunspots, from 1645 to 1715, previously noticed by Gustav Friedrich Wilhelm Spörer in 1887–89. This interruption in the normal sunspot cycle is now referred to as the Maunder Minimum.
1925	Edward Victor Appleton, and his research student M. A. F. Barnett, use radio transmissions in the United Kingdom to verify the existence of the electrically-conducting ionosphere, or the Kennelly-Heaviside layer as it was subsequently known. This was confirmed in 1926 by Gregory Breit and M. A. Tuve in America. A height of about 100,000 meters was inferred for the radio reflecting layer by measuring the time delay between the transmission of the radio signal and the reception of its echo.
1929	William M. H. Greaves and Harold W. Newton distinguish between great, non-recurrent geomagnetic storms and smaller ones with a 27-day recurrence.
1931–40	Sydney Chapman and Vincent C. A. Ferraro propose that a magnetic storm is caused when an electrically neutral plasma cloud ejected from the Sun envelops the Earth.

1931, 1943 George Ellery Hale and H. W. Newton present evidence that great geomagnetic storms are associated with solar flares observed with the spectrohelioscope.

1935–37 J. Howard Dellinger suggests that the sudden ionosphere disturbances that interfere with short-wave radio signals have a solar origin.

1946, 1950 Scott E. Forbush and his colleagues describe flare-associated transient increases in the cosmic ray intensity at the Earth's surface, and attribute them to very energetic charged particles from the Sun.

1951–52 Herbert Friedman and his colleagues at the U. S. Naval Research Laboratory use instruments aboard sounding rockets to show that the Sun emits enough X-ray and ultraviolet radiation to create the ionosphere.

1951–57 Ludwig F. Biermann argues that a continuous flow of solar corpuscles is required to push comet ion tails into straight paths away from the Sun, correctly inferring solar wind speeds as high as 0.5 to 1.0 million meters per second.

1954 Philip Morrison proposes that magnetized clouds of gas, emitted by the active Sun, account for world-wide decreases in the cosmic-ray intensity observed at Earth, lasting for days and correlated roughly with geomagnetic storms.

1954–58 Scott E. Forbush demonstrates the inverse correlation between the intensity of cosmic rays arriving at Earth and the number of sunspots over two 11-year solar activity cycles.

1955–66 Cesare Emiliani shows that cyclic oxygen isotopic variations in deep-sea sediments can record the ebb and flow of the ice ages over the past 700 thousand years. Although he initially thought the data recorded temperature variations, Nicholas J. Shackleton subsequently showed that it records changes in the global ice volume.

1958 Eugene N. Parker suggests that a perpetual supersonic flow of electric corpuscles, that he called the solar wind, naturally results from the expansion of a very hot corona. He also demonstrates that the solar magnetic field will be pulled into interplanetary space, modulating the amount of cosmic rays reaching Earth and attaining a spiral shape in the plane of the Sun's equator due to the combined effects of the radial solar wind flow and the Sun's rotation.

1958–81 Stellar evolution theory is used by several different authors, including Douglas O. Gough, C. B. Haselgrove, Fred Hoyle, Martin Schwarzschild, and Roger K. Ulrich, to reliably predict that the solar luminosity has risen steadily by about 30 percent during the 4.5 billion-year life of the Sun. Schwarzschild speculates in 1958 that the changing solar brightness might have detectable geological and geophysical consequences.

1958–59 The first American satellite, Explorer 1, was launched into orbit on 1 February 1958, followed by Explorer 3 on 26 March 1958. Instruments aboard these spacecraft, provided by James A. Van Allen and colleagues, discovered belts of charged particles that girdle the Earth's equator.

1958, 1965 Wolfgang Gleissberg finds an 80-year cycle in the record of the number of sunspots, and subsequently in the frequency of auroras.

1959 Thomas Gold suggests that geomagnetic storms are caused by a shock front associated with magnetic clouds ejected from the Sun, and coins the term magnetosphere for the region in the vicinity of the Earth in which the Earth's magnetic field dominates all dynamical processes involving charged particles.

1959 Edward P. Ney argues that the 11-year, solar-cycle modulation of cosmic rays may produce a climate effect in the Earth's lower atmosphere, the troposphere, by producing enhanced ionization, more stormy weather, greater cloud cover and reduced ground temperatures at times of activity minimum and *vice versa*.

1961	James W. Dungey proposes a mechanism for transmitting solar wind energy to the magnetosphere by direct magnetic linkage or merging between the interplanetary and terrestrial magnetic fields in a process now known as magnetic reconnection. In 1946–48 Ronald G. Giovanelli developed a theory of solar flares involving magnetic neutral points, and in 1953 Dungey had shown how magnetic reconnection might energize solar flares.
1961	William Ian Axford and Colin O. Hines raise the possibility that the magnetosphere is energized by fluid friction at its boundary with the solar wind.
1961	A comparison is made by Minze Stuiver between variations in sunspot activity and fluctuations in the radiocarbon, or carbon 14, concentration during the past 13 centuries, with suggestions of some correspondence between the two. The larger the number of sunspots the greater the depression of cosmic-ray intensity in the higher atmosphere with a corresponding decrease in radiocarbon.
1962–67	Mariner 2 was launched on 7 August 1962. Using data obtained during Mariner's voyage to Venus, Marcia Neugebauer and Conway W. Snyder demonstrate that a low-speed solar wind plasma is continuously emitted by the Sun, and discover high-speed wind streams that recur with a 27-day period within the orbital plane of the planets.
1962–64	An interplanetary shock associated with solar activity is detected using instruments aboard the Mariner 2 spacecraft in 1962, reported by Charles P. Sonett, David S. Colburn, Leverett Davis, Jr., Edward J. Smith and Paul J. Coleman, Jr. in 1964.
1964	Syun-Ichi Akasofu develops the notion of a magnetosphere substorm.
1964–68	Norman F. Ness and colleagues use instruments aboard NASA's Interplanetary Monitoring Platform 1, or IMP-1, launched on November 27, 1963, to detect a large bow shock formed in the solar wind ahead of the magnetosphere, and a long magnetic tail on the night side of the Earth. An extended magnetotail was previously measured by Charles P. Sonett in 1960 and Edward J. Smith in 1962 using instruments on Explorer 6 and 10.
1965	Hans E. Suess discovers an approximate 200-year periodicity in the amount of radioactive carbon 14 within tree rings back to 5,000 years BC.
1966	Donald H. Fairfield and Larry J. Cahill, Jr. argue that geomagnetic activity is greatest when the interplanetary magnetic field is southward, leading to an enhanced merging rate between it and the geomagnetic fields. Also see the 1969 paper by Joan Hirshberg and David S. Colburn and the 1975 paper by Rande K. Burton, Robert L. Mc Pherron and Christopher T. Russell.
1966–78	Olin C. Wilson measures the long-term magnetic activity of roughly 100 Sun-like stars by way of variations in their violent emission lines of singly ionized calcium – the H and K lines at 393.4 and 396.8 nanometers.
1972	Carl Sagan and George Mullen show that the faint brightness of the young Sun is in conflict with the temperature history of the Earth, and suggest compensating terrestrial action by modification of the Earth's early atmosphere. With an unchanging atmosphere, the Earth's oceans would have frozen over about 2 billion years ago, which is in conflict with geological evidence for liquid water on Earth more than 3 billion years ago.
1973	James E. Lovelock and Lynn Margulis suggest the *Gaia* hypothesis in which plants and animals have developed the capability to control their environment, keeping it comfortable for living things in spite of threatening changes.
1974	John Thomas Gosling and colleagues report observations of coronal mass ejections, or CMEs, then called coronal disturbances or coronal transients, with the coronagraph aboard Skylab. They found that many CMEs are not associated with solar flares and that some CMEs have the high outward speed of up to a million meters per second needed to produce interplanetary shocks.

1975–77 A meeting held at the California Institute of Technology to discuss the solar constant and the Earth's atmosphere stimulates interest in spacecraft measurements of the solar irradiance of Earth. The results of these discussions were published in a book edited by Oran R. White in 1977, and contributed to the inclusion of the Active Cavity Radiometer Irradiance Monitor (ACRIM) aboard the Solar Maximum Mission (SMM) satellite.

1976 John A. Eddy shows that the prolonged sunspot minimum from 1645 to 1715 coincided with a decrease in solar activity, as characterized by a marked absence of terrestrial auroras, an abnormally high terrestrial carbon 14 abundance, and exceptionally low temperatures on Earth.

1976 Edward J. Smith and John H. Wolfe use Pioneer 10 and 11 data to investigate Co-rotating Interaction Regions, or CIRs, in the interplanetary medium. They were first detected during studies of the geomagnetic field by Charles P. Sonett and colleagues about 1970. CIRs are the probable source of the 27-day repetition period of the so-called substorm, but there is still controversy about this conclusion.

1976 James D. Hays, John Imbrie and Nicholas J. Shackleton use an analysis of oxygen isotopes in deep-sea sediments to show that the major ice ages during the past half million years recurred at intervals of 19, 23, 41 and 100 thousand years, with a dominant 100-thousand-year recurrence. This supported the idea that the ice ages are caused by a variation in the intensity and distribution of solar energy arriving at the Earth. The double wobble period of 19 and 23 thousand years was explained by refined astronomical calculations by André Berger in the following year.

1977 Michael J. Newman and Robert T. Rood suggest that the faint-young-Sun paradox could be resolved by a strong greenhouse effect in the past. This was considered by Carl Sagan and Christopher Chyba in 1997. Also see 1972.

1980, 1984 Minze Stuiver and Paul D. Quay attribute the changing atmospheric carbon 14 to a variable Sun, and Charles P. Sonett provides further evidence for the 200-year periodicity in the radiocarbon data.

1981 Richard C. Willson, Samuel Gulkis, Michael Janssen, Hugh S. Hudson, and Gary A. Chapman report high-precision measurements of variations in the total solar irradiance of the Earth, or the solar constant, with an amplitude of up to 0.2 percent, made with the Active Cavity Radiometer Irradiance Monitor (ACRIM) on the Solar Maximum Mission satellite.

1981 Jo Ann Joselyn and Patrick S. Mc Intosh make a convincing case that large magnetic storms can occasionally be associated with disappearing solar filaments. Filament disruptions were previously found to be associated with coronal mass ejections.

1985–99 Ice cores drilled in Greenland and Antarctica are used to determine the local temperature and atmospheric carbon dioxide and methane content in the polar regions for up to 420,000 years ago, including four glacial-interglacial cycles. This data confirmed the evidence from deep-sea sediments for initiation of the major ice ages every 100 thousand years by changes in the intensity and distribution of sunlight arriving at Earth. The ice-core data also showed that the climate repeatedly warms and cools when the amounts of carbon dioxide and methane increase or decrease, respectively. (See papers with first authors of J. Chappellaz, C. Genthon, C. Lorius, J. R. Petit, and D. Raynaud.)

1986–88 Louis A. Frank and his colleagues report images of the entire aurora oval from space using the Dynamics Explorer 1 satellite.

1987 Hilary V. Cane, Neil R. Sheeley Jr. and Russell A. Howard show that strong interplanetary shocks are associated with fast coronal mass ejections that move at speeds greater than 500 thousand meters per second. They used data from the third International Sun-Earth Explorer (ISEE-3) low-frequency (< 1 MHz) radio instrument and the Solwind coronagraph.

1987 Bruce T. Tsurutani and Walter D. Gonzalez show that Alfvén waves in the high-speed (coronal hole) solar wind streams can cause periodic magnetic reconnection, frequent magnetic substorms and aurora activity during the declining phase of the 11-year solar cycle, near solar minimum. They used data gathered in 1978–79 by the third International Sun-Earth Explorer (ISEE-3).

1987, 1991 George C. Reid demonstrates that the globally averaged sea surface temperature over the past 130 years shows a very significant correlation with the envelope of the 11-year sunspot cycle. He explained the correlation by a variation in the Sun's total irradiance of Earth (the solar constant) over the same time interval.

1988–93 Karin Labitzke and Harold Van Loon discover an association between the variability of the Sun and winds in the middle atmosphere. The winter storms in the stratosphere follow an 11-year pattern of low-pressure systems over the North Atlantic Ocean, matching the solar cycle in both period and phase.

1988–92 Richard C. Willson and Hugh S. Hudson use radiometric data taken with the Active Cavity Radiometer Irradiance Monitor (ACRIM) instrument on the Solar Maximum Mission (SMM) to demonstrate that the total solar irradiance of Earth (the solar constant) varies in step with the 11-year cycle of solar activity. It has a total decline and rise of about 0.1 percent. They also used ACRIM data to show that the intense magnetic fields of sunspots produce a short-term (days) decrease of several tenths of a percent in the solar constant when the sunspots cross the visible solar disk. Douglas V. Hoyt, John R. Hickey and their colleagues confirmed this solar cycle variation in the solar constant using data from the Earth Radiation Budget (ERB) experiment on the Nimbus 7 satellite from 1978 to 1991.

1990 Peter V. Foukal and Judith Lean show that the observed changes of the total solar irradiance of the Earth (the solar constant) during the previous 11-year cycle of magnetic activity can be explained by the increased emission of photosphere magnetic structures, called faculae. Their increased area and brightening near solar activity maximum exceed the irradiance reductions caused by dark sunspots whose numbers also increase in step with solar activity.

1990 Richard R. Radick, G. W. Lockwood and Sallie L. Baliunas show that main-sequence stars similar to the Sun become brighter as their magnetic activity level increases. Baliunas and Robert Jastrow speculate that the long-term brightness changes of Sun-like stars indicate that the Sun could undergo substantial luminosity variations on time scales of centuries. They could exceed by a factor of four or five the 0.1 percent change in the Sun's total irradiance of the Earth (the solar constant) observed during the previous 11-year solar cycle.

1990–93 John T. Gosling and his colleagues argue that large, non-recurrent geomagnetic storms are caused by interplanetary disturbances driven by coronal mass ejections.

1991, 1994 Eigil Friis-Christensen and Knud Lassen find a high correlation between the variable period of the "11-year" sunspot cycle and the long-term variations of the land air temperature in the Northern Hemisphere for the past 130 years, and perhaps for the past five centuries.

1992, 1993 1995 Steven W. Kahler and John T. Gosling reason that large, non-recurrent geomagnetic storms, interplanetary shock waves and energetic interplanetary particle events are all mainly due to coronal mass ejections rather than solar flares. Hugh Hudson, Bernhard Haisch and Keith T. Strong argue that both flares and coronal mass ejections result from solar eruptions that can have terrestrial consequences.

B. Ulysses Principal Investigators and Their Institutions

	Principal Investigator	Institution
COSPIN	R. Bruce McKibben	University of Chicago, Illinois
DUST	Eberhard Grün	Max-Planck-Institut für Kernphysik, Heidelberg, Germany
EPAC	Erhard Keppler	Max-Planck-Institut für Aeronomie, Katlenburg-Lindau, Germany
GRB	Kevin C. Hurley	University of California, Berkeley
GWE	Bruno Bertotti	Universitá di Pavia, Italy
HISCALE	Louis J. Lanzerotti	Bell Laboratories, Lucent Technologies, Murray Hill, New Jersey
SCE	Mike K. Bird	Universität Bonn, Germany
SWICS	Johannes Geiss	University of Bern, Switzerland
	George Gloeckler	University of Maryland
SWOOPS	Dave J. McComas	Los Alamos National Laboratory, New Mexico
URAP	Robert J. MacDowall	Goddard Space Flight Center, Greenbelt, Maryland
VHM/FGM	Andre Balogh	Imperial College, London, England

Internet Addresses

I. Fundamental Data

Fundamental physical constants and physical reference data can be found at the URL:

http://phyiscs.nist.gov/

Abstracts and complete articles for many of the references found in this book can be located at the NASA Astrophysics Data System at the URL:

http://adswww.harvard.edu/index.html

II. Information About SOHO, Ulysses and Yohkoh

Information about SOHO can be obtained from the World Wide Web at the URL:

http://sohowww.nascom.nasa.gov/

Collections of images from each of twelve SOHO instrument teams can be obtained (and freely used with appropriate acknowledgment) at the URL:

http://sohowww.nascom.nasa.gov/gallery/

The NASA home page for the Ulysses mission can be found at the URL:

http://ulysses.jpl.nasa.gov/

The European Space Agency, or ESA, home page for Ulysses can be found at the URL:

http://helio.estec.esa.nl/ulysses/welcome.html

The Yohkoh mission, and especially its Soft X-ray Telescope, or SXT, are discussed at the URLs:

http://www.lmsal.com/SXT/

http://isass1.solar.isas.ac.jp/

http://ydac.mssl.ucl.ac.uk/ydac/

http://solar.physics.montana.edu/

An educational Yohkoh Public Outreach Project has the URL:

http://www.lmsal.com/YPOP/

III. Information About Other Solar Missions

The home page for the Transition Region And Coronal Explorer, or TRACE, mission has the URL:

http://vestige.lmsal.com/TRACE/

The home page for the High Energy Solar Spectroscopic Imager, or HESSI, is located at:

http://hesperia.gsfc.nasa.gov/hessi/

or

http://hessi.ssl.berkeley.edu/

IV. Information About NASA's Sun-Earth Connection

Information about NASA research in solar and space physics can be found at the NASA Space Physics Division Sun-Earth Connections home page at the URL:

http://sec.gsfc.nasa.gov/

The National Space Science Data Center maintains information about past, current and future solar missions at the URL:

http://nssdc.gsfc.nasa.gov/solar/

Links to all current, past and future flight projects in space physics, including the interplanetary medium, the Earth's magnetosphere, and related studies of Venus, Jupiter and Saturn can be found at the URL:

http://nssdc.gsfc.nasa.gov/space/

V. Information About Helioseismology

Comprehensive educational outreach information about helioseismology is provided at:

http://bigcat.obs.aau.dk/helio_outreach/english/

The Global Oscillations Network Group, or GONG, project is described at the URL:

http://www.gong.noao.edu/

An on-line educational description of some aspects of solar astronomy, including helioseismology, is provided at the URL:

http://solar-center.stanford.edu/

The home page for the Solar Oscillations Investigation (SOI) Michelson Doppler Imager (MDI) on SOHO is at:

http://soi.stanford.edu/

VI. Other Solar Topics

A review of solar flares is given in

http://hesperia.gsfc.nasa.gov/hessi/

The National Solar Observatory (NSO) home page is located at:

http://www.nso.noao.edu/

The NASA Marshall Space Flight Center Solar Physics Branch is located at:

http://science.msfc.nasa.gov/ssl/PAD/SOLAR/default.htm

VII. Information About Cosmic and Heliospheric Physics

A thorough cosmic and heliospheric learning center is available at:

http://helios.gsfc.nasa.gov/

VIII. Information About Solar Activity and Space Weather

The Space Environment Center (SEC) at the National Oceanic and Atmospheric Administration (NOAA) provides real-time monitoring and forecasting of solar and geophysical events at the URL:

http://solar.sel.noaa.gov/

Information about the Earth's climate can be found at the National Climate Data Center with the URL:

http://www.noaa.gov/

References*

Acton, L. W., et al. (1982): Chromospheric evaporation in a well-observed compact flare. Astrophysical Journal **263**, 409–422.
Acton, L. W., et al. (1992): The Yohkoh mission for high-energy solar physics. Science **258**, 618–625.
Acton, L. W., et al. (1992): The morphology of 20×10^6 K plasma in large non-impulsive solar flares. Publications of the Astronomical Society of Japan **44**, L71–L75.
Acuna, M. H., et al. (1995): The global geospace science program and its investigations. Space Science Reviews **71**, 5–21.
Adhemár, J. A.: Révolutions de la Mer. Paris: Deluges Périodiques 1842.
Aelig, M. R., et al.: Solar wind iron charge states observed with high time resolution with SOHO/CELIAS/CTOF. In: Proceedings of the fifth SOHO workshop. The corona and solar wind near minimum activity. ESA SP-101. Noordwijk, The Netherlands: ESA Publications Division 1997, pp. 157–161.
Akasofu, S.-I. (1964): The development of the auroral substorm. Planetary and Space Science **12**, 273–282.
Akasofu, S.-I.: Polar and magnetospheric substorms. Dordrecht, Holland: Reidel 1968.
Akasofu, S.-I.: Physics of magnetospheric substorms. Dordrecht, Holland: Reidel 1977.
Akasofu, S.-I. (1989): Substorms. EOS **70**, 529–532.
Akasofu, S.-I., Chapman, S.: Solar-terrestrial physics. Oxford, England: Oxford University Press 1972.
Alexander, P. (1992): History of solar coronal expansion studies. EOS, Transactions of the American Geophysical Union **73**, No. 41, 433, 438.
Alfvén, H. (1947): Granulation, magneto-hydrodynamic waves, and the heating of the solar corona. Monthly Notices of the Royal Astronomical Society **107**, 211–219.
Alfvén, H.: Cosmical electrodynamics. Oxford, England: Clarendon Press 1950.
Alfvén, H.: Electric current structure of the magnetosphere. In: Physics of the Hot Plasma in the Magnetosphere (Eds. B. Hultqvist and L. Stenflo). New York: Plenum 1975, pp. 1–22.
Alfvén, H. (1977): Electric currents in cosmic plasmas. Reviews of Geophysics and Space Physics **15**, 271–284.
Alfvén, H., Carlqvist, P. (1967): Currents in the solar atmosphere and a theory of solar flares. Solar Physics **1**, 220–228.
Alfvén, H, Herlofson, N. (1950): Cosmic radiation and radio stars. Physical Review **78**, 616. Reproduced in: A source book in astronomy and astrophysics 1900–1975 (Eds. K. R. Lang and O. Gingerich). Cambridge, Massachusetts: Harvard University Press 1977, pp. 779–781.
Allen, C. W. (1947): Interpretation of electron densities from corona brightness. Monthly Notices of the Royal Astronomical Society **107**, 426–432.
Altschuler, M. D., Newkirk, G. (1969): Magnetic fields and the structure of the solar corona. I. Methods in calculating coronal fields. Solar Physics **9**, 131–149.
Anderson, C. D. (1932): Energies of cosmic-ray particles. Physical Review **41**, 405–421.
Anderson, C. D. (1932): The apparent existence of easily deflectable positives. Science **76**, 238–239.

*Abstracts and complete articles for many of the references found in this book can be located at the NASA Astrophysics Data System at the URL:

http://adswww.harvard.edu/index.html

Anderson, C. D. (1933): The positive electron. Physical Review **43**, 491–494. Reproduced in Hillas (1972).
Anderson, C. D., Neddermeyer, S. H. (1936): Cloud chamber observations of cosmic rays at 4300 meters elevation and near sea-level. Physical Review **50**, 263–271.
Anderson, K. A., Winckler, J. R. (1962): Solar flare X-ray burst on September 28, 1961. Journal of Geophysical Research **67**, 4103–4117.
Anderson, L. S., Athay, R. G. (1989): Chromospheric and coronal heating. Astrophysical Journal **336**, 1089–1091.
Anderson, L. S., Athay, R. G. (1989): Model solar chromosphere with prescribed heating. Astrophysical Journal **346**, 1010–1018.
Anderson, P. C, et al. (1998): Energetic auroral electron distributions derived from global X-ray measurements and comparison with in-situ particle measurements. Geophysical Research Letters **25**, 4105–4108 (1998).
Ando, H., Osaki, Y. (1975): Nonadiabatic nonradial oscillations – an application to the five-minute oscillation of the Sun. Publications of the Astronomical Society of Japan **27**, 581–603.
Antia, H. M. (1998): Estimate of solar radius from f-mode frequencies. Astronomy and Astrophysics **330**, 336–340.
Antia, H. M., Chitre, S. M. (1995): Helioseismic bounds on the central temperature of the Sun. Astrophysical Journal **442**, 434–445.
Antia, H. M., Chitre, S. M., Kale, D. M. (1978): Overstabilization of acoustic modes in a polytropic atmosphere. Solar Physics **56**, 275–292.
Antonucci, E., Dennis, B. R. (1983): Observations of chromospheric evaporation during the Solar Maximum Mission. Solar Physics **86**, 67–77.
Antonucci, E., et al. (1982): Impulsive phase of flares in soft X-ray emission. Solar Physics **78**, 107–123.
Antonucci, E., Gabriel, A. H., Dennis, B. R. (1984): The energetics of chromospheric evaporation in solar flares. Astrophysical Journal **287**, 917–925.
Appleton, E. V. (1932): Wireless studies of the ionosphere. Proceedings of the Institute of Electrical Engineers **71**, 642–650.
Appleton, E. V., Barnett, M. A. F. (1925): Local reflection of wireless waves from the upper atmosphere. Nature **115**, 333–334.
Appleton, E. V., Barnett, M. A. F. (1925): On some direct evidence for downward atmospheric reflection of electric rays. Proceedings of the Royal Society (London) **A109**, 621–641.
Appleton, E., Hey, J. S. (1946): Solar radio noise. Philosophical Magazine **37**, 73–84.
Armstrong, A. H., Harrison, F. B., Heckman, H. H., Rosen, L. (1961): Charged particles in the inner Van Allen radiation belt. Journal of Geophysical Research **66**, 351–361.
Arndt, M. B., Habbal, S. R., Karovska, M. (1994): The discrete and localized nature of the variable emission from active regions. Solar Physics **150**, 165–178.
Arnoldy, R. L. (1971): Signature in the interplanetary medium for substorms. Journal of Geophysical Research **76**, 5189–5201.
Arnoldy, R. L., Kane, S. R., Winckler, J. R. (1967): A study of energetic solar flare X-rays. Solar Physics **2**, 171–178.
Arnoldy, R. L., Kane, S. R., Winckler, J. R. (1968): Energetic solar flare X-rays observed by satellite and their correlation with solar radio and energetic particle emission. Astrophysical Journal **151**, 711–736.
Arrhenius, S. (1896): On the influence of carbonic acid in the air upon the temperature of the ground. Philosophical Magazine and Journal of Science **41**, 237–268.
Aschwanden, M. J., Benz, A. O. (1997): Electron densities in solar flare loops, chromospheric evaporation upflows, and acceleration sites. Astrophysical Journal **480**, 825–839.
Aschwanden, M. J., et al. (1996): Electron time-of-flight measurements during the Masuda flare, 1992 January 13. Astrophysical Journal **464**, 985–998.

Aschwanden, M. J., et al. (1996): Electron time-of-flight distances and flare loop geometries compared from CGRO and Yohkoh observations. Astrophysical Journal **468**, 398–417.

Aschwanden, M. J., et al. (1998): The scaling law between electron time-of-flight distances and loop lengths in solar flares. Astrophysical Journal **470**, 1198–1217.

Aschwanden, M. J., et al. (2000): Time variability of the "quiet" sun observed with TRACE. II. Physical parameters, temperature evolution, and energetics of euv nanoflares. Astrophysical Journal **535**, 1047–1065.

Athay, R. G., Moreton, G. E. (1961): Impulsive phenomena of the solar atmosphere I. Some optical events associated with flares showing explosive phase. Astrophysical Journal **133**, 935–945.

Athay, R. G., White, O. R. (1978): Chromospheric and coronal heating by sound waves. Astrophysical Journal **226**, 1135–1139.

Athay, R. G., White, O. R. (1979): Chromospheric oscillations observed with OSO 8 IV. Power and phase spectra for CIV. Astrophysical Journal **229**, 1147–1162.

Avignon, Y., Martres, M. J., Pick, M. (1966): Etude de la "composante lentement variable" en relation avec la structure des centres d'activité solaire associés. Annales d'Astrophysique **29**, 33–42.

Axford, W. I. (1960): The modulation of galactic cosmic rays in the interplanetary medium. Planetary and Space Science **13**, 115–130.

Axford, W. I. (1962): The interaction between the solar wind and the Earth's magnetosphere. Journal of Geophysical Research **67**, 3791–3796.

Axford, W. I. (1972): The interaction of the solar wind with the interstellar medium. In Solar Wind: Proceedings of the Second International Conference. (Eds. C. P. Sonnett, P. J. Coleman, Jr. and J. M. Wilcox). NASA SP-308., p. 609.

Axford, W. I. (1980): Very hot plasmas in the solar system. Highlights of Astronomy **5**, 351–359.

Axford, W. I. (1985): The solar wind. Solar Physics **100**, 575–586.

Axford, W. I. (1994): The good old days. Journal of Geophysical Research **99**, 19,199–19, 212.

Axford, W. I., Dessler, A. J., Gottlieb, B. (1963): Termination of solar wind and solar magnetic field. Astrophysical Journal **137**, 1268–1278.

Axford, W. I., Hines, C. O. (1961): A unifying theory of high-latitude geophysical phenomena and geomagnetic storms. Canadian Journal of Physics **39**, 1433–1464.

Axford, W. I., Mc Kenzie, J. F.: Solar Wind. In: Cosmic winds and the heliosphere (Eds. J. R. Jokipii, C.P. Sonett, M. S. Giampapa). Tucson, Arizona: University of Arizona Press 1997, pp. 31–66.

Babcock, H. W. (1947): Zeeman effect in stellar spectra. Astrophysical Journal **105**, 105–119.

Babcock, H. W. (1961): The topology of the Sun's magnetic field and the 22-year cycle. Astrophysical Journal **133**, 572–587.

Babcock, H. W., Babcock, H. D. (1955): The sun's magnetic field, 1952–1954. Astrophysical Journal **121**, 349–366.

Bahcall, J. N. (1978): Solar neutrino experiments. Review of Modern Physics **50**, 881–903.

Bahcall, J. N. (1996): Solar neutrinos: Where we are, where we are going. Astrophysical Journal **467**, 475–484.

Bahcall, J. N., Basu, S., Kumar, P. (1997): Localized helioseismic constraints on solar structure. Astrophysical Journal (Letters) **485**, L91-L94.

Bahcall, J. N., Bethe, H. A. (1990): A solution to the solar neutrino problem. Physical Review Letters **65**, 2233–2235.

Bahall, J. N., Davis, R., Wolfenstein, L. (1988): Solar neutrinos: A field in transition. Nature **334**, 487–493.

Bahcall, J. N., et al. (1995): Progress and prospects in neutrino astrophysics. Nature **375**, 29–34.

Bahcall, J. N., et al. (1997): Are standard solar models reliable? Physical Review Letters **78**, 171–174.

Baker, D. N., Carovillano, R. (1997): IASTP and solar-terrestrial physics. Advances in Space Research **20**, 531–538.

Baliunas, S. L.: The past, present and future of solar magnetism: Stellar magnetic activity. In: The sun in time (Eds. C. P. Sonett, M. S. Giampapa, and M. S. Matthews). Tucson, Arizona: The University of Arizona Press 1991, pp. 809–831.

Baliunas, S. L., et al. (1996): A dynamo interpretation of stellar activity cycles. Astrophysical Journal **460**, 848–854.

Baliunas, S. L., Jastrow, R. (1990): Evidence for long-term brightness changes of solar-type stars. Nature **348**, 520–523.

Baliunas, S. L., Soon, W. (1995): Are variations in the length of the activity cycle related to changes in brightness in solar-type stars? Astrophysical Journal **450**, 896–901.

Baliunas, S. L., Vaughan, A. H. (1985): Stellar activity cycles. Annual Review of Astronomy and Astrophysics **23**, 379–412.

Balogh, A. et al. (1995): The heliospheric magnetic field over the south polar region of the sun. Science **268**, 1007–1010.

Bame, S. J., et al. (1974): The quiet corona: Temperature and temperature gradient. Solar Physics **35**, 137–152.

Bame, S. J., et al. (1975): Solar wind heavy ion abundances. Solar Physics **43**, 463–473.

Bame, S. J., et al. (1976): Solar cycle evolution of high-speed solar wind streams. Astrophysical Journal **207**, 977–980.

Bame, S. J., et al. (1977): A search for a general gradient in the solar wind speed at low solar latitudes. Journal of Geophysical Research **82**, 173–176.

Bame, S. J., et al. (1977): Evidence for a structure-free state at high solar wind speeds. Journal of Geophysical Research **82**, 1487–1492.

Bame, S. J., et al. (1993): Ulysses observations of a recurrent high speed solar wind stream and the heliomagnetic streamer belt. Geophysical Research Letters **20**, No. 21, 2323–2326.

Bard, E., et al. (1997): Solar modulation of cosmogenic nuclide production over the last millennium: comparison between ^{14}C and ^{10}Be records. Earth and Planetary Science Letters **150**, 453–462.

Barnes, A. (1975): Plasma processes in the expansion of the solar wind and in the interplanetary medium. Reviews of Geophysics and Space Physics **13**, No. 3, 1049–1053.

Barnes, A. (1992): Acceleration of the solar wind. Reviews of Geophysics **30**, 43–55.

Barnes, A., Gazis, P. R., Phillips, J. L. (1995): Constraints on solar wind acceleration mechanisms from Ulysses plasma observations: The first polar pass. Geophysical Research Letters **22**, No. 23, 3309–3311.

Barnett, T. P. (1989) A solar-ocean relation: fact or fiction? Geophysical Research Letters **16**, 803–806.

Barnola, J. M., et al. (1987): Vostok ice core provides 160,000-year record of atmospheric CO_2. Nature **329**, 408–414.

Bartels, J. (1932): Terrestrial-magnetic activity and its relations to solar phenomena. Terrestrial Magnetism and Atmospheric Electricity **37**, 1–52.

Bartels, J. (1934): Twenty-seven day recurrences in terrestrial magnetic and solar activity, 1923–1933. Terrestrial Magnetism and Atmospheric Electricity **39**, 201–202.

Bartels, J. (1940): Solar radiation and geomagnetism. Terrestrial Magnetism and Atmospheric Electricity **45**, 339–343.

Bartels, J. (1963): Discussion of time-variations of geomagnetic activity indices Kp and Ap, 1932–1961. Annales de Géophysique **19**, 1–20.

Bastian, T. S., Benz, A. O., Gary, D. E. (1998): Radio emission from solar flares. Annual Review of Astronomy and Astrophysics **36**, 131–188.

Basu, S., et al. (1996): The Sun's hydrostatic structure from LOWL data. Astrophysical Journal **460**, 1064–1070.
Beer, J., et al. (1988): Information on past solar activity and geomagnetism from ^{10}Be in the Camp Century ice core. Nature **331**, 675–679.
Beer, J., et al. (1990): Use of ^{10}Be in polar ice to trace the 11-year cycle of solar activity: Information on cosmic ray history. Nature **347**, 164–166.
Belcher, J. W., Davis, L. Jr. (1971): Large-amplitude Alfvén waves in the interplanetary medium, 2. Journal of Geophysical Research **76**, 3534–3563.
Belcher, J. W., Davis, L. Jr., Smith, E. J. (1969): Large-amplitude Alfvén waves in the interplanetary medium: Mariner 5. Journal of Geophysical Research **74**, 2303–2308.
Bell, B., Glazer, H. (1958): Some sunspot and flare statistics. Smithsonian Contributions to Astrophysics **3**, 25–38.
Bentley, R. D., et al. (1994): The correlation of solar flare hard X-ray bursts with Doppler blueshifted soft X-ray flare emission. Astrophysical Journal (Letters) **421**, L55–L58.
Bentley, R. D., Mariska, J. T. (eds.): Magnetic reconnection in the solar atmosphere. San Francisco: Astronomical Society of the Pacific Conference Series **111**, 1996.
Benz, A. O. (1986): Millisecond radio spikes. Solar Physics **104**, 99–110.
Benz, A. O., Bernold, T. E. X., Dennis, B. R. (1983): Radio blips and hard X-rays in solar flares. Astrophysical Journal **271**, 355–366.
Benz, A. O., Csillaghy, A., Aschwanden, M. J. (1996): Metric spikes and electron acceleration in the solar corona. Astronomy and Astrophysics **309**, 291–300.
Benz, A. O., et al. (1981): Solar radio blips and X-ray kernals. Nature **291**, 210–211.
Benz, A. O., et al. (1992): Electron beams in the low corona Solar Physics **141**, 335–346.
Benz, A. O., et al. (1994): Particle acceleration in flares. Solar Physics **153**, 33–53.
Benz, A. O., Krucker, S. (1998): Energy distribution of heating processes in the quiet solar corona. Astrophysical Journal (Letters) **501**, L213–L216.
Benz, A. O., Krucker, S. (1998): Heating events in the quiet solar corona. Solar Physics **182**, 349–363.
Benz, A. O., Krucker, S. (1999): Heating events in the quiet solar corona: multiwavelength correlations. Astronomy and Astrophysics **341**, 286–295.
Berger, A. (1977): Support for the astronomical theory of climatic change. Nature **268**, 44–45.
Berger, A. (1978): Long-term variations of daily insolation and Quaternary climatic changes. Journal of Atmospheric Science **35**, No. 2, 2362–2367.
Berger, A. (1978): Long-term variations of caloric insolation resulting from the earth's orbital elements. Quaternary Research **9**, 139–167.
Berger, A. (1980): The Milankovitch astronomical theory of paleoclimates: A modern review. Vistas in Astronomy **24**, 103–122.
Berger, A. (1988): Milankovitch theory and climate. Review of Geophysics **26**, 624–657.
Berger, A.: Long-term history of climate ice ages and Milankovitch periodicity. In: The Sun in time (Eds. C. P. Sonett, M. S. Giampapa, and M. S. Mathews). Tucson, Arizona: The University of Arizona Press 1991, pp. 498–510.
Bertaux, J. L., et al. (1997): First results from SWAN Lyman-α solar wind mapper on SOHO. Solar Physics **175**, 737–770. Reprinted in: The first results from SOHO (Eds. B. Fleck and Z. Svestka). Boston: Kluwer Academic Publishers 1997.
Bertaux, J. L., et al.: The first 1.5 year of observations from SWAN Lyman-alpha solar wind mapper on SOHO. In: Proceedings of the fifth SOHO workshop. The corona and solar wind near minimum activity. ESA SP-404. Noorwidjk, The Netherlands: ESA Publications Division 1997, pp. 29–36.
Biermann, L. F. (1948): Über die Ursache der chromosphärischen Turbulenz und des UV-Exzesses der Sonnenstrahlung. Zeitschrift für Astrophysik **25**, 161–177.
Biermann, L. F. (1951): Kometenschweife und solare Korpuskularstrahlung. Zeitschrift für Astrophysik **29**, 274–286.

Biermann, L. F. (1953): Physical processes in comet tails and their relation to solar activity. La physique des cometes. IAU Colloquium No. 4, 251–262.

Biermann, L. F. (1957): Solar corpuscular radiation and the interplanetary gas. Observatory **77**, 109–110. Reproduced in: A source book in astronomy and astrophysics, 1900–1975 (Eds. K. R. Lang and O. Gingerich) Cambridge, Massachusetts: Harvard University Press 1979, pp. 147–148.

Biermann, L. F., Haxel, O., Schlüter, A. (1951): Neutrale ultrastrahlung von der sonne. Zeitschrift für Naturforschung **A6**, 47–48.

Bigelow, F. H.: The solar corona discussed by spherical harmonics. Washington, DC: The Smithsonian Institution 1889.

Bigelow, F. H. (1890): The Solar Corona. Sidereal Messenger **9**, 93.

Billings, D. E. (1959): Distribution of matter with temperature in the emission corona. Astrophysical Journal **130**, 961–971.

Birch, A. C., Kosovichev, A. G. (1998): Latitudinal variation of solar subsurface rotation inferred from p-mode frequency splittings measured with SOI-MDI and GONG. Astrophysical Journal (Letters) **503**, L187–L190.

Birkeland, K. (1896): Sur les rayons cathodiques sons l'action de forces magnetiques. Archives des Sciences Physiques et Naturelles **1**, 497.

Birkeland, K.: The Norwegian aurora polaris expedition, 1902–1903, Vol. I., On the cause of magnetic storms and the origin of terrestrial magnetism. Christinania, Denmark: H. Aschehoug & Co. 1908, 1st Section, 1913, 2nd Section.

Blackett, P. M. S., Occhialini, G. P. S. (1933): Some photographs of the tracks of penetrating radiation. Proceedings of the Royal Society (London) **139A**, 699–718.

Blackwell, D. E. (1960): The zodiacal light and its interpretation. Endeavor **19**, 14–19.

Blackwell, D. E., Ingham, M. F. (1961): Observations of the zodiacal light from a very high altitude station. Monthly Notices of the Royal Astronomical Society **122**, 129–141.

Bochsler, P., Geiss, J., Maeder, A. (1990): The abundance of ^{3}He in the solar wind – A constraint for models of solar evolution. Solar Physics **128**, 203–215.

Bohlin, J. D., Sheeley, N. R. Jr. (1978): Extreme ultraviolet observations of coronal holes. Solar Physics **56**, 125–151.

Boischot, A. (1957): Caracteres d'un type d'émission hertzienne associé a certaines éruptions chromosphériques. Comptes Rendus de l'Academie des Sciences **244**, 1326–1329.

Boischot, A. (1958): Etude du rayonnement radioélectrique solaire sur 169 MHz a l'aide d'un grand interférometre a réseau. Annales d'Astrophysique **21**, 273–344.

Boischot, A., Denisse, J.-F. (1957): Les émissions de type IV et l'origine des rayons cosmiques associés aux éruptions chromosphériques. Comptes Rendus de l'Academie des Sciences **245**, 2194–2197.

Bond, G., et al. (1993): Correlations between climate records from north Atlantic sediments and Greenland ice. Nature **365**, 143–147.

Bondi, H. (1952): On spherically symmetrical accretion. Monthly Notices of the Royal Astronomical Society **112**, 195–204.

Bondi, H, Hoyle, F. (1944): On the mechanism of accretion by stars. Monthly Notices of the Royal Astronomical Society **104**, 273–282.

Bone, N.: The aurora, sun-earth interactions. New York: Ellis Norwood 1991.

Bonetti, A., et al. (1963): Explorer 10 plasma measurements. Journal of Geophysical Research **68**, 4017–4062.

Borrini, G., et al. (1982): Helium abundance enhancements in the solar wind. Journal of Geophysical Research **87**, 7370–7378.

Bothe, W., Kolhörster, W. (1929): The nature of the high-altitude radiation. Zeitschrift für Physik **56**, 751–777. Reproduced in English in Hillas (1972).

Bothmer, V., et al. (1996): Ulysses observations of open and closed magnetic field lines within a coronal mass ejection. Astronomy and Astrophysics **316**, 493–498.

Bothmer, V., et al.: Solar energetic particle events and coronal mass ejections: New insights from SOHO. In: 31st ESLAB Symposium. The Netherlands: ESA/ESTEC 1997, pp. 207–216.
Bothmer, V., Schwenn, R. (1998): The structure and origin of magnetic clouds in the solar wind. Annales Geophysicae **16**, 1–24.
Bradley, R. S., Jones, P. D. (1993): "Little Ice Age" summer temperature variations: their nature and relevance to recent global warming trends. Holocene **3**, 367–376.
Bradt, H. L., Peters, B. (1948): Investigation of the primary cosmic radiation with nuclear photographic emulsions. Physical Review **74**, 1828–1837.
Bradt, H. L., Peters, B. (1950): The heavy nuclei of the primary cosmic radiation. Physical Review **77**, 54–70.
Brandt, J. C., et al. (1969): Interplanetary gas. A calculation of angular momentum of the solar wind. Astrophysical Journal **156**, 1117–1124.
Braun, D. C., Duvall, T. J. Jr., La Bonte, B. J. (1987): Acoustic absorption by sunspots. Astrophysical Journal (Letters) **319**, L27–L31.
Braun, D. C., Duvall, T. J. Jr., La Bonte, B. J. (1988): The absorption of high-degree p-mode oscillations in and around sunspots. Astrophysical Journal **335**, 1015–1025.
Bravo, S., Stewart, G. A. (1997): Fast and slow wind from solar coronal holes. Astrophysical Journal **489**, 992–999.
Bray, R. J., Loughhead, R. E., Durrant, C. J.: The solar granulation. New York: Cambridge University Press 1967, 1984.
Breen, A. R., et al.: Ground and space-based studies of solar wind acceleration. In: The corona and solar wind near minimum activity. Proceedings of the fifth SOHO workshop. ESA SP-404. Noordwijk, The Netherlands: ESA Publications 1997, pp. 223–228.
Breit, G., Tuve, M. A. (1926): A test of the existence of the conducting layer. Physical Review **28**, 554–575.
Brekke, P., Hassler, D. M., Wilhelm, K. (1997): Doppler shifts in the quiet-Sun transition region and corona observed with SUMER on SOHO. Solar Physics **175**, 349–374. Reprinted in: The first results from SOHO (Eds. B. Fleck and Z. Svestka) Boston: Kluwer Academic Publishers 1997, pp. 349–374.
Brekke, P., Hassler, D. M., Wilhelm, K.: Systematic redshifts in the quiet sun transition region and corona observed with SUMER on SOHO. In: The corona and solar wind near minimum activity. Proceedings of the fifth SOHO workshop. ESA SP-404. Noordwijk, The Netherlands: ESA Publications Division 1997, pp. 229–234.
Bridge, H. S. et al. (1962): Direct observations of the interplanetary plasma. Journal of the Physical Society of Japan **17**, Supplement A-II, 553–559.
Broecker, W. S., Denton, G. H. (1990): What drives glacial cycles? Scientific American **262**, 49–56 – January.
Brown, J. C. (1971): The deduction of energy spectra of non-thermal electrons in flares from the observed dynamic spectra of hard X-ray bursts. Solar Physics **18**, 489–502.
Brown, J. C. (1972): The decay characteristics of models of solar hard X-ray bursts. Solar Physics **25**, 158–177.
Brown, J. C. (1972): The directivity and polarization of thick target X-ray bremsstrahlung from solar flares. Solar Physics **26**, 441–459.
Brown, J. C. (1973): Thick target X-ray bremsstrahlung from partially ionized targets in solar flares. Solar Physics **28**, 151–158.
Brown, J. C.: The interpretation of spectra, polarization, and directivity of solar hard X-rays. In: Solar gamma-, X-, and EUV radiation. Proceedings of IAU symposium no. 68 (Ed. Sharad R. Kane). Boston: D. Reidel 1975, pp. 245–282.
Brown, J. C. (1991): Energetic particles in solar flares: theory and diagnostics. Philosophical Transactions of the Royal Society (London) **A336**, 413–424.
Brown, J. C., Emslie, A. G. (1989): Self-similar Lagrangian hydrodynamics of beam-heated solar flare atmospheres. Astrophysical Journal **339**, 1123–1131.

Brown, J. C., et al. (1990): Beam heating in solar flares: electrons or protons? Astrophysical Journal Supplement **73**, 343–348.
Brown, T. M. (1985): Solar rotation as a function of depth and latitude. Nature **317**, 591–594.
Brown, T. M., Christensen-Dalsgaard, J. (1998): Accurate determination of the solar photospheric radius. Astrophysical Journal (Letters) **500**, L195–L198.
Brown, T. M., et al. (1989): Inferring the Sun's internal angular velocity from observed p-mode frequency splittings. Astrophysical Journal **343**, 526–546.
Brown, T. M., Morrow, C. A. (1987): Depth and latitude dependence of solar rotation. Astrophysical Journal (Letters) **314**, L21–L26.
Brueckner, G. E., Bartoe, J.-D. F. (1983): Observations of high-energy jets in the corona above the quiet sun, the heating of the corona, and the acceleration of the solar wind. Astrophysical Journal **272**, 329–348.
Bruner, E. C., Jr. (1978): Dynamics of the solar transition zone. Astrophysical Journal **226**, 1140–1146.
Bruno, R., et al. (1986): In-situ observations of the latitudinal gradients of the solar wind parameters during 1976 and 1977. Solar Physics **104**, 431–445.
Bruzek, A. (1964): On the association between loop prominences and flares. Astrophysical Journal **140**, 746–759.
Budyko, M. I. (1969): Effect of solar radiation variations on the climate of Earth. Tellus **21**, 611–620.
Bumba, V. (1958): Relation between chromospheric flares and magnetic fields of sunspot groups. Crimean Astrophysical Observatory, Bulletin **19**, 105–114.
Bumba, V., Howard, R. (1965): Large-scale distribution of solar magnetic fields. Astrophysical Journal **141**, 1502–1512.
Bunsen, R. (1859): Letter to H. E. Roscoe in November 1859. Quoted by Roscoe in: The life and experiences of Sir Henry Enfield Roscoe, London 1906, p. 71. Reproduced by A. J. Meadows in: The origins of astrophysics, article in The general history of astronomy, vol. 1. astrophysics and twentieth-century astronomy to 1950, part A (ed. O. Gingerich). New York: Cambridge University Press 1984, p. 5.
Bürgi, A., Geiss, J. (1986): Helium and minor ions in the corona and solar wind: dynamics and charge states. Solar Physics **103**, 347–383.
Burkepile, J. T., St. Cyr, O. C.: A revised and expanded catalogue of mass ejections observed by the solar maximum mission coronagraph. NCAR/TN-369+STR. Boulder, Colorado: National Center for Atmospheric Research 1993.
Burlaga, L. F. (1971): Hydromagnetic waves and discontinuities in the solar wind. Space Science Reviews **12**, 600–657.
Burlaga, L. F. (1983): Understanding the heliosphere and its energetic particles. Proceedings of the 18th International Conference of Cosmic Rays, **12**, 21–60.
Burlaga, L. F. (1988): Magnetic clouds and force-free fields with constant alpha. Journal of Geophysical Research **93**, 7217–7224.
Burlaga, L. F.: Magnetic clouds. In: Physics of the inner heliosphere II. particles, waves and turbulence. (Eds. R. Schwenn, E. Marsch). New York: Springer-Verlag 1990, pp. 1–22.
Burlaga, L. F., Behannon, K. W., Klein, L. W. (1987): Compound streams, magnetic clouds and major magnetic storms. Journal of Geophysical Research **92**, 5725–5734.
Burlaga, L. F. et al. (1978): Sources of magnetic fields in recurrent interplanetary streams. Journal of Geophysical Research **83**, 4177–4185.
Burlaga, L. F., et al. (1981).: Magnetic loop behind an interplanetary shock: Voyager, Helios, and IMP 8 observations. Journal of Geophysical Research **86**, 6673–6684.
Burlaga, L. F., King, J. H. (1979): Intense interplanetary magnetic fields observed by geocentric spacecraft during 1963–1975. Journal of Geophysical Research **84**, 6633–6640.
Burlaga, L. F., Lepping, R. P. (1977): The causes of recurrent geomagnetic storms. Planetary and Space Science **25**, 1151–1160.

Burnight, T. R. (1949): Soft X-radiation in the upper atmosphere. Physical Review **76**, 165.

Burton, M. E., et al. (1996): Ulysses out-of-ecliptic observations of interplanetary shocks. Astronomy and Astrophysics **316**, 313–322.

Burton, R. K., Mc Pherron, R. L., Russell, C. T. (1975): An empirical relationship between interplanetary conditions and Dst. Journal of Geophysical Research **80**, 4204–4214.

Byram, E. T., Chubb, T. A., Friedman, H. (1954): Solar X-ray emission. Physical Review **96**, 860.

Byram, E. T., Chubb, T. A., Friedman, H. (1956): The solar X-ray spectrum and the density of the upper atmosphere. Journal of Geophysical Research **61**, 251–263.

Cahill, L. J., Patel, V. L. (1967): The boundary of the geomagnetic field, August to November 1961. Planetary and Space Science **15**, 997–1033.

Cane, H. V.: The current status of our understanding of energetic particles, coronal mass ejections and flares. In: Coronal mass ejections. geophysical monograph 99 (Eds. N. Crooker, J. A. Joselyn and J. Feynman). Washington, DC: American Geophysical Union 1997, pp. 205–215.

Cane, H. V., Mc Guire, R. E., Rosenvinge, T. T. Von (1986): Two classes of solar energetic particle events associated with impulsive and long duration soft X-ray events. Astrophysical Journal **301**, 448–459.

Cane, H. V., Sheeley, N. R. Jr., Howard, R. A. (1987): Energetic interplanetary shocks, radio emission, and coronal mass ejections. Journal of Geophysical Research **92**, 9869–9874.

Canfield, R. C., Hudson, H. S., Mc Kenzie, D. E. (1999): Sigmoidal morphology and eruptive solar activity. Geophysical Research Letters **26**, No. 6, 627–630.

Canuto, V. M., et al. (1983): The young Sun and the atmosphere and photochemistry of the early Earth. Nature **305**, 281–286.

Cargill, P. J., Klimchuk, J. A. (1997): A nanoflare explanation for the heating of coronal loops observed by Yohkoh. Astrophysical Journal **478**, 799–806.

Cargill, P. J., Priest, E. R. (1983): The heating of post-flare loops. Astrophysical Journal **266**, 383–389.

Carlqvist, P. (1969): Current limitation and solar flares. Solar Physics **7**, 377–392.

Carmichael, H.: A process for flares. In: AAS-NASA symposium on the physics of solar flares NASA SP-50 (ed. W. N. Hess). Washington, DC: National Aeronautics and Space Administration 1964, pp. 451–456.

Carrington, R. C. (1858): On the distribution of the solar spots in latitude since the beginning of the year 1854. Monthly Notices of the Royal Astronomical Society **19**, 1–3. Reproduced in: Early solar physics (Ed. A. J. Meadows). Oxford, England: Pergamon Press 1970, pp. 169–172.

Carrington, R. C. (1860): Description of a singular appearance seen in the Sun on September 1, 1859. Monthly Notices of the Royal Astronomical Society **20**, 13–15. Reproduced in: Early solar physics (Ed. A. J. Meadows). Oxford, England: Pergamon Press 1970, pp. 181–183.

Carrington, R. C.: Observations of spots on the sun. London: Williams and Norgate 1863.

Celsius, A. (1747): Bemerkungen über der Magnetnadel Stündliche Veränderungen in ihrer Abweichung. Svenska Ventensk. Handl. **8**, 296.

Chae, J., et al. (1998): Chromospheric upflow events associated with transition region explosive events. Astrophysical Journal (Letters) **504**, L123–L126.

Chamberlin, T. C. (1899): An attempt to frame a working hypothesis of the cause of the glacial periods on an atmospheric basis. The Journal of Geology **7**, 545–584.

Chandra, S. (1991): The solar UV related changes in total ozone from a solar rotation to a solar cycle. Geophysical Research Letters **18**, 837–840.

Chapman, G. A. (1984): On the energy balance of solar active regions. Nature **308**, 252–254.

Chapman, G. A. (1987): Variations of solar irradiance due to magnetic activity. Annual Review of Astronomy and Astrophysics **25**, 633–667.

Chapman, G. A., Cookson, A. M., Dobias, J. J. (1996): Variations in total solar irradiance during solar cycle 22. Journal of Geophysical Research **101**, 13,541–13,548.

Chapman, G. A., Cookson, A. M., Dobias, J. J. (1997): Solar variability and the relation of facular to sunspot areas during solar cycle 22. Astrophysical Journal **482**, 541–545.

Chapman, G. A., et al. (1984): Solar luminosity fluctuations and active region photometry. Astrophysical Journal (Letters) **282**, L99–L101.

Chapman, S. (1918): An outline of a theory of magnetic storms. Proceedings of the Royal Society (London) **A95**, 61–83.

Chapman, S. (1918): The energy of magnetic storms. Monthly Notices of the Royal Astronomical Society **79**, 70–83.

Chapman, S. (1929): Solar streams of corpuscles – their geometry, absorption of light and penetration. Monthly Notices of the Royal Astronomical Society **89**, 456–470.

Chapman, S. (1950): Corpuscular influences upon the upper atmosphere. Journal of Geophysical Research **55**, 361–372. Reproduced in: A source book in astronomy and astrophysics, 1900–1975 (Eds. K. R. Lang and O. Gingerich). Cambridge, Massachusetts: Harvard University Press 1979, pp. 125–131.

Chapman, S. (1954): The viscosity and thermal conductivity of a completely ionized gas. Astrophysical Journal **120**, 151–155.

Chapman, S. (1957): Notes on the solar corona and the terrestrial atmosphere. Smithsonian Contributions to Astrophysics **2**, No. 1, 1–14.

Chapman, S. (1958): Thermal diffusion in ionized gases. Proceedings of the Physical Society of London **72**, 353–362.

Chapman, S. (1959): Interplanetary space and the earth's outermost atmosphere. Proceedings of the Royal Society (London) **A253**, 462–481.

Chapman, S. (1959): The outermost ionosphere. Journal of Atmospheric and Terrestrial Physics **15**, 43–47.

Chapman, S., Bartels, J.: Geomagnetism. Oxford, England: At the Clarendon Press 1940.

Chapman, S., Ferraro, V. C. A. (1929): The electrical state of solar streams of corpuscles. Monthly Notices of the Royal Astronomical Society **89**, 470–479.

Chapman, S., Ferraro, V. C. A. (1931): A new theory of magnetic storms, I. The initial phase. Terrestrial Magnetism and Atmospheric Electricity **36**, 77–97, 171–186; **37**, 147–156, 421–429 (1932).

Chapman, S., Ferraro, V. C. A. (1933): A new theory of magnetic storms, II. The main phase. Terrestrial Magnetism and Atmospheric Electricity **38**, 79–86.

Chapman, S., Ferraro, V. C. A. (1940): A theory of the first phase of geomagnetic storms. Terrestrial Magnetism and Atmospheric Electricity **45**, 245–268.

Chappellaz, J., et al. (1990): Ice-core record of atmospheric methane over the past 160,000 years. Nature **345**, 127–131.

Chappellaz, J., et al. (1993): Synchronous changes in atmospheric CH_4 and Greenland climate between 40 and 8 kyr BP. Nature **366**, 443–445.

Charlson, R. J., Wigley, T. M. L. (1994): Sulfate aerosol and climate change. Scientific American **270**, 48–57 – February.

Cheng, C.-C., Doschek, G. A., Feldman, U. (1979): The dynamical properties of the solar corona from intensities and line widths of EUV forbidden lines. Astrophysical Journal **227**, 1037–1046.

Cheng, C.-C., et al. (1981): Spatial and temporal structures of impulsive bursts from solar flares observed in UV and hard X-rays. Astrophysical Journal (Letters) **248**, L39–L43.

Cheng, C.-C., Pallavicini, R. (1987): Analysis of ultraviolet and X-ray observations of three homologous solar flares from SMM. Astrophysical Journal **318**, 459–473.

Chiang, W. H., Foukal, P. V. (1985): The influence of faculae on sunspot heat blocking. Solar Physics **97**, 9–20.
Chitre, S. M., Gokhale, M. H. (1975): The five-minute oscillations in the solar atmosphere. Solar Physics **43**, 49–55.
Christensen-Dalsgaard, J., et al. (1985): Speed of sound in the solar interior. Nature **315**, 378–382.
Christensen-Dalsgaard, J., et al. (1996): The current state of solar modeling. Science **272**, 1286–1292.
Christensen-Dalsgaard, J., Gough, D. O., Thompson, M. J. (1991): The depth of the solar convection zone. Astrophysical Journal **378**, 413–437.
Christensen-Dalsgaard, J., Gough, D. O., Toomre, J. (1985): Seismology of the Sun. Science **229**, 923–931.
Christiansen, W. N.: The first decade of solar radio astronomy in Australia. In: The early years of radio astronomy (Ed. W. T. Sullivan, III). New York: Cambridge University Press 1984.
Chubb, T. A., Friedman, H., Kreplin, R. W. (1960): Measurements made of high-energy X-rays accompanying three class 2+ solar flares. Journal of Geophysical Research **65**, 1831–1832.
Chupp, E. L. (1971): Gamma ray and neutron emissions from the Sun. Space Science Reviews **12**, 486–525.
Chupp, E. L.: Gamma-ray astronomy: Dordrecht, Holland: D. Reidel 1976.
Chupp, E. L. (1984): High energy neutral radiation from the Sun. Annual Review of Astronomy and Astrophysics **22**, 359–387.
Chupp, E. L. (1990): Emission characteristics of three intense solar flares observed in cycle 21. Astrophysical Journal Supplement **73**, 213–226.
Chupp, E. L., Benz, A. O, Eds. (1994): Particle acceleration phenomena in astrophysical plasmas. Proceedings of the international astronomical union (IAU) colloquium 142. Astrophysical Journal Supplement **90**, 511–983.
Chupp, E. L., et al. (1973): Solar gamma ray lines observed during the solar activity of August 2 to August 11, 1972. Nature **241**, 333–334.
Chupp, E. L., et al. (1982): A direct observation of solar neutrons following the 0118 UT flare on 1980 June 21. Astrophysical Journal (Letters) **263**, L95–L99.
Chupp, E. L., et al. (1987): Solar neutron emissivity during the large flare on 1982 June 3. Astrophysical Journal **318**, 913–929.
Cini Castagoli, G., et al. (1984): Solar cycles in the last centuries in ^{10}Be and σ^{18} in polar ice and in thermoluminescence signals of a sea sediment. Nuovo Cimento **7C**, 235–244.
Cini Castagoli, G., Lal, D. (1980): Solar modulation effects in terrestrial production of carbon-14. Radiocarbon **22**, 133–158.
Claverie, A., et al. (1979): Solar structure from global studies of the five-minute oscillation. Nature **282**, 591–594.
Clay, J. (1927): Penetrating radiation. Koninklijke Nederlandse Akademie van Wetenschappen te Amsterdam, Proceedings of the Section of Sciences **30**, 1115.
Clay, J. (1932): The Earth-magnetic effect and the corpuscular nature of (cosmic) ultra-radiation. IV. Koninklijke Nederlandse Akademie van Wetenschappen te Amsterdam, Proceedings of the Section of Sciences **35**, 1282–1290. Reproduced in Hillas (1972).
Cline, T. L., Mc Donald, F. B. (1968): Relativistic electrons from solar flares. Solar Physics **5**, 507–530.
Cliver, E. W. (1994): Solar activity and geomagnetic storms: The first 40 years. EOS **75**, 569, 574–575.
Cliver, E. W. (1994): Solar activity and geomagnetic storms: The corpuscular hypothesis. EOS **75**, 609, 612–613.

Cliver, E. W. (1995): Solar activity and geomagnetic storms: From M regions and flares to coronal holes and CMEs. EOS **76**, 75, 83–84.

Cliver, E. W. (1995): Solar flare nomenclature. Solar Physics **157**, 285–293.

Cliver, E. W., Boriakoff, V., Feynman, J. (1998): Solar variability and climate change: Geomagnetic aa index and global surface temperature. Geophysical Research Letters **25**, 1035–1038.

Cliver, E. W., et al. (1989): Solar flare nuclear gamma-rays and interplanetary proton events. Astrophysical Journal **343**, 953–970.

Cocconi, G., et al. (1958): The cosmic ray flare effect. Nuovo Cimento Supplement Series **8**, No. 2, 161–168.

Coleman, P. J., Jr., et al. (1966): Measurements of magnetic fields in the vicinity of the magnetosphere and in interplanetary space: Preliminary results from Mariner 4. Space Research **6**, 907–928.

Coles, W. A., et al. (1980): Solar cycle changes in the polar solar wind. Nature **286**, 239–241.

Coles, W. A., Rickett, B. (1976): IPS observations of solar wind speed out of the ecliptic. Journal of Geophysical Research **77**, 4797–4799.

Compton, A. H. (1932): Variation of the cosmic rays with latitude. Physical Review **41**, 111–113.

Compton, A. H. (1933): A geographic study of the cosmic rays. Physical Review **43**, 387–403.

Corti, G., et al.: Physical parameters in plume and interplume regions from UVCS observations. In: The corona and solar wind near minimum activity. Proceedings of the fifth SOHO workshop. ESA SP-404. Noordwijk, The Netherlands: ESA Publications Division 1997, pp. 289–294.

Covington, A. E. (1951): Some characteristics of 10.7 cm solar noise. Journal of the Royal Astronomical Society of Canada **45**, 15–22.

Covington, A. E., Harvey, G. A. (1958): Impulsive and long-enduring sudden enhancements of solar radio emission at 10-cm wave-length. Journal of the Royal Astronomical Society of Canada **52**, 161–166.

Cox, A. (1969): Geomagnetic reversals. Science **163**, 237–245.

Cox, A. N., Livingston, W. C., Matthews, M. S. (eds.): Solar interior and atmosphere. Tucson, Arizona: The University of Arizona Press 1991.

Craig, I. J. D., Mc Clymont, A. N., Underwood, J. H. (1978): The temperature and density structure of active region coronal loops. Astronomy and Astrophysics **70**, 1–11.

Croll, J.: Climate and time in their geological relations. New York: Appleton 1875.

Crommelynck, D., et al. (1995): First realization of the space absolute radiometric reference (SARR) during the ATLAS 2 flight period. Advances in Space Research **16**, No. 8, 17–23.

Crooker, N. U., Cliver, E. W. (1994): Postmodern view of M-regions. Journal of Geophysical Research **99**, 23,383–23,390.

Crooker, N., Joselyn, J. A., Feynman, J. (Eds.): Coronal mass ejections, geophysical monograph 99. Washington, DC: American Geophysical Union, 1997.

Crosby, N. B., Aschwanden, M. J., Dennis, B. R. (1993): Frequency distributions and correlations of solar X-ray flare parameters. Solar Physics **143**, 275–299.

Crowley, T. J. (1983): The geologic record of climatic change. Review of Geophysics and Space Physics **21**, 828–877.

Crowley, T. J., Kim, K.-Y. (1996): Comparison of proxy records of climate change and solar forcing. Geophysical Research Letters **23**, 359–362.

Crowley, T. J., Kim, K.-Y. (1999): Modeling the temperature response to forced climate change over the last six centuries. Geophysical Research Letters **26**, 1901–1904.

Crutzen, P. J., Isaksen, I. S. A., Reid, G. C. (1975): Solar proton events: Stratospheric sources of nitric oxide. Science **189**, 457–459.

Cubasch, U., et al. (1997): Simulation of the influence of solar radiation variations on the global climate with an ocean-atmosphere general circulation model. Climate Dynamics **13**, 757–767.

Culhane, J. L., Jordan, C. (Eds.): The Physics of Solar Flares. London: The Royal Society 1991.

Cummings, A. C., Stone, E. C., Webber, W. R. (1993): Estimate of the distance to the solar wind termination shock from gradients of anomalous cosmic ray oxygen. Journal of Geophysical Research **98**, 15,165–15,168.

Currie, R. G. (1974): Solar cycle signal in surface air temperature. Journal of Geophysical Research **79**, 5657–5660.

Cushman, G. W., Rense, W. A. (1976): Evidence of outward flow of plasma in a coronal hole. Astrophysical Journal (Letters) **207**, L61–L62.

Damon, P. E., Jirikowic, J. L.: Solar forcing of global climate change. In: The sun as a variable star (Eds. J. Pap, H. Hudson, and S. Solanki). New York: Cambridge University Press 1994.

Damon, P. E., Peristykh, A. N. (1999): Solar cycle length and twentieth century northern hemisphere warming. Geophysical Research Letters **26**, 2469–2472.

Damon, P. E., Sonett, C. P.: Solar and terrestrial components of the atmospheric ^{14}C variation spectrum. In: The Sun in time (Eds. C. P. Sonett, M. S. Giampapa and M. S. Matthews). Tucson, Arizona: University of Arizona Press 1991, pp. 360–388.

Dansgaard, W., et al. : Climatic record revealed by the Camp Century ice core. In: The late-glacial ages (Ed. K. Turekian). New Haven, Connecticut: Yale University Press 1980, pp. 37–46.

Dansgaard, W. et al.: North Atlantic climate oscillations revealed by deep Greenland ice cores. In: Climate processes and climate sensitivity. American geophysical union geophysical monograph 29 (Eds. J. E. Hansen and T. Takahashi). Washington, D.C.: American Geophysical Union 1984, pp. 288–298.

Dansgaard, W. et al. (1989): A new Greenland deep ice core. Science **218**, 1273–1277.

David, C., Gabriel, A. H., Bely-Dubau, F.: Temperature structure in coronal holes. In: The corona and solar wind near minimum activity Proceedings of the fifth SOHO workshop. ESA SP-404. Noordwijk, The Netherlands: ESA Publications Division 1997, pp. 319–322.

Davis, L. Jr. (1955): Interplanetary magnetic fields and cosmic rays. Physical Review **100**, 1440–1444.

Davis, L. Jr.: The interplanetary magnetic field. In: Solar wind: The proceedings of a conference sponsored by the national aeronautics and space administration. NASA SP-308. Washington, DC: NASA 1972. pp. 73–103.

Davis, R., Harmer, D. S., Hoffman, K. C. (1968): Search for neutrinos from the sun. Physical Review Letters **20**, 1205–1209. Reproduced in: A source book in astronomy and astrophysics 1900–1975 (Eds. K. R. Lang and O. Gingerich). Cambridge, Massachusetts: Harvard University Press 1979, pp. 389–395.

De Forest, C. E., et al. (1997): Polar plume anatomy: Results of a coordinated observation. Solar Physics **175**, 393–410. Reproduced in: The first results from SOHO (Eds. B. Fleck and Z. Svestka). Dordrecht, Holland: Kluwer Academic Publishers 1997, pp. 393–410.

De Forest, C. E., Gurman, J. B. (1998): Observation of quasi-periodic compressive waves in solar polar plumes. Astrophysical Journal (Letters) **501**, L217–L220.

De Jager, C. (1986): Solar flares and particle acceleration. Space Science Reviews **44**, 43–90.

De Jager, C., Kundu, M. R. (1963): A note on bursts of radio emission and high energy (> 20 keV) X-rays from solar flares. Space Research **3**, 836–838.

De Jong, A. F. M., Mook, W. G., Becker, B. (1979). Confirmation of the Suess wiggles. Nature **280**, 48–49.

De Mairan, J. J. D.: Traité physique et historique de l'Aurore Boréale. Paris: Imprimerie Royale 1733.

De Vorkin, D. H.: Science with a vengeance: How the military created the US space sciences after world war II. New York: Springer-Verlag 1992.

De Vries, H. L. (1958): Variation in concentration of radiocarbon with time and location on Earth. Proceeding Konikl. Ned. Akad. Wetenschap. 94–102.

Dearborn, D. S. P., Blake, J. R. (1980): Is the sun constant? Astrophysical Journal **237**, 616–619.

Dearborn, D. S. P., Blake, J. R. (1980): Magnetic fields and the solar constant. Nature **287**, 365–366.

Dearborn, D. S. P., Blake, J. R. (1982): Surface magnetic fields and the solar luminosity. Astrophysical Journal **257**, 896–900.

Debrunner, H., et al. (1983): The solar cosmic ray neutron event on June 3, 1982. Proceedings 18th International Cosmic Ray Conference **4**, 75–78.

Debrunner, H., Flückiger, E. O., Lockwood, J. A. (1990): Signature of the solar cosmic-ray event on 1982 June 3. Astrophysical Journal Supplement **73**, 259–262.

Dellinger, J. H. (1935): A new radio transmission phenomenon. Physical Review **48**, 705.

Dellinger, J. H. (1937): Sudden ionospheric disturbances. Terrestrial Magnetism and Atmospheric Electricity **42**, 49–53.

Delmas, R. J., et al. (1992): 1000 years of explosive volcanism recorded at South Pole. Tellus **44B**, 335–350.

Denisse, J. F.: The early years of radio astronomy in France. In: The early years of radio astronomy (Ed. W. T. Sullivan, III). New York: Cambridge University Press 1984, pp. 303–316.

Denisse, J. F., Boischot, A., Pick, M.: Propiétes des éruptions chromospheriques associées a la production de rayons cosmiques par le soliel. In: Space Research, Proceedings of the First International Space Science Symposium. Amsterdam: North Holland Publishing Co. 1960. pp. 637–648.

Dennis, B. R. (1985): Solar hard X-ray bursts. Solar Physics **100**, 465–490.

Dennis, B. R. (1988): Solar flare hard X-ray observations. Solar Physics **118**, 49–94.

Dennis, B. R., Zarro, D. M. (1993): The Neupert effect: what can it tell us about the impulsive and gradual phases of solar flares? Solar Physics **146**, 177–190.

Dennison, P. A., Hewish, A. (1967): The solar wind outside the plane of the ecliptic. Nature **213**, 343–346.

Dere, K. P. (1994): Explosive events, magnetic reconnection, and coronal heating. Advances in Space Research **14**, No. 4, 13–22.

Dere, K. P., et al. (1997): EIT and LASCO observations of the initiation of a coronal mass ejection. Solar Physics **175**, 601–612. Reproduced in: The first results from SOHO (Eds. B. Fleck, Z. Svestka). Dordrecht, Holland: Kluwer Academic Publishers 1997, pp. 601–612.

Dessler, A. J. (1967): Solar wind and interplanetary magnetic field. Reviews of Geophysics **5**, 1–41.

Deubner, F.-L. (1975): Observations of low wave number nonradial eigenmodes of the sun. Astronomy and Astrophysics **44**, 371–375.

Deubner, F.-L., Gough, D. (1984): Helioseismology – oscillations as a diagnostic of the solar interior. Annual Review of Astronomy and Astrophysics **22**, 593–619.

Deubner, F.-L., Ulrich, R. K., Rhodes, E. J. Jr. (1979): Solar p-mode oscillations as a tracer of radial differential rotation. Astronomy and Astrophysics **72**, 177–185.

Dicke, R. H. (1978): Is there a chronometer hidden deep in the Sun? Nature **276**, 676–680.

Dickinson, R. W., Cicerone, R. J. (1986): Future global warming from atmospheric trace gases. Nature **319**, 109–115.

Dietrich, W. F. (1973): The differential energy spectra of solar-flare ^{1}H, ^{3}He and ^{4}He. Astrophysical Journal **180**, 955–973.

Dilke, F. W. W., Gough, D. O. (1972): The solar spoon. Nature **240**, 262–264, 293, 294.

Dobson, G. M. B. (1968): Forty years' research on atmospheric ozone at Oxford: a history. Applied Optics **7**, 387–405.

Dodson, H. W., Hedeman, E. R. (1970): Major Hα flares in centers of activity with very small or no spots. Solar Physics **13**, 401–419.

Dodson, H. W., Hedeman, E. R., Owren, L. (1953): Solar flares and associated 200 Mc/sec radiation. Astrophysical Journal **118**, 169–196.

Donnelly, R. F. (1967): The solar flare radiations responsible for sudden frequency deviations. Journal of Geophysical Research **72**, 5247–5256.

Doschek, G. A. (1983): Solar instruments on the P78-1 spacecraft. Solar Physics **86**, 9–17.

Doschek, G. A. (1983): Solar flare X-ray spectra from the P78-1 spacecraft. Solar Physics **86**, 49–58.

Doschek, G. A. (1991): High-temperature plasma in solar flares. Philosophical Transactions of the Royal Society (London) **A336**, 451–460.

Doschek, G. A., et al. (1980): High-resolution X-ray spectra of solar flares III. General spectral properties of X1-X5 type flares. Astrophysical Journal **239**, 725–737.

Doschek, G. A., et al. (1993): The 1992 January 5 flare at 13.3 UT: Observations from Yohkoh. Astrophysical Journal **416**, 845–856.

Doschek, G. A., Feldman, U., Bohlin, J. D. (1976): Doppler wavelength shifts of transition zone lines measured in Skylab solar spectra. Astrophysical Journal (Letters) **205**, L177–L180.

Doschek, G. A., Kreplin, R. W., Feldman, U. (1979): High-resolution solar flare X-ray spectra. Astrophysical Journal (Letters) **233**, L157–L160.

Doschek, G. A., Mariska, J. T., Sakao, T. (1996): Soft X-ray flare dynamics. Astrophysical Journal **459**, 823–835.

Doschek, G. A., Strong, K. T., Tsuneta, S. (1995): The bright knots at the tops of soft X-ray flare loops. Quantitative results from Yohkoh. Astrophysical Journal **440**, 370–385.

Dowdy, J. F., Jr., Rabin, D., Moore, R. L. (1986): On the magnetic structure of the quiet transition region. Solar Physics **105**, 35–45.

Dröge, F. (1977): Millisecond fine-structures of solar burst radiation in the range 0.2–1.4 GHz. Astronomy and Astrophysics **57**, 285–290.

Dryer, M. (1982): Coronal transient phenomenon. Space Science Reviews **33**, 233–275.

D'Silva, S. (1998): Computing travel time in time-distance helioseismology. Astrophysical Journal (Letters) **498**, L79-L82.

Duijveman, A., Hoyng, P., Machado, M. E. (1982): X-ray imaging of three flares during the impulsive phase. Solar Physics **81**, 137–157.

Dulk, G. A. (1985): Radio emission from the sun and other stars. Annual Review of Astronomy and Astrophysics **23**, 169–224.

Dungey, J. W. (1953): Conditions for the occurrence of electrical discharges in astrophysical systems. Philosophical Magazine **44**, 725–738.

Dungey, J. W.: The neutral point discharge theory of solar flares. A reply to Cowling's criticism. In: Electromagnetic phenomena in cosmical physics. Proceedings of IAU symposium no. 6 (ed. B. Lehnert). Cambridge, England: Cambridge at the University Press 1958. pp. 135–140.

Dungey, J. W. (1961): Interplanetary magnetic field and the auroral zones. Physical Review Letters **6**, 47–48.

Dungey, J. W. (1979): First evidence and early studies of the Earth's bow shock. Nuovo Cimento **C2**, 655–660.

Dungey, J. W. (1994): Memories, maxims, and motives. Journal of Geophysical Research **99**, 19,189–19,197.

Duvall, T. L. Jr. (1979): Large-scale solar velocity fields. Solar Physics **63**, 3–15.

Duvall, T. L. Jr. (1982): A dispersion law for solar oscillations. Nature **300**, 242–243.

Duvall, T. L. Jr., et al. (1984): Internal rotation of the Sun. Nature **310**, 22–25.
Duvall, T. L. Jr., et al. (1993): Time-distance helioseismology. Nature **362**, 430–432.
Duvall, T. L. Jr., et al. (1996): Downflows under sunspots detected by helioseismic tomography. Nature **379**, 235–237.
Duvall, T. L. Jr., et al. (1997): Time-distance helioseismology with the MDI instrument. Solar Physics **170**, 63–73. Reprinted in: The first results from SOHO (Eds. B. Fleck and Z. Svestka). Boston: Kluwer Academic Publishers 1997, pp. 63–73.
Duvall, T. L. Jr., Harvey, J. W. (1983): Observations of solar oscillations of low and intermediate degree. Nature **302**, 24–27.
Duvall, T. L. Jr., Harvey, J. W. (1984): Rotational frequency splitting of solar oscillations. Nature **310**, 19–22.
Duvall, T. L. Jr., Harvey, J. W., Pomerantz, M. A. (1986): Latitude and depth variation of solar rotation. Nature **321**, 500–501.

Earl, J. A. (1961): Cloud-chamber observations of primary cosmic-ray electrons. Physical Review Letters **6**, 125–128.
Eather, R. H.: Majestic lights. The aurora in science, history, and the arts. Washington, DC: American Geophysical Union 1980.
Eddy, J. A. (1976): The Maunder minimum. The reign of Louis XIV appears to have been a time of real anomaly in the behavior of the sun. Science **192**, 1189–1202.
Eddy, J. A. (1977): Climate and the changing sun. Climate Change **1**, 173–190.
Eddy, J. A. (1977): The case of the missing sunspots. Scientific American **236**, 80–95 – May.
Eddy, J. A.: Historical evidence for the existence of the solar cycle. In: The solar output and its variation (ed. Oran R. White). Boulder, Colorado: Colorado Associated University Press 1977, pp. 51–71.
Eddy, J. A.: A new sun: The solar results from Skylab. Washington, DC: National Aeronautics and Space Administration SP-402 1979.
Eddy, J. A.: Keynote address: An historical review of solar variability, weather, and climate. In: Weather and climate responses to solar variations (ed. B. M. Mc Cormac). Boulder, Colorado: Colorado Associated University Press 1983, pp. 1–23.
Eddy, J. A. (1990): Some thoughts on Sun-weather relations. Philosophical Transactions of the Royal Society (London) **A330**, 543–545.
Eddy, J. A., Gilliland, R. L., Hoyt, D. V. (1982): Changes in the solar constant and climatic effects. Nature **300**, 689–693.
Eddy, J. A., Stephenson, F. R., Yau, K. K. C. (1989): On pre-telescopic sunspot records. Quarterly Journal of the Royal Astronomical Society **30**, 60–73.
Edlén, B. (1941): An attempt to identify the emission lines in the spectrum of the solar corona. Arkiv för Matematik, Astronomi och Fysik **28B**, 1–4. Reproduced in: A source book in astronomy and astrophysics, 1900–1975 (Eds. K. R. Lang and O. Gingerich). Cambridge, Massachusetts: Harvard University Press 1975, pp. 120–124.
Edlén, B. (1942): Die Deutung der Emissionslinien im Spektrum der Sonnenkorona. Zeitschrift für Astrophysik **22**, 30–64.
Edlén, B. (1945): The identification of the coronal lines. Monthly Notices of the Royal Astronomical Society **105**, 323–333.
Elliott, J. R., Kosovichev, A. G. (1998): The adiabatic exponent in the solar core. Astrophysical Journal (Letters) **500**, L199–L202.
Ellison, M. A. (1942): Some studies of the motions of hydrogen flocculi by Doppler displacements of the Hα line. Monthly Notices of the Royal Astronomical Society **102**, 11–21.
Ellison, M. A. (1946): Visual and spectrographic observations of a great solar flare 1946 July 25. Monthly Notices of the Royal Astronomical Society **106**, 500–508.
Ellison, M. A. (1948): Distinction between flares and prominences. Observatory **68**, 69–70.

Ellison, M. A. (1949): Characteristic properties of chromospheric flares. Monthly Notices of the Royal Astronomical Society **109**, 1–27.

Elphinstone, R. E., Murphree, J. S., Cogger, L. L. (1996): What is a global auroral substorm? Reviews of Geophysics **34**, 169–232.

Elsworth, Y. P., et al. (1990): Variation of low-order acoustic solar oscillations over the solar cycle. Nature **345**, 322–324.

Elsworth, Y. P., et al. (1995): Slow rotation of the Sun's interior. Nature **376**, 669–672.

Emiliani, C. (1966). Isotopic paleotemperatures. Science **154**, 851–857.

Evans, J. V. (1982): The sun's influence on the earth's atmosphere and interplanetary space. Science **216**, 467–474.

Evenson, P., Meyer, P., Pyle, K. R. (1983): Protons from the decay of solar flare neutrons. Astrophysical Journal **274**, 875–882.

Evershed, J. (1948): Spectrum lines in chromospheric flares. Observatory **68**, 67–68.

Fabian, P., Pyle, J. A., Wells, R. J. (1979): The August 72 proton event and the atmospheric ozone layer. Nature **277**, 458–460.

Fainberg, J., Stone, R. G. (1974): Satellite observations of type III radio bursts at low frequencies. Space Science Reviews **16**, 145–188.

Fairfield, D. H., Cahill, L. J., Jr. (1966): Transition region magnetic field and polar magnetic disturbances. Journal of Geophysical Research **71**, 155–169.

Falconer, D. A., et al. (1997): Neutral-line magnetic shear and enhanced coronal heating in solar active regions. Astrophysical Journal **482**, 519–534.

Falconer, D. A., Moore, R. L., Porter, J. G., Hathaway, D. H. (1998): Network coronal bright points: Coronal heating concentrations found in the solar magnetic network. Astrophysical Journal **501**, 386–396.

Farrugia, C. J., et al. (1993): The earth's magnetosphere under continued forcing: Substorm activity during the passage of an interplanetary magnetic cloud. Journal of Geophysical Research **98**, 7657–7671.

Feldman, U., et al. (1980): High-resolution X-ray spectra of solar flares. IV. General spectral properties of M-type flares. Astrophysical Journal **241**, 1175–1185.

Feldman, U., et al. (1994): The morphology of the 10^7 plasma in solar flares I. Nonimpulsive flares. Astrophysical Journal **424**, 444–458.

Feldman, U., et al. (1996): Electron temperature, emission measure, and X-ray flux in A2 to X2 X-ray class solar flares. Astrophysical Journal **460**, 1034–1041.

Feldman, W. C., et al. (1976): High-speed solar wind parameters at 1 AU. Journal of Geophysical Research **81**, 5054–5060.

Feldman, W. C., et al.: Plasma and magnetic fields from the Sun. In: The solar output and its variations (Ed. O. R. White). Boulder, Colorado: Colorado Associated University Press 1977, pp. 351–381.

Feldman, W. C., et al. (1981): The solar origins of solar wind interstream flows: Near-equatorial coronal streamers. Journal of Geophysical Research **86**, A7, 5408–5416.

Feldman, W. C., et al. (1996): Constraints on high-speed solar wind structure near its coronal base: a Ulysses perspective. Astronomy and Astrophysics **316**, 355–367.

Fichtel, C. E., Guss, D. E. (1961): Heavy nuclei in solar cosmic rays. Physical Review **6**, 495–497.

Fireman, E. L.: Solar activity during the past 10,000 years from radionuclides in lunar samples. In: The ancient sun: Fossil record in the earth, moon and meteorites (Eds. R. O. Pepin, J. A. Eddy, and R. B. Merrill). New York: Pergamon Press 1980, pp. 365–386.

Fischer, H., et al. (1999): Ice core records of atmospheric CO_2 around the last three glacial terminations. Science **283**, 1712–1714.

Fisk, L. A., Kozlovsky, B, Ramaty, R. (1974): An interpretation of the observed oxygen and nitrogen enhancements in low-energy cosmic rays. Astrophysical Journal (Letters) **190**, L35–L37.

Fitzgerald, G. F. (1892): Sunspots and magnetic storms. The Electrician **30**, 48.
Fitzgerald, G. F. (1900): Sunspots, magnetic storms, comets' tails, atmospheric electricity and aurorae. The Electrician **46**, 249, 287–288.
Fleck, B., Domingo, V., Poland, A. I., Eds. (1995): The SOHO mission. Solar Physics **162**, No. 1, 2. Reprinted by Boston: Kluwer Academic Publishers 1995.
Fleck, B., Svestka, Z., Eds. (1997): The first results from SOHO. Solar Physics **170**, No. 1, **175**, No. 2. Reprinted by Boston: Kluwer Academic Publishers 1997.
Fletcher, L., Huber, M. C. E.: O^{5+} acceleration by turbulence in polar coronal holes. In: The corona and solar wind near minimum activity. Proceedings of the fifth SOHO workshop. ESA SP-404. Noorwidjk, The Netherlands: ESA Publications Division 1997, pp. 379–384.
Foley, C. R., Culhane, J. L., Acton, L. W. (1997): Yohkoh soft X-ray determination of plasma parameters in a polar coronal hole. Astrophysical Journal **491**, 933–938.
Folland, C. K., Karl, T. R., Vinikov, K. Ya. : Observed climate variations and change. In: Climate change, the IPCC scientific assessment (Eds. J. T. Houghton, G. J. Jenkins and J. J. Ephraums). Cambridge, England: Cambridge University Press 1990, pp. 195–218.
Forbes, T. G. (1986): Fast-shock formation in line-tied magnetic reconnection models of solar flares. Astrophysical Journal **305**, 553–563.
Forbes, T. G., Acton, L. W. (1996): Reconnection and field line shrinkage in solar flares. Astrophysical Journal **459**, 330–341.
Forbes, T. G., Malherbe, J. M., Priest, E. R. (1989): The formation of flare loops by magnetic reconnection and chromospheric ablation. Solar Physics **120**, 285–307.
Forbes, T. G., Priest, E. R. (1987): A comparison of analytical and numerical models for steadily driven reconnection. Review of Geophysics **25**, 1583–1607.
Forbush, S. E. (1937): On diurnal variation in cosmic-ray intensity. Terrestrial Magnetism and Atmospheric Electricity **42**, 1–16.
Forbush, S. E. (1938): On cosmic-ray effects associated with magnetic storms. Terrestrial Magnetism and Atmospheric Electricity **43**, 203–218.
Forbush, S. E. (1938): On the world-wide changes in cosmic ray intensity. Physical Review **54**, 975–988.
Forbush, S. E. (1946): Three unusual cosmic-ray increases possibly due to charged particles from the sun. Physical Review **70**, 771–772
Forbush, S. E. (1954): World-wide cosmic-ray variations, 1937–1952. Journal of Geophysical Research **59**, 525–542.
Forbush, S. E. (1958): Cosmic-ray intensity variations during two solar cycles. Journal of Geophysical Research **63**, 651–669.
Forbush, S. E., Stinchcomb, T. B., Schein, M. (1950): The extraordinary increase of cosmic-ray intensity on November 19, 1949. Physical Review **79**, 501–504.
Forrest, D. J., Chupp, E. L. (1983): Simultaneous acceleration of electrons and ions in solar flares. Nature **305**, 291–292.
Forrest, D. J., et al. (1985): Neutral pion production in solar flares. Proceedings of the 19th International Cosmic Ray Conference **4**, 146–149.
Forsyth, R. J., et al. (1996): The heliospheric magnetic field at solar minimum: Ulysses observations from pole to pole. Astronomy and Astrophysics **316**, 287–295.
Fossat, E., Grec, G., Pomerantz, M. A. (1981): Solar pulsations observed from the geographic south pole – initial results. Solar Physics **74**, 59–63.
Foukal, P. (1990): The variable Sun. Scientific American **262**, 34–41 – February.
Foukal, P. (1990): Solar luminosity variations over timescales of days to the past few solar cycles. Philosophical Transactions of the Royal Society (London) **A330**, 591–599.
Foukal, P. (1994): Stellar luminosity variations and global warming. Science **264**, 238–239.
Foukal, P., Lean, J. (1986): The influence of faculae on total irradiance and luminosity. Astrophysical Journal **302**, 826–835.

Foukal, P., Lean, J. (1988): Magnetic modulation of solar luminosity by photospheric activity. Astrophysical Journal **328**, 347–357.
Foukal, P., Lean, J. (1990): An empirical model of total solar irradiance variation between 1874 and 1988. Science **247**, 556–558.
Frank, L. A., Craven, J. D. (1988): Imaging results from Dynamics Explorer 1. Review of Geophysics and Space Physics **26**, 249–283.
Frank, L. A., et al. (1986): The theta aurora. Journal of Geophysical Research **91**, 3177–3224.
Fraunhofer, J. (1814–15): Denkschriften (Munich Academy of Sciences) **5**, 195, 202. Auszug davon in Gilb. Ann. Physik **56**, 264 (1817)
Frazier, E. N. (1968): A spatio-temporal analysis of velocity fields in the solar photosphere. Zeitschrift für Astrophysik **68**, 345–358.
Freier, P., et al. (1948): Evidence for heavy nuclei in the primary cosmic radiation. Physical Review **74**, 213–217.
Freier, P., et al. (1948): The heavy component of primary cosmic rays. Physical Review **74**, 1818–1827.
Freier, P., Ney, E. P., Winckler, J. R. (1959): Balloon observations of solar cosmic rays on march 26, 1958. Journal of Geophysical Research **64**, 685–688.
Frick, P., et al. (1997): Wavelet analysis of stellar chromospheric activity variations. Astrophysical Journal **483**, 426–434.
Friedlander, M. W.: Cosmic rays. Cambridge, Massachusetts: Harvard University Press 1989.
Friedman, H.: X-ray and extreme ultraviolet observations of the Sun. In: Space research II (Eds. H. C. Van De Hulst, C. De Jager and A. F. Moore). Amsterdam: North-Holland Pub. Co. 1961. Reproduced in: A source book in astronomy and astrophysics 1900–1975 (Eds. K. R. Lang and O. Gingerich). Cambridge, Massachusetts: Harvard University Press 1979, pp. 56–61.
Friedman, H.: Solar X-ray emission. In: The solar corona. Proceedings of international astronomical union symposium No. 16 (Ed. J. W. Evans). New York: Academic Press 1963, pp. 45–48.
Friedman, H.: Sun and Earth. New York: Scientific American Library 1986.
Friedman, H., Lichtman, S. W., Byram, E. T. (1951): Photon counter measurements of solar X-rays and extreme ultraviolet light. Physical Review **83**, 1025–1030.
Friis-Christensen, E., Lassen, K. (1991): Length of the solar cycle: An indicator of solar activity closely associated with climate. Science **254**, 698–700.
Friis-Christensen, E., Lassen, K.: Solar activity and global temperature. In: The sun as a variable star: Solar and stellar luminosity variations (Eds. J. Pap, C. Fröhlich, H. Hudson, S. K. Solanki). New York: Cambridge University Press 1994, pp. 339–347.
Fröhlich, C. (1987): Variability of the solar "constant" on time scales of minutes to years. Journal of Geophysical Research **92**, 796–800.
Fröhlich, C., et al. (1995): VIRGO: Experiment for helioseismology and solar irradiance monitoring. Solar Physics **162**, 101–128.
Fröhlich, C., et al. (1997): First results from VIRGO, the experiment for helioseismology and solar irradiance monitoring in SOHO. Solar Physics **170**, 1–25. Reprinted in: The first results from SOHO (Eds. B. Fleck and Z. Svestka). Boston: Kluwer Academic Publishers 1997, pp. 1–25.
Frölich, C., et al. (1997): In-flight performances of VIRGO solar irradiance instruments on SOHO. Solar Physics **175**, 267–286.
Frölich, C., Lean, J. (1998): The sun's total irradiance: Cycles, trends and related climate change uncertainties since 1976. Geophysical Research Letters **25**, 4377–4380.
Frölich, C., Lean, J.: Total solar irradiance variations: The construction of a composite and its comparison with models. In: New eyes to see inside the sun and stars. IAU symposium 185. (Eds. F. L. Deubner, J. Christensen-Dalsgaard and D. Kurtz). Dordrecht, Holland: Kluwer Academic Publications 1998, pp. 89–102.

Fukuda, Y., et al. (1998): Evidence for oscillation of atmospheric neutrinos. Physical Review Letters **81**, 1562–1567.

Fürst, E., Benz, A. O., Hirth, W. (1982): About the relation between radio and soft X ray emission in case of very weak solar activity. Astronomy and Astrophysics **107**, 178–185.

Gabriel, A. H. (1971): Measurements on the Lyman alpha corona. Solar Physics **21**, 392–400.

Gabriel, A. H. (1976): A magnetic model of the solar transition region. Philosophical Transactions of the Royal Society (London) **A281**, 339–352.

Gabriel, A. H.: Structure of the quiet chromosphere and corona. In: The energy balance and hydrodynamics of the solar chromosphere and corona. Proceedings of IAU colloquium no. 36 (Eds. R.-M. Bonnet and Ph. Delache). Paris: G. de Bussac Clermont-Ferrand 1977, pp. 375–399.

Gabriel, A. H., et al. (1971): Rocket observations of the ultraviolet solar spectrum during the total eclipse of 1970 March 7. Astrophysical Journal **169**, 595–614.

Gabriel, A. H. et al. (1997): Performance and early results from the GOLF instrument flown in the SOHO mission. Solar Physics **175**, 207–226. Reprinted in: The first results from SOHO (Eds. B. Fleck and Z. Svestka). Boston: Kluwer Academic Publishers 1997, pp. 207–226.

Galileo Galilei: Istoria e dimostrazioni intorno alle macchie solari e loro accidenti. Rome, 1613.

Galloway, D. J., Weiss, N. O. (1981): Convection and magnetic fields in stars. Astrophysical Journal **243**, 945–953.

Gary, G. A., et al. (1987): Nonpotential features observed in the magnetic field of an active region. Astrophysical Journal **314**, 782–794.

Gauss, C. F.: Allgemeine theorie des erdmagnetismus, resultate aus den beobachtungen des magnetischen verein im jarhre 1838. Translated by Mrs. Sabine, revised by Sir John Herschel in Scientific Memoirs Selected From Transactions of Foreign Academies and Learned Societies and From Foreign Journals **2**, 184–251 (1841).

Gavaghan, H.: Something new under the sun: Satellites and the beginning of the space age. New York: Springer-Verlag 1998.

Geiss, J., et al. (1970): Apollo 11 and 12 solar wind composition experiments.: Fluxes of He and Ne isotopes. Journal of Geophysical Research **75**: 5972–5979.

Geiss, J., et al. (1995): The southern high-speed stream – results from the SWICS instrument on Ulysses. Science **268**, 1033–1036.

Geiss, J., Gloeckler, G., Von Steiger, R. (1995): Origin of the solar wind from composition data. Space Science Reviews **72**, 49–60.

Geiss, J., Reeves, H. (1972): Cosmic and solar system abundances of deuterium and helium-3. Astronomy and Astrophysics **18**, 126–132.

Geiss, J., Witte, M. (1996): Properties of the interstellar gas inside the heliosphere. Space Science Reviews **78**, 229–238.

Genthon, C., et al. (1987): Vostok ice core: climatic response to CO_2 and orbital forcing changes over the last climatic cycle. Nature **329**, 414–418.

Gilbert, W.: De magnete, magneticisque corporibus, et de magno magnete tellure: physiologia nova, plurimis and argumentis, and experimentis demonstrata London 1600. English translation by P. Fleury Mottelay, William Gilbert of colchester ... on the great magnet of the earth. Ann Arbor 1893 and Silvanus P. Thompson, reprinted from the 1900 edition by Basic Books, New York 1958.

Giles, P. M., et al. (1997): A subsurface flow of material from the Sun's equator to its pole. Nature **390**, 52–54.

Gille, J. C., Smythe, C. M., Heath, D. F. (1984): Observed ozone response to variations in solar ultraviolet radiation. Science **225**, 315–317.

Gilliland, R. L. (1980): Solar luminosity variations. Nature **286**, 838–839.

Gilliland, R. L. (1981): Solar radius variations over the past 265 years. Astrophysical Journal **248**, 1144–1155.

Gilliland, R. L. (1982): Modeling solar variability. Astrophysical Journal **253**, 399–405.

Gilliland, R. L. (1989): Solar evolution. Palaeogeography, Palaeoclimatology, Palaeocology **75**, 35–55.

Ginzburg, V. L. (1946): On solar radiation in the radio spectrum. Comptes Rendus (Doklady) de l'Académie des Sciences de l' URSS **52**, 487.

Ginzburg, V. L. (1956): The nature of cosmic radio emission and the origin of cosmic rays. Nuovo Cimento Supplement **3**, 38–48. Reproduced in: A source book in astronomy and astrophysics 1900–1975. Cambridge, Massachusetts: Harvard University Press 1977, pp. 677–684.

Giovanelli, R. G. (1939): The relations between eruptions and sunspots. Astrophysical Journal **89**, 555–567.

Giovanelli, R. G. (1940): Solar eruptions. Astrophysical Journal **91**, 344–349.

Giovanelli, R. G. (1946): A theory of chromospheric flares. Nature **158**, 81–82.

Giovanelli, R. G. (1947): Magnetic and electric phenomena in the Sun's atmosphere associated with sunspots. Monthly Notices of the Royal Astronomical Society **107**, 338–355.

Giovannelli, R. G. (1948): Chromospheric flares. Monthly Notices of the Royal Astronomical Society **108**, 163–176.

Giovanelli, R. G., Mc Cabe, M. K. (1958): The flare-surge event. Australian Journal of Physics **11**, 191–200.

Gleeson, L. J., Axford, W. I. (1968): Solar modulation of galactic cosmic rays. Astrophysical Journal **154**, 1011–1019.

Gleissberg, W. (1943): Predictions for the coming sunspot-cycle. Terrestrial Magnetism and Atmospheric Electricity **48**, 243–244.

Gleissberg, W. (1958): The eighty-year sunspot cycle. Journal of the British Astronomical Association **68**, 148–152.

Gleissberg, W. (1965): The 80-year solar cycle in auroral frequency number. Journal of the British Astronomical Association **75**, 227–231.

Gleissberg, W. (1966): Ascent and descent in the eighty-year cycles of solar activity. Journal of the British Astronomical Association **76**: 265–270.

Gloeckler, G., et al. (1989): Heavy ion abundances in coronal hole solar wind flows. EOS **70**, 424.

Gloeckler, G., Geiss, J. (1996): Abundance of ^{3}He in the local interstellar cloud. Nature **381**, 210–212.

Gold, T.: Discussion of shock waves and rarefied gases. In: Gas dynamics of cosmic clouds (Eds. J. C. Van De Hulst, J. M. Burgers). New York: North-Holland 1955, p. 103.

Gold, T. (1959): Magnetic field in the solar system. Nuovo Cimento Supplemento **13**, 318–323.

Gold, T. (1959): Plasma and magnetic fields in the solar system. Journal of Geophysical Research **64**, 1665–1674.

Gold, T. (1960): Energetic particle fluxes in the solar system and near the Earth. Astrophysical Journal Supplement **4**, 406–426.

Gold, T. (1962): Magnetic storms. Space Science Review **1**, 100–114.

Gold, T.: Magnetic energy shedding in the solar atmosphere. In: AAS-NASA symposium on the physics of solar flares NASA SP-50 (ed. W. N. Hess). Washington, DC: National Aeronautics and Space Administration 1964, pp. 389–395.

Gold, T., Hoyle, F. (1960): On the origin of solar flares. Monthly Notices of the Royal Astronomical Society **120**, 89–105.

Goldreich, P., Keeley, D. A. (1977): Solar seismology. I. The stability of the solar p-modes. Astrophysical Journal **211**, 934–942.

Goldreich, P., Keeley, D. A. (1977): Solar seismology. II. The stochastic excitation of the solar p-modes by turbulent convections. Astrophysical Journal **212**, 243–247.

Goldreich, P., Kumar, P. (1990): Wave generation by turbulent convection. Astrophysical Journal **363**, 694–704.

Goldstein, B. E. (1993): The solar wind as we know it today. EOS. Transactions of the American Geophysical Union **74**, No. 20, 229.

Goldstein, B. E., et al. (1996): Ulysses plasma parameters: latitudinal, radial, and temporal variations. Astronomy and Astrophysics **316**, 296–303.

Golub, L., et al. (1980): Magnetic fields and coronal heating. Astrophysical Journal **238**, 343–348.

Golub, L., Krieger, A. S., Silk, J. K., Timothy, A. F., Vaiana, G. S. (1974): Solar X-ray bright points. Astrophysical Journal (Letters) **189**, L93–L97.

Golub, L., Pasachoff, J. M.: The solar corona. New York: Cambridge University Press 1997.

Gonzalez, W. D. et al. (1994): What is a geomagnetic storm? Journal of Geophysical Research **99**, 5771–5792.

Gonzalez, W. D., Tsurutani, B. T. (1987): Criteria of interplanetary parameters causing intense magnetic storms (Dst < -100 nT). Planetary and Space Science **35**, 1101–1108.

Goode, P. R. et al. (1991): What we know about the Sun's internal rotation from oscillations. Astrophysical Journal **367**, 649–657.

Gopalswamy, N., Hanaoka, Y. (1998): Coronal dimming associated with a giant prominence eruption. Astrophysical Journal (Letters) **498**, L179–L182.

Gorney, D. J. (1990): Solar cycle effects on the near-earth space environment. Reviews of Geophysics **28**, 315–336.

Gosling, J. T. (1993): The solar flare myth. Journal of Geophysical Research **98**, 18,937–18,949.

Gosling, J. T.: The solar flare myth in solar-terrestrial physics. In: Solar system plasmas in space and time. geophysical monograph 84 (Eds. J. L. Burch and J. H. Waite, Jr.) Washington, DC: American Geophysical Union 1994. pp. 65–69.

Gosling, J. T.: Coronal mass ejections – an overview. In: Coronal mass ejections. geophysical monograph 99 (Eds. N. Crooker, J. A. Joselyn and J. Feynman). Washington, DC: American Geophysical Union 1997, pp. 9–16.

Gosling, J. T., et al. (1974): Mass ejections from the Sun: A view from Skylab. Journal of Geophysical Research **79**, 4581–4587.

Gosling, J. T., et al. (1975): Direct observations of a flare related coronal and solar wind disturbance. Solar Physics **40**, 439–448.

Gosling, J. T., et al. (1976): The speeds of coronal mass ejection events. Solar Physics **48**, 389–397.

Gosling, J. T., et al. (1976): Solar wind speed variations 1962–1974. Journal of Geophysical Research **81**, 5061–5070.

Gosling, J. T., et al. (1981): Coronal streamers in the solar wind at 1 AU. Journal of Geophysical Research **86**, A7, 5438–5448.

Gosling, J. T., et al: (1987): Bidirectional solar wind electron heat flux events. Journal of Geophysical Research **92**, 8519–8535.

Gosling, J. T., et al. (1990): Coronal mass ejections and large geomagnetic storms. Geophysical Research Letters **17**, 901–904.

Gosling, J. T., et al. (1991): Geomagnetic activity associated with Earth passsage of interplanetary shock disturbances and coronal mass ejections. Journal of Geophysical Research **96**, 7831–7839.

Gosling, J. T., et al. (1995): Coronal mass ejections at high heliographic latitudes: Ulysses. Space Science Reviews **72**, 133–136.

Gosling, J. T., Hundhausen, A. J. (1977): Waves in the solar wind. Scientific American **236**, 36–43 – March.

Gosling, J. T., Pizzo, V., Bame, S. J. (1973): Anomalously low proton temperatures in the solar wind following interplanetary shock waves – evidence for magnetic bottles? Journal of Geophysical Research **78**, 2001–2009.

Gough, D. O.: Random remarks on solar hydrodynamics. In: The energy balance and hydrodynamics of the solar chromosphere and corona. Proceedings of the international astronomical union colloquium no. 36 (Eds. R.-M. Bonnet and Ph. Delache). Paris: G. De Bussac, Clermont-Ferrand 1976. pp. 3–36.

Gough, D. O. (1981): Solar interior structure and luminosity variations. Solar Physics **74**, 21–34.

Gough, D. O. (1989): Deep roots of solar cycles. Nature **336**, 618–619.

Gough, D. O., et al. (1996): Perspectives in helioseismology. Science **272**, 1281–1284.

Gough, D. O., Toomre, J. (1991): Seismic observations of the solar interior. Annual Review of Astronomy and Astrophysics **29**, 627–684.

Graham, G. (1724): An account of observations made of the variation of the horizontal needle at London, in the latter part of the year 1722 and beginning of 1723. Philosophical Transactions of the Royal Society (London) **32**, No. 383, 96–107.

Grall, R. R., et al. (1996): Rapid acceleration of the polar solar wind. Nature **379**, 429–432.

Gray, D. F., Livingston, W. C. (1997): Monitoring the solar temperature: Spectroscopic temperature variations of the Sun. Astrophysical Journal **474**, 802–809.

Greaves, W. M. H., Newton, H. W. (1929): On the recurrence of magnetic storms. Monthly Notices of the Royal Astronomical Society **89**, 641–646.

Grec, G., Fossat, E., Pomerantz, M. A. (1980): Solar oscillations – full disk observations from the geographic south pole. Nature **288**, 541–544.

Grec, G., Fossat, E., Pomerantz, M. A. (1983): Full-disk observations of solar oscillations from the geographic south pole – latest results. Solar Physics **82**, 55–66.

Greenland Ice-Core Project (GRIP) Members (1993): Climate instability during the last interglacial period recorded in the GRIP ice core. Nature **364**, 203–207.

Gringauz, K. I. (1961): Some results of experiments in interplanetary space by means of charged particle traps on soviet space probes. Space Research **2**, 539–553.

Gringauz, K. I., et al. (1960): A study of the interplanetary ionized gas, high-energy electrons, and corpuscular radiation from the sun by means of the three-electrode trap for charged particles on the second soviet cosmic rocket. Soviet Physics (Doklady) **5**, 361–364.

GRIP Members (1993): Climate instability during the last interglacial period recorded in the GReenland Ice-core Project (GRIP) ice core. Nature **364**, 203–207.

Grotrian, W. (1934): Über das Fraunhofersche Spektrum der Sonnenkorona. Zeitschrift für Astrophysik **8**, 124–146.

Grotrian, W. (1939): On the question of the significance of the lines in the spectrum of the solar corona. Naturwissenschaften **27**, 214. English translation in: A source book in astronomy and astrophysics 1900–1975 (Eds. K. R. Lang and O. Gingerich). Cambridge, Massachusetts: Harvard University Press 1979, pp. 120–122.

Guenther, D. B., Demarque, P. (1997): Seismic tests of the Sun's interior structure, composition, and age, and implications for solar neutrinos. Astrophysical Journal **484**, 937–959.

Guenther D. B., Jaffe, A., Demarque, P. (1989): The standard solar model: Composition, opacities, and seismology. Astrophysical Journal **345**, 1022–1033.

Guhathakurta, M., Fisher, R. (1998): Solar wind consequences of a coronal hole density profile: Spartan 201–03 coronagraph and Ulysses observations from 1.15 $R_\odot$ to 4 AU. Astrophysical Journal (Letters) **499**, L215–L218.

Gurnett, D. A., et al. (1993): Radio emission from the heliopause triggered by an interplanetary shock. Science **262**, 199–203.

Habbal, S. R., et al. (1997): Origins of the slow and the ubiquitous fast solar wind. Astrophysical Journal (Letters) **489**, L103–L106.

Hagenaar, H. J., Schrijver, C. J., Title, A. M. (1997): The distribution of cell sizes of the solar chromospheric network. Astrophysical Journal **481**, 988–995.

Haigh, J. D. (1994): The role of stratospheric ozone in modulating the solar radiative forcing of climate. Nature **370**, 544–546.

Haigh, J. D. (1996): The impact of solar variability on climate. Science **272**, 981–984.

Haisch, B. M., Rodono, M. (1989): Solar and stellar flares: Proceedings of IAU colloquium no. 104. Solar Physics **121**, Nos. 1, 2. Reprinted Boston: Kluwer 1989.

Haisch, B., Strong, K. T., Rodono, M. (1991): Flares on the sun and other stars. Annual Review of Astronomy and Astrophysics **29**, 275–324.

Hale, G. E. (1892): A remarkable solar disturbance. Astronomy and Astrophysics **11**, 611–613.

Hale, G. E. (1892): On the condition of the sun's surface in June and July, 1892, as compared with the record of terrestrial magnetism. Astronomy and Astrophysics **11**, 917–925.

Hale, G. E. (1908): Solar vortices. Astrophysical Journal **28**, 100–116.

Hale, G. E. (1908): On the probable existence of a magnetic field in sun-spots. Astrophysical Journal **28**, 315–343. Reproduced in: A source book in astronomy and astrophysics 1900–1975 (Eds. K. R. Lang and O. Gingerich). Cambridge, Massachusetts: Harvard University Press 1979, pp. 96–105.

Hale, G. E. (1908): The Zeeman effect in the sun. Publications of the Astronomical Society of the Pacific **20**, 287–288.

Hale, G. E. (1926): Visual observations of the solar atmosphere. Proceedings of the National Academy of Science **12**, 286–295.

Hale, G. E. (1929): The spectrohelioscope and its work. Part I. History, instruments, adjustments, and methods of observation. Astrophysical Journal **70**, 265–311.

Hale, G. E. (1931): The spectrohelioscope and its work. Part III. Solar eruptions and their apparent terrestrial effects. Astrophysical Journal **73**, 379–412.

Hale, G. E., et al. (1919): The magnetic polarity of sun-spots. Astrophysical Journal **49**, 153–178.

Halley, E. (1716): An account of the late surprising appearance of the lights seen in the air. Philosophical Transactions of the Royal Society (London) **29**, 406–428.

Hammer, C. U. (1977): Past volcanism revealed by Greenland ice sheet impurities. Nature **270**, 482–486.

Hanaoka, Y. (1994): A flare caused by interacting coronal loops. Astrophysical Journal (Letters) **420**, L37–L40.

Hanaoka, Y. (1996): Flares and plasma flow caused by interacting coronal loops. Solar Physics **165**, 275–301.

Hanaoka, Y. (1997): Double-loop configuration of solar flares. Solar Physics **173**, 319–346.

Hansen, J. E., Lacis, A. A. (1990): Sun and dust versus greenhouse gases: an assessment of their relative roles in global climate change. Nature **346**, 713–719.

Hansen, J. E., Lebedeff, S. (1987): Global trends of measured surface air temperature. Journal of Geophysical Research **92**, 13,345–13,372.

Hansen, J. E., Lebedeff, S. (1988): Global surface air temperatures: Update through 1987. Geophysical Research Letters **15**, 323–326.

Hansen, R. T., Garcia, C. J., Hansen, S. F., Yasukawa, E. (1974): Abrupt depletions of the inner corona. Publications of the Astronomical Society of the Pacific **86**, 500–515.

Hara, H., et al. (1994): Temperatures of coronal holes observed with Yohkoh SXT. Publications of the Astronomical Society of Japan **46**, 493–502.

Harrison, R. A. (1991): Coronal mass ejection. Philosophical Transactions of the Royal Society (London) **A336**, 401–412.

Harrison, R. A. (1994): A statistical study of the coronal mass ejection phenomenon. Advances in Space Research **14**, No. 4, 23–28.

Harrison, R. A.: CME onset studies. In: The corona and solar wind near minimum activity. Proceedings of the fifth SOHO workshop ESA SP-404. Noordwijk, The Netherlands: ESA Publications Division 1997, pp. 85–91.

Harrison, R. A. (1997): EUV blinkers – the significance of variations in the extreme ultraviolet quiet Sun. Solar Physics **175**, 467–485. Reprinted in: The first results from SOHO (Eds. B. Fleck and Z. Svestka). Boston: Kluwer Academic Publishers 1997, pp. 467–485.

Harrison, R. A.: One year of CDS: Highlights from observations using the coronal diagnostic spectrometer on SOHO. In: The corona and solar wind near minimum activity. Proceedings of the fifth SOHO workshop. ESA SP-404. Noorwijk, The Netherlands: ESA Publications Division 1997. pp. 7–16.

Harrison, R. A., et al. (1990): The launch of solar coronal mass ejections: Results from the coronal mass ejection onset program. Journal of Geophysical Research **95**, 917–937.

Hart, A. B. (1954): Motions in the Sun at the photospheric level. IV. The equatorial rotation and possible velocity fields in the photosphere. Monthly Notices of the Royal Astronomical Society **114**, 17–38.

Hart, A. B. (1956): Motions in the Sun at the photopsheric level. VI. Large scale motions in the equatorial region. Monthly Notices of the Royal Astronomical Society **116**, 38–55.

Hartle, R. E., Sturrock, P. A. (1968): Two-fluid model of the solar wind. Astrophysical Journal **151**, 1155–1170.

Hartmann, L. W., Noyes, R. W. (1987): Rotation and magnetic activity in main-sequence stars. Annual Review of Astronomy and Astrophysics **25**, 271–301.

Hartz, T. R. (1964): Solar noise observations from the alouette satellite. Annales d'Astrophysique **27**, 831–836.

Hartz, T. R. (1969): Type III solar radio noise bursts at hectometer wavelengths. Planetary and Space Science **17**, 267–287.

Harvey, J. W. (1995): Helioseismology. Physics Today **48**, 32–38 – October.

Harvey, J. W., et al. (1996): The global oscillation network group (GONG) project. Science **272**, 1284–1286.

Harvey, J. W., Kennedy, J. R., Leibacher, J. W. (1987): GONG – to see inside our sun. Sky and Telescope **74**, 470–476 – November.

Harvey, K. L., Martin, S. F. (1973): Ephemeral active regions. Solar Physics **32**, 389–402.

haselgrove, B., Hoyle, F. (1959): Main-sequence stars. Monthly Notices of the Royal Astronomical Society **119**, 112–120.

Hassler, D. M., et al. (1997): Observations of polar plumes with the SUMER instrument on SOHO. Solar Physics **175**, 375–391. Reprinted in: The first results from SOHO (Eds. B. Fleck and Z. Svestka). Boston: Kluwer Academic Publishers 1997, pp. 375–391.

Hassler, D. M., et al. (1998): Solar wind outflow and the chromospheric magnetic network. Science **283**, 810–813.

Hathaway, D. H., et al. (1996): GONG observations of solar surface flows. Science **272**, 1306–1309.

Hays, J. D., Imbrie, J., Shackleton, N. J. (1976): Variations in the Earth's orbit: Pacemaker of the ice ages. Science **194**, 1121–1132.

Heath, D. F., Krueger, A. J., Crutzen, J. (1977): Solar proton events: Influence on stratospheric ozone. Science **197**: 888–889.

Heaviside, O. (1902). In: Encyclopedia Britannica **113**, 215.

Henderson-Sellers, A. (1979): Clouds and the long term stability of the earth's atmosphere and climate. Nature **279**, 786–788.

Hess, V. F. (1912): Concerning observations of penetrating radiation on seven free balloon flights. Physikalishe Zeitschrift **13**, 1084–1091. English translation in: A source book in astronomy and astrophysics 1900–1975 (Eds. K. R. Lang and O. Gingerich). Cambridge, Massachusetts: Harvard University Press 1979, pp. 13–20. Also reproduced in English in Hillas (1972).

Hewish, A. (1955): The irregular structure of the outer regions of the solar corona. Proceedings of the Royal Society (London) **228A**, 238–251.

Hewish, A. (1958): The scattering of radio waves in the solar corona. Monthly Notices of the Royal Astronomical Society **118**, 534–546.

Hey, J. S. (1946): Solar radiations in the 4–6 metre radio wavelength band. Nature **157**, 47–48.

Hey, J. S.: The evolution of radio astronomy. New York: Science History Publications, Neale Watson Academic Publications 1973.

Heyvaerts, J., Priest, E. R. (1984): Coronal heating by reconnection in DC current systems. A theory based on Taylor's hypothesis. Astronomy and Astrophysics **137**, 63–78.

Heyvaerts, J., Priest, E. R., Rust, D. M. (1977): An emerging flux model for the solar flare phenomenon. Astrophysical Journal **216**, 123–137.

Hick, P., et al. (1995): Synoptic IPS and Yohkoh soft X-ray observations. Geophysical Research Letters **22**, 643–646.

Hickey, J. R., et al.: Solar variability indications from Nimbus 7 satellite data. In: Variations of the solar constant (ed. S. Sofia). Washington, DC: NASA CP-2191 1981, pp. 59–72.

Hickey, J. R., et al. (1988): Observation of total solar irradiance variability from Nimbus satellites. Advances in Space Research **9**, No. 7, 5–10.

Hickey, J. R., et al. (1988): Total solar irradiance measurements by ERB/Nimbus-7: A review of nine years. Space Science Reviews **48**, 321–342.

Hiei, E., Hundhausen, A. J., Sime, D. G. (1993): Reformation of a coronal helmet streamer by magnetic reconnection after a coronal mass ejection. Geophysical Research Letters **20**, 2785–2788.

Hildner, E., et al. (1975): The sources of material comprising a mass ejection coronal transient. Solar Physics **45**, 363–376.

Hill, T. W., Dessler, A. J. (1991): Plasma motions in planetary magnetospheres. Science **252**, 410–415 – April.

Hillas, A. M.: Cosmic Rays. New York: Pergamon Press 1972.

Hines, C. O. (1974): A possible mechanism for the production of sun-weather correlations. Journal of Atmospheric Science **31**, 589–591.

Hirayama, T. (1974): Theoretical model of flares and prominences. I. Evaporating flare model. Solar Physics **34**, 323–338.

Hirshberg, J., Bame, S. J., Robbins, D. E. (1972): Solar flares and solar wind helium enrichments: July 1965–July 1967. Solar Physics **23**, 467–486.

Hirshberg, J., Colburn, D. S. (1969): Interplanetary field and geomagnetic variations – a unified view. Planetary and Space Science **17**, 1183–1206.

Hodgson, R. (1860): On a curious appearance seen in the Sun. Monthly Notices of the Royal Astronomical Society **20**, 15–16. Reproduced in: Early solar physics (Ed. A. J. Meadows). Oxford: Pergamon Press 1970, p. 185.

Hoeksema, J. T., Scherrer, P. H. (1987): Rotation of the coronal magnetic field. Astrophysical Journal **318**, 428–436.

Hoeksema, J. T., Wilcox, J. M., Scherrer, P. H. (1982): Structure of the heliospheric current sheet in the early portion of sunspot cycle 21. Journal of Geophysical Research **87A**, 10,331–10,338.

Hollweg, J. V. (1972): Alfvénic motions in the solar atmosphere. Astrophysical Journal **177**, 255–259.

Hollweg, J. V. (1973): Transverse Alfvén waves in the solar wind. Wave pressure, Poynting flux, and angular momentum. Journal of Geophysical Research **78**, 3643–3652.
Hollweg, J. V. (1975): Waves and instabilities in the solar wind. Reviews of Geophysics **13**, 263–289.
Hollweg, J. V. (1978): Some physical processes in the solar wind. Reviews of Geophysics **16**, 689–720.
Hollweg, J. V. (1984): Resonances of coronal loops. Astrophysical Journal **277**, 392–403.
Holzer, T. E. (1977): Effects of rapidly diverging flow, heat addition, and momentum addition in the solar wind and stellar winds. Journal of Geophysical Research **82**, 23–35.
Holzer, T. E. (1989): Interaction between the solar wind and the interstellar medium Annual Review of Astronomy and Astrophysics **27**, 199–234.
Holzer, T. E., Axford, W. I. (1970): The theory of stellar winds and related flows. Annual Review of Astronomy and Astrophysics **8**, 31–60.
Holzer, T. E., Leer, E.: Coronal hole structure and the high speed solar wind. In: The corona and solar wind near minimum activity. Proceedings of the fifth SOHO workshop. ESA SP-404. Noordwijk, The Netherlands: ESA Publications Division 1997, pp. 65–74.
Hood, L. L. (1987): Solar ultraviolet radiation induced variations in the stratosphere and mesosphere. Journal of Geophysical Research **92**, 876–888.
Howard, R. A., et al. (1982): The observation of a coronal transient directed at earth. Astrophysical Journal (Letters) **263**, L101–L104.
Howard, R. A., et al. (1985): Coronal mass ejections: 1979–1981. Journal of Geophysical Research **90**, 8173–8191.
Howard, R. A., et al.: Observations of CMEs from SOHO/LASCO. In: Coronal mass ejections. geophysical monograph 99 (Eds. N. Crooker, J. A. Joselyn and J. Feynman). Washington, DC: American Geophysical Union 1997, pp. 17–26.
Howard, R. A., Koomen, M. J. (1974): Observation of sectored structure in the outer solar corona: Correlation with interplanetary magnetic field. Solar Physics **37**, 469–475.
Howard, R. F. (1974): Studies of solar magnetic fields. Solar Physics **38**, 283–299.
Howard, R. F. (1985): Eight decades of solar research at Mount Wilson. Solar Physics **100**, 171–187.
Howard, R. F., La Bonte, B. J. (1980): The Sun is observed to be a torsional oscillator with a period of 11 years. Astrophysical Journal (Letters) **239**, L33–L36.
Howe, R., et al. (2000): Dynamic variations at the base of the solar convection zone. Science **287**, 2456–2460
Hoyle, F.: Some recent researches in solar physics. Cambridge, England: Cambridge at the University Press 1949.
Hoyle, F.: Remarks on the computation of stellar evolution tracks. In: Stellar populations (Ed. J. K. O'Connell). Vatican City: Specola Vaticana 1958, pp. 223–226.
Hoyle, F., Bates, D. R. (1948): The production of the E-layer. Terrestrial Magnetism and Atmospheric Electricity **53**, 51–62.
Hoyng, P., et al. (1981): Origin and location of the hard X-ray emission in a two-ribbon flare. Astrophysical Journal (Letters) **246**, L155–L159.
Hoyt, D. V, Eddy, J. A., Hudson, H. S. (1983): Sunspot areas and solar irradiance variations during 1980. Astrophysical Journal **275**, 878–888.
Hoyt, D. V. et al. (1992): The Nimbus 7 solar total irradiance: a new algorithm for its derivation. Journal of Geophysical Research **97**, 51–63.
Hoyt, D. V., Schatten, K. H. (1993): A discussion of plausible solar irradiance variations, 1700–1992. Journal of Geophysical Research **98**, 18,895–18,906.
Hoyt, D. V., Schatten, K. H.: The role of the sun in climate change. New York: Oxford University Press 1997.

Hoyt, D. V., Schatten, K. H., Nesmes-Ribes, E. (1994): The one hundredth year of Rudolf Wolf's death: Do we have the correct reconstruction of solar activity? Geophysical Research Letters **21**, 2067–2070.

Hsieh, K. C., Simpson, J. A. (1970): The relative abundances and energy spectra of ^{3}He and ^{4}He from solar flares. Astrophysical Journal (Letters) **162**, L191–L196.

Hudson, H. S. (1972): Thick-target processes and white-light flares. Solar Physics **24**, 414–428.

Hudson, H. S. (1987): Solar flare discovery. Solar Physics **113**, 1–12.

Hudson, H. S. (1988): Observed variability of the solar luminosity. Annual Review of Astronomy and Astrophysics **26**, 473–508.

Hudson, H. S. (1991): Solar flares, microflares, nanoflares and coronal heating. Solar Physics **133**, 367–369.

Hudson, H. S.: The solar antecedents of geomagnetic storms. In: Magnetic storms (Eds. B. T. Tsurutani, W. D. Gonzales, Y. Kamide). Washington, DC: American Geophysical Union 1997.

Hudson, H. S., Acton, L. W., Freeland, S. L. (1996): A long-duration solar flare with mass ejection and global consequences. Astrophysical Journal **470**, 629–635.

Hudson, H. S., et al. (1982): The effects of sunspots on solar irradiance. Solar Physics **76**, 211–219.

Hudson, H. S., et al. (1994): Impulsive behavior in solar soft X-radiation. Astrophysical Journal (Letters) **422**, L25–L27.

Hudson, H. S., et al. (1998): X-ray coronal changes during halo CMEs. Geophysical Research Letters **25**, 2481–2484.

Hudson, H. S., Haisch, B. M., Strong, K. T. (1995): Comment on 'The solar flare myth' by J. T. Gosling. Journal of Geophysical Research **100**, 3473–3477.

Hudson, H. S., Ryan, J. (1995): High-energy particles in solar flares. Annual Review of Astronomy and Astrophysics **33**, 239–282.

Hudson, H. S., Silva, S., Woodard, M. (1982): The effect of sunspots on solar irradiance. Solar Physics **76**, 211–219.

Hudson, H. S., Webb, D. F.: Soft X-ray signatures of coronal ejections. In: Coronal mass ejections. geophysical monograph 99 (Eds. N. Crooker, J. A. Joselyn and J. Feynman). Washington, DC: American Geophysical Union 1997, pp. 27–38.

Hufbauer, K.: Exploring the sun: Solar science since Galileo. Baltimore, Maryland: Johns Hopkins University Press 1991.

Hulburt, E. O. (1938): Photoelectric ionization in the ionosphere. Physical Review **53**, 344–351.

Humboldt, F. W. H. A. von: Voyage aux régions équinoxiales du Nouveau Continent, fait en 1799, 1800, 1801, 1802, 1803, et 1804 par Al [exandre] de Humboldt et A [imé] Bonpland. Paris., 1805–1834.

Humboldt, F. W. H. A. von: Kosmos. 1845, 1847.

Hundhausen, A. J.: Coronal expansion and solar wind. New York: Springer-Verlag 1972.

Hundhausen, A. J.: Interplanetary shock waves and the structure of solar wind disturbances. In: Solar Wind: NASA SP-308 (Eds. C. P. Sonett, P. J. Coleman, J. M. Wilcox). Washington, DC: NASA 1972, pp. 393–417.

Hundhausen, A. J.: An interplanetary view of coronal holes. In: Coronal holes and high speed wind streams (ed. J. Zirker). Boulder, Colorado: Colorado Associated University Press 1977, pp. 225–329.

Hundhausen, A. J. (1993): Sizes and locations of coronal mass ejections: SMM observations from 1980 and 1984–1989. Journal of Geophysical Research **98**, 13,177–13,200.

Hundhausen, A. J.: An introduction. In: Coronal mass ejections: geophysical monograph 99 (Eds. N. Crooker, J. A. Joselyn and J. Feynman). Washington, DC: American Geophysical Union 1997, pp. 1–7.

Hundhausen, A. J.: Coronal mass ejections. In: Cosmic winds and the heliosphere (Eds. J. R. Jokipii, C. P. Sonett, and M. S. Giampapa). Tucson, Arizona: University of Arizona Press 1997, pp. 259–296.

Hundhausen, A. J., Bame, S. J., Montgomery, M. D. (1970): Large-scale characteristics of flare-associated solar wind disturbances. Journal of Geophysical Research **75**, 4631–4642.

Hundhausen, A. J., Burkepile, J. T., St. Cyr, O. C. (1994): Speeds of coronal mass ejections: SMM observations from 1980 and 1984–1989. Journal of Geophysical Research **99**, 6543–6552.

Hundhausen, A. J., Stanger, A. L., Serbicki, S. A.: Mass and energy contents of coronal mass ejections: SMM results from 1980 and 1984–1989. In: Solar dynamical phenomena and solar wind consequences. Proceedings of the third SOHO workshop. ESA SP-373. Noordwijk, The Netherlands: ESA Publications Division 1994, pp. 409–412.

Imbrie, J. (1982): Astronomical theory of the Pleistocene ice ages. A brief historical. Icarus **50**, 408–432.

Imbrie, J., et al.: The orbital theory of Pleistocene climate: Support from a revised chronology of the marine δ^{18}O record. In: Milankovitch and climate, part 1 (Eds. A. L. Berger et. al.). The Netherlands: Reidel 1984, pp. 269–305.

Imbrie, J., et al. (1992): On the structure and origin of major glaciation cycles. 1. Linear responses to Milankovich forcing. Paleoceanography **7**, 701–738.

Imbrie, J., Imbrie, J. Z. (1980): Modeling the climatic response to orbital variations. Science **207**, 943–953.

Imbrie, J., Imbrie, K. P.: Ice ages – solving the mystery. Short Hills, New Jersey: Enslow Publishers 1979.

Imbrie, J., Imbrie, K. P.: Ice ages – solving the mystery. Second edition. Cambridge, Massachusetts: Harvard University Press 1986

Innes, D. E, Inhester, B., Axford, W. I., Wilhelm, K. (1997): Bi-directional plasma jets produced by magnetic reconnection on the Sun. Nature **386**, 811–813.

Isenberg, P. (1991). The solar wind. Geomagnetism **4**, 1–85.

James, I. N., James, P. M. (1989): Ultra-low-frequency variability in a simple atmospheric circulation model. Nature **342**, 53–55.

Janssen, P. J. C. (1872): Observations of the solar eclipse of 12 December 1871. Nature **5**, 249. Reproduced in: Early solar physics (ed. A. J. Meadows). New York: Pergamon Press 1970, pp. 223–224.

Jockers, K. (1970): Solar wind models based on exospheric theory. Astronomy and Astrophysics **6**, 215–239.

Johnson, T. H. (1938): Nature of primary cosmic radiation. Physical Review **54**, 385–387.

Jokipii, J. R. (1971): Propagation of cosmic rays in the solar wind. Review of Geophysics and Space Physics **9**, 27–87.

Jokipii, J. R., et al. (1995): Interpretation and consequences of large-scale magnetic variances observed at high heliographic latitude. Geophysical Research Letters **22**, No. 23, 3385–3388.

Jokipii, J. R., KÓta, J. (1989): The polar heliospheric magnetic field. Geophysical Research Letters **16**, 1–4.

Jokipii, J. R., Mc Donald, F. B. (1995): Quest for the limits of the heliosphere. Scientific American **272**, 58–63 – April.

Jokipii, J. R., Sonett, C. P., Giampapa, M. S. (eds.): Cosmic winds and the heliosphere. Tucson, Arizona: University of Arizona Press 1997.

Jones, P. D., Wigley, T. M. L., Wright, P. B. (1986): Global temperature variations between 1861 and 1984. Nature **322**, 430–434.

Joselyn, J. A., Mc Intosh, P. S. (1981): Disappearing solar filaments: A useful predictor of geomagnetic activity. Journal of Geophysical Research **86**, 4555–4564.

Jouzel, J., et al. (1987): Vostok ice core: a continuous isotope temperature record over the last climatic cycle (160,000 years). Nature **329**, 402–408.

Jouzel, J., et al. (1993): Extending the Vostok ice core record of palaeoclimate to the penultimate glacial period. Nature **364**, 407–412.

Joy, A. H., Humason, M. L. (1949): Observations of the faint dwarf star L726-8. Publications of the Astronomical Society of the Pacific **61**, 133–134.

Kahler, S. W. (1977): The morphological and statistical properties of solar X-ray events with long decay times. Astrophysical Journal **214**, 891–897.

Kahler, S. W. (1982): The role of the big flare syndrome in correlations of solar energetic proton fluxes and associated microwave burst parameters. Journal of Geophysical Research **87**, 3439–3448.

Kahler, S. W. (1987): Coronal mass ejections. Reviews of Geophysics **25**, 663–675.

Kahler, S. W. (1992): Solar flares and coronal mass ejections. Annual Review of Astronomy and Astrophysics **30**, 113–141.

Kahler, S. W., et al. (1986): Solar filament eruptions and energetic particle events. Astrophysical Journal **302**, 504–510.

Kahler, S. W., Sheeley, N. R., Jr., Liggett, M. (1989): Coronal mass ejections and associated X-ray flare durations. Astrophysical Journal **344**, 1026–1033.

Kahn, F. D. (1961): Sound waves trapped in the solar atmosphere. Astrophysical Journal **134**, 343–346.

Kakinuma, T.: Observations of interplanetary scintillation: solar wind velocity measurements. In: Study of travelling interplanetary phenomena (Eds. M. A. Shea, D. F. Smart and S. T. Wu.) Dordrecht, Holland: D. Reidel 1977, pp. 101–118.

Kallenrode, M.-B.: Space physics. An introduction to plasmas and particles in the heliosphere and magnetospheres. New York: Springer-Verlag 1998.

Kanbach, G., et al. (1993): Detection of a long-duration solar gamma-ray flare on June 11, 1991 with EGRET on COMPTON-GRO. Astronomy and Astrophysics Supplement Series **97**, 349–353.

Kane, S. R.: Impulsive (flash) phase of solar flares: Hard X-ray, microwave, EUV and optical emissions. In: Coronal disturbances. Proceedings of IAU symposium no. 57 (Ed. G. Newkirk, Jr.) Boston: D. Reidel 1974. pp. 105–141.

Kane, S. R. (Ed.): Solar gamma-, X-, and EUV radiation. Proceedings of IAU symposium no. 68. Boston: D. Reidel 1975.

Kane, S. R., et al.: Impulsive phase of solar flares. In: Solar flares: A monograph from Skylab solar workshop II (Ed. P. A. Sturrock). Boulder, Colorado: Colorado Associated University Press 1980. pp. 187–229.

Kane, S. R., et al. (1986): Rapid acceleration of energetic particles in the 1982 February 8 solar flare. Astrophysical Journal (Letters) **300**, L95–L98.

Kane, S. R., et al. (1995): Energy release and dissipation during giant solar flares. Astrophysical Journal (Letters) **446**, L47–L50.

Kano, R., Tsuneta, S. (1995): Scaling law of solar coronal loops obtained with Yohkoh. Astrophysical Journal **454**, 934–944.

Kano, R., Tsuneta, S. (1996): Temperature distributions and energy scaling law of solar coronal loops obtained with Yohkoh. Publications of the Astronomical Society of Japan **48**, 535–543.

Karlén, W., Kuylenstierna, J: Evidence from the Scandinavian tree lines since the last ice age. In: The global warming debate (ed. J. Emsley). London: European Science and Environment Forum, 1996. pp. 192–204.

Kasting, J. F., Ackerman, T. P. (1986): Climatic consequences of very high carbon dioxide levels in Earth's early atmosphere. Science **234**, 1383–1385.

Kasting, J. F., Grinspoon, D. H.: The faint young sun problem. In: The Sun in time (Eds. C. P. Sonett, M. S. Giampapa, M. S. Matthews). Tucson, Arizona: The University of Arizona Press 1991, pp. 447–462.

Kawabata, K. (1960): The relationship between post-burst increases of solar microwave radiation and sudden ionospheric disturbances. Report of Ionosphere and Space Research in Japan **14**, 405–426.

Keating, G. M., et al. (1986): Detection of stratospheric HNO_3 and NO_2 response to short-term solar ultraviolet variability. Nature **322**, 43–46.

Keating, G. M., et al. (1987): Response of middle atmosphere to short-term ultraviolet variations 1. Observations. Journal of Geophysical Research **92**, 889–902.

Kellogg, P. J. (1962): Flow of plasma around the earth. Journal of Geophysical Research **67**, 3805–3811.

Kelly, P. M. (1977): Solar influence on North Atlantic mean sea level pressure. Nature **269**, 320–322.

Kelly, P. M., Wigley, T. M. L. (1990): The influence of solar forcing trends on global mean temperature since 1861. Nature **347**, 460–462.

Kelly, P. M., Wigley, T. M. L. (1992): Solar cycle length, greenhouse forcing and global climate. Nature **360**, 328–330.

Kelvin, Lord: see W. Thomson.

Kennelly, A. E. (1902): Electrical World **39**, 473.

Kiepenheuer, K. O. (1950): Cosmic rays and radio emission from our galaxy. Physical Review **79**, 738–739. Reproduced in: A source book in astronomy and astrophysics 1900–1975 (Eds. K. R. Lang and O. Gingerich). Cambridge, Massachusetts: Harvard University Press 1977, pp. 677–679.

Kirchhoff, G. (1861): On the chemical analysis of the solar atmosphere. Philosophical Magazine and Journal of Science **21**, 185–188. Reproduced in: Early solar physics (Ed. A. J. Meadows). New York: Pergamon Press 1970. pp. 103–106.

Kirchhoff, G., Bunsen, R. (1860): Chemical analysis of spectrum – observations. Philosophical Magazine and Journal of Science **20**, 89–109, **22**, 329–349, 498–510 (1861).

Kivelson, M. G., Russell, C. T. (Eds).: Introduction to space physics. Cambridge, England: Cambridge University Press 1997.

Klein, J., et al. (1980): Radiocarbon concentrations in the atmosphere: 8000 year record of variations in tree rings. Radiocarbon **22**, 950–961.

Klimchuk, J. A., Gary, D. E. (1995): A comparison of active region temperatures and emission measures observed in soft X-rays and microwaves and implications for coronal heating. Astrophysical Journal **448**, 925–937.

Klimchuk, J. A., Porter, L. J. (1995): Scaling of heating rates in solar coronal loops. Nature **377**, 131–133 (1995).

Ko, Y-K., et al. (1997): An empirical study of the electron temperature and heavy ion velocities in the south polar coronal hole. Solar Physics **171**, 345–361.

Kohl, J. L., et al. (1980): Measurement of coronal temperatures from 1.5 to 3 solar radii. Astrophysical Journal (Letters) **241**, L117–L121.

Kohl, J. L., et al. (1995): Spartan 201 coronal spectroscopy during the polar passes of Ulysses. Space Science Reviews **72**, 29–38.

Kohl, J. L., et al. (1997): First results from the SOHO ultraviolet coronagraph spectrometer. Solar Physics **175**, 613–644. Reprinted in: The first results from SOHO (Eds. B. Fleck and Z. Svestka). Boston: Kluwer Academic Publishers 1997, pp. 613–644.

Kohl, J. L., et al. (1998): UVCS/SOHO empirical determinations of anisotropic velocity distributions in the solar corona. Astrophysical Journal (Letters) **501**, L127–L131.

Kohl, J. L., Withbroe, G. L. (1982): EUV spectroscopic plasma diagnostics for the solar wind acceleration region. Astrophysical Journal **256**, 263–270.

Kojima, M., Kakinuma, T. (1987): Solar cycle evolution of solar wind speed structure between 1973 and 1985 observed with the interplanetary scintillation method. Journal of Geophysical Research **92**, 7269–7279.

Kolhörster, W. (1913): Messungen der durchdringenden Strahlung im Freiballon im größeren Höhen. Physikalishe Zeitschrift **14**, 1153–1156.

Kominz, M. A., Pisias, N. G. (1979): Pleistocene climate: Deterministic or stochastic. Science **204**, 171–173.

Kopp, R. A., Holzer, T. E. (1976): Dynamics of coronal hole regions 1.: Steady polytropic flows with multiple critical points. Solar Physics **49**, 43–56.

Kopp, R. A., Pneuman, G. W. (1976): Magnetic reconnection in the corona and the loop prominence phenomenon. Solar Physics **50**, 85–98.

Kosovichev, A. G., et al. (1997): Structure and rotation of the solar interior: Initial results from the MDI medium-L program. Solar Physics **170**, 43–61. Reprinted in: The first results from SOHO (Eds. B. Fleck and Z. Svestka). Boston: Kluwer Academic Publishers 1997, pp. 43–61.

Kosovichev, A. G., Schou, J. (1997): Detection of zonal shear flows beneath the Sun's surface from f-mode frequency splitting. Astrophysical Journal (Letters) **482**, L207–L210.

Kosovichev, A. G., Zharkova, V. V. (1998): X-ray flare quakes the Sun. Nature **393**, 317.

Kosugi, T., et al. (1992): The hard X-ray telescope (HXT) onboard Yohkoh: Its performance and some initial results. Publications of the Astronomical Society of Japan **44**, L45–L49.

Kozlovsky, B, Ramaty, R. (1977): Narrow lines from alpha-alpha reactions. Astrophysical Letters **19**, 19–24.

Kreplin, R. W. (1961): Solar X-rays. Annales de Géophysique **17**, 151–161.

Kreplin, R. W., Chubb, T. A., Friedman, H. (1962): X-ray and Lyman-alpha emission from the Sun as measured from the NRL SR-1 satellite. Journal of Geophysical Research **67**, 2231–2253.

Krieger, A. S., et al.: X-ray observations of coronal holes and their relation to high velocity solar wind streams. In: Solar Wind Three (Ed. C. T. Russell). Los Angeles, California: Institute of Geophysics and Planetary Physics, UCLA 1974.

Krieger, A. S., Timothy, A. F., Roelof, E. C. (1973): A coronal hole and its identification as the source of a high velocity solar wind stream. Solar Physics **29**, 505–525.

Krucker, S., Benz, A. O. (1998): Energy distribution of heating processes in the quiet solar corona. Astrophysical Journal (Letters) **501**, L213–L216.

Krucker, S., Benz, A. O., Bastian, T. S., Acton, L. W. (1997): X-ray network flares of the quiet sun. Astrophysical Journal **488**, 499–505.

Krüger, A.: Introduction to solar radio astronomy and radio physics. Dordrecht: D. Reidel 1979.

Kuhn, J. R., Libbrecht, K. G., Dicke, R. H. (1988): The surface temperature of the Sun and changes in the solar constant. Science **242**, 908–911.

Kuhn, J. R., Kasting, J. F. (1983): The effects of increased CO_2 concentrations of surface temperature of the early earth. Nature **301**, 53–55.

Kukla, G. (1975): Missing link between Milankovitch and climate. Nature **253**, 600–603.

Kukla, G., et al. (1981): Orbital signature of interglacials. Nature **290**, 295–300.

Kundu, M. R. (1961): Bursts of centimeter-wave emission and the region of origin of X-rays from solar flares. Journal of Geophysical Research **66**, 4308–4312.

Kundu, M. R.: Solar radio astronomy. New York: Wiley Interscience 1965.

Kundu, M. R. (1982): Advances in solar radio astronomy. Reports on Progress in Physics **45**, 1435–1541.

Kundu, M. R., et al. (1995): Microwave and hard X-ray observations of footpoint emission from solar flares. Astrophysical Journal **454**, 522–530.

Kundu, M. R., Lang, K. R. (1985): The sun and nearby stars: Microwave observations at high resolution. Science **228**, 9–15.

Kundu, M. R., Vlahos, L. (1982): Solar microwave bursts – a review. Space Science Reviews **32**, 405–462.

Kundu, M. R., Woodgate, B., Schmahl, E. J. (Eds.): Energetic phenomena on the sun. Boston: Kluwer Academic Publishers 1989.

Kuperus, M., Ionson, J. A., Spicer, D. S. (1981): On the theory of coronal heating mechanisms. Annual Review of Astronomy and Astrophysics **19**, 7–40.

Kyle, H. L., Hoyt, D. V., Hickey, J. R.: A review of the Nimbus 7 ERB solar data set. In: The sun as a variable star: Solar and stellar luminosity variations (Eds. J. M. Pap, C. Fröhlich, H. S. Hudson, S. K. Solanki). New York: Cambridge University Press 1994, pp. 9–12.

Labitzke, K. (1987): Sunspots, the QBO and the stratospheric temperature in the north polar region. Geophysical Research Letters **14**, 535–537.

Labitzke, K., Van Loon, H. (1988): Associations between the 11-year solar cycle, the QBO, and the atmosphere. Part 1. The troposphere and stratosphere in the northern hemisphere winter. Journal of Atmospheric and Terrestrial Physics **50**, 197–206.

Labitzke, K., Van Loon, H. (1990): Associations between the 11-year solar cycle, the quasi-biennial oscillation and the atmosphere: a summary of recent work. Philosophical Transactions of the Royal Society (London) **A330**, 557–589.

Labitzke, K., Van Loon, H. (1993): Some recent studies of probable connections between solar and atmospheric variability. Annales Geophysicae **11**, 1084–1094.

Lacis, A. A., Carlson, B. E. (1992): Global warming: Keeping the Sun in proportion. Nature **360**, 297.

Lamb, H. H. (1965): The early Medieval warm epoch and its sequel. Palaeogeography, Palaeoclimatology, Palaeoecology **1**, 13–37.

Lamb, H. H.: Climate: present, past and future: Climate history and the future. London: Methuen 1977.

Lamb, H. H.: Climate history and the modern world. London: Methuen 1982.

Lane, J. H. (1870): On the theoretical temperature of the sun; under the hypothesis of a gaseous mass maintaining its volume by its internal heat, and depending on the laws of gases as known to terrestrial experiment. American Journal of Science and Arts (2nd series) **50**, 57–74. Reproduced in: Early Solar Physics (Ed. A. J. Meadows). New York: Pergamon Press 1970, pp. 257–276.

Lang, K. R. (1994): Radio evidence for non-thermal particle acceleration on stars of late spectral type. Astrophysical Journal Supplement **90**, 753–764.

Lang, K. R. (1996): Unsolved mysteries of the sun – Part 1, 2. Sky and Telescope **92**, No. 2, 38–42 – August, **92**, No. 3, 24–28 – September.

Lang, K. R. (1997): SOHO reveals the secrets of the sun. Scientific American **276**, No. 3, 32–47 – March. Updated in Magnificent Cosmos, a Scientific American Publication (1998) – March.

Lang, K. R.: Astrophysical formulae. vol. I. Radiation, gas processes and high energy astrophysics. New York: Springer Verlag 1999.

Lang, K. R.: Astrophysical formulae. vol. II. Space, time, mass and cosmology. New York: Springer Verlag 1999.

Lang, K. R.: The sun. In: The new solar system (Eds. J. Kelly Beatty, C. C. Petersen, A. Chaikin). New York: Cambridge University Press 1999, pp. 23–38.

Lang, K. R., et al. (1993): Magnetospheres of solar active regions inferred from spectral-polarization observations with high spatial resolution. Astrophysical Journal **419**, 398–417.

Lang, K. R., Gingerich, O.: A source book in astronomy and astrophysics 1900–1975. Cambridge, Massachusetts: Harvard University Press 1979.

Langer, S. H., Petrosian, V. (1977): Impulsive solar X-ray bursts. III. Polarization, directivity, and spectrum of the reflected and total bremsstrahlung radiation from a beam of electrons directed toward the photosphere. Astrophysical Journal **215**, 666–676.

Lassen, K., Friis-Christensen, E. (1995): Variability of the solar cycle length during the past five centuries and the apparent association with terrestrial climate. Journal of Atmospheric and Terrestrial Physics **57**, 835–845.

Lassen, K., Friis-Christensen, E.: A long-term comparison of sunspot cycle length and temperature change from Zurich observatory. In: The global warming debate (Ed. J. Emsley). London: European Science and Environment Forum, 1996, pp. 224–232.
Lattes, C. M. G., et al. (1947): Processes involving charged mesons. Nature **159**, 694–697.
Lattes, C. M. G., Occhialini, G. P. S., Powell, C. F. (1947): Observations on the tracks of slow mesons in photographic emulsions. Nature **160**, 453–456, 492. Reproduced in Hillas (1972).
Lean, J. (1987): Solar uv irradiance variation: A review. Journal of Geophysical Research **92**, 839–868.
Lean, J. (1989): Contribution of ultraviolet irradiance variations to changes in the sun's total irradiance. Science **244**, 197–200.
Lean, J. (1991): Variations in the sun's radiative output. Reviews of Geophysics **29**, 505–535.
Lean, J. (1997): The sun's variable radiation and its relevance for earth. Annual Review of Astronomy and Astrophysics **35**, 33–67.
Lean, J., Beer, J., Bradley, R. (1995): Reconstruction of solar irradiance since 1610: Implications for climate change. Geophysical Research Letters **22**, 3195–3198.
Lean, J., et al. (1995): Correlated brightness variations in solar radiative output from the photosphere to the corona. Geophysical Research Letters **22**, 655–658.
Lean, J., et al. (1998): Magnetic sources of the solar irradiance cycle. Astrophysical Journal **492**, 390–401.
Lean, J., Foukal, P. (1988): A model of solar luminosity modulation by magnetic activity between 1954 and 1984. Science **240**, 906–908.
Lean, J., Rind, D. (1994): Solar variability: implications for global change. EOS **75**, No. 1, 1–6.
Lean, J., Skumanich, A., White, O. (1992): Estimating the Sun's radiative output during the Maunder Minimum. Geophysical Research Letters **19**, 1591–1594.
Lee, R. B., et al. (1995): Long-term solar irradiance variability during sunspot cycle 22. Journal of Geophysical Research **100**, 1667–1675.
Leer, E., Holzer, T. E. (1980): Energy addition in the solar wind. Journal of Geophysical Research **85**, 4681–4688.
Leibacher, J., Stein, R. F. (1971): A new description of the solar five-minute oscillation. Astrophysical Letters **7**, 191–192.
Leighton, R. B. (1961): Considerations on localized velocity fields in stellar atmospheres: Prototype – The solar atmosphere. In: Aerodynamic phenomena in stellar atmospheres. Proceedings of the fourth symposium on cosmical gas dynamics. Supplemento del Nuovo Cimento **22**, 321–325.
Leighton, R. B. (1963): The solar granulation. Annual Review of Astronomy and Astrophysics **1**, 19–40.
Leighton, R. B., Noyes, R. W., Simon, G. W. (1962): Velocity fields in the solar atmosphere I. Preliminary report. Astrophysical Journal **135**, 474–499.
Lepping, R. P., et al. (1991): The interaction of a very large interplanetary magnetic cloud with the magnetosphere and with cosmic rays. Journal of Geophysical Research **96**, 9425–9438.
Letaw, J. R., Silberberg, R., Tsao, C. H. (1987). Radiation hazards on space missions. Nature **330**, 709–710.
Le Treut, H, , Ghil, M. (1983): Orbital forcing, climatic interactions, and glaciation cycles. Journal of Geophysical Research **99**, 5167–5190.
Levine, J. S., Hays, P. B., Walker, J. C. G. (1979): The evolution and variability of atmospheric ozone over geological time. Icarus **39**, 295–309.
Levine, R. H. (1974): Acceleration of thermal particles in collapsing magnetic regions. Astrophysical Journal **190**, 447–456.
Levine, R. H., Altschuler, M. D., Harvey, J. W. (1977): Solar sources of the interplanetary magnetic field and solar wind. Journal of Geophysical Research **82**, 1061–1065.

Libbrecht, K. G. (1989): Solar p-mode frequency splittings. Astrophysical Journal **336**, 1092–1097.
Libbrecht, K. G., Woodard, M. F. (1990): Solar-cycle effects on solar oscillation frequencies. Nature **345**, 779–782.
Libby, W. F.: Radiocarbon dating. Chicago, Illinois: The University of Chicago Press 1955.
Lin, R. P. (1987): Solar particle acceleration and propagation. Reviews of Geophysics **25**, 676–684.
Lin, R. P., et al. (1981): A new component of hard X-rays in solar flares. Astrophysical Journal (Letters) **251**, L109–L114.
Lin, R. P., et al. (1984): Solar hard X-ray microflares. Astrophysical Journal **283**, 421–425.
Lin, R. P., Evans, L. G., Fainberg, J. (1973): Simultaneous observations of fast solar electrons and type III radio burst emission near 1 au. Astrophysical Letters **14**, 191–198.
Lindemann, F. A. (1919): Note on the theory of magnetic storms. Philosophical Magazine **38**, 669–684.
Lindsey, C., Braun, D. C. (2000): Seismic images of the far side of the Sun. Science **287**, 1799–1801.
Lingenfelter, R. E. (1969): Solar flare optical, neutron, and gamma-ray emission. Solar Physics **8**, 341–347.
Lingenfelter, R. E., et al. (1965): High-energy solar neutrons 1. Production in flares. Journal of Geophysical Research **70**, 4077–4086.
Lingenfelter, R. E., Ramaty, R.: High energy nuclear reactions in solar flares. In: High energy nuclear reactions in astrophysics (Ed. B. Shen). New York: W. A. Benjamin 1967. pp. 99–158.
Linsky, J. L. (1980): Stellar chromospheres. Annual Review of Astronomy and Astrophysics **18**, 439–488.
Lites, B. W., Hansen, E. R. (1977): Ultraviolet brightenings in active regions as observed from OSO-8. Solar Physics **55**, 347–358.
Livingston, W., Wallace, L., White, O. R. (1988): Spectrum line intensity as a surrogate for solar irradiance variations. Science **240**, 1765–1767.
Lockwood, G. W., et al. (1984): The photometric variability of solar-type stars. IV. Detection of rotational modulation among Hyades stars. Publications of the Astronomical Society of the Pacific **96**, 714–722.
Lockwood, G. W., et al. (1992): Long-term solar brightness changes estimated from a survey of sun-like stars. Nature **360**, 653–655.
Lockwood, G. W., Skiff, B. A., Radick, R. R. (1997): The photometric variability of sun-like stars: Observations and results, 1984–1995. Astrophysical Journal **485**, 789–811.
Lockyer, J. N. (1869): Spectroscopic observations of the sun. III, IV. Proceedings of the Royal Society **17**, 350–356, 415–418. Reproduced in: Early solar physics (Ed. A. J. Meadows) New York: Pergamon Press 1970, pp. 193–202, 233–236.
Lockyer, J. N.: Contributions to solar physics. London: Macmillan 1874.
Lodge, O. (1900): Sun spots, magnetic storms, comet tails, atmospheric electricity, and aurorae. The Electrician **46**, 249–250, 287–288.
Lomb, N. R., Andersen, A. P. (1980): The analysis and forecasting of the Wolf sunspot numbers. Monthly Notices of the Royal Astronomical Society **190**, 723–732.
Loomis, E. (1860): On the geographical distribution of auroras in the northern hemisphere. American Journal of Science and Arts **30**, 89.
Loomis, E.: The aurora borealis, or polar light: Its phenomena and laws. Smithsonian Institute Annual Report 1864.
Loomis, E. (1866–1871): Notices of auroras extracted from the meteorological journals of Reverend Ezra Stiles. Transactions of the American Academy of Arts and Sciences **1**, 155.

Lorius, C., et al. (1985): A 150,000-year climatic record from Antarctic ice. Nature **316**, 591–596.

Lorius, C., et al. (1988): Antarctic ice core: CO_2 and climatic change over the last climatic cycle. EOS **69**, 681, 683–684.

Lorius, C., et al. (1990): The ice-core record: climate sensitivity and future greenhouse warming. Nature **347**, 139–147.

Lovelock, J. E.: Gaia, a new look at life on Earth. Oxford, England: Oxford University Press, 1979.

Lovelock, J. E.: The ages of gaia. New York: Norton 1988.

Lovelock, J. E., Margulis, L. (1973): Atmospheric homeostasis by and for the biosphere: The gaia hypothesis. Tellus **26**, 1–9.

Lovelock, J. E., Whitfield, M. (1982): Life span of the biosphere. Nature **296**, 561–563.

Low, B. C. (1996): Solar activity and the corona. Solar Physics **167**, 217–265.

Lüst, R., Schlüter, A. (1954): Kraftfreie magneticfelder. Zeitschrift für Astrophysik **34**, 263–282.

Lyons, L. R. (1992): Formation of auroral arcs via magnetosphere-ionosphere coupling. Reviews of Geophysics **30**, 93–112.

Lyot, B. (1930): La couronne solair etudie en dehors des eclipses. Comptes Rendus de l'Academie des Sciences Paris **191**, 834.

Machado, M. E., et al. (1988): The observed characteristics of flare energy release I. Magnetic structure at the energy release site. Astrophysical Journal **326**, 425–450.

Macklin, R. J. Jr., Neugebauer, M. M. (Eds.): The solar wind: Proceedings of a conference held at the california institute of technology, pasadena, california, April 1–4, 1964, and sponsored by the jet propulsion laboratory. Oxford, England: Pergamon Press 1966.

Mac Queen, R. M. (1980): Coronal transients: A summary. Philosophical Transactions of the Royal Society (London) **A297**, 605–620.

Mac Queen, R. M., et al. (1974): The outer solar corona as observed from Skylab: Preliminary results. Astrophysical Journal (Letters) **187**, L85–L88.

Mac Queen, R. M., et al. (1976): Initial results from the high altitude observatory white light coronagraph on Skylab – a progress report. Philosophical Transactions of the Royal Society (London) **A281**, 405–414.

Maher, K. A., Stevenson, D. J. (1988): Impact frustration of the origin of life. Nature **331**, 612–614.

Mairan, J. J.: Traité physique et historique de l'aurorae boréale, Paris: L'Imprimerie Royale 1733 (1st edition), 1754 (2nd revised edition).

Mall, U., Fichtner, H., Rucinski, D. (1996): Interstellar atom and pick-up ion fluxes along the Ulysses flight-path. Astronomy and Astrophysics **316**, 511–518.

Mann, M. E., Bradley, R. S., Hughes, M. K. (1998): Global-scale temperature patterns and climate forcing over the past six centuries. Nature **392**, 779–787.

Mann, M. E., Bradley, R. S., Hughes, M. K. (1999): Northern hemisphere temperatures during the past millennium: Inferences, uncertainties, and limitations. Geophysical Research Letters **26**, 759–762.

Mann, M. E., Park, J., Bradley, R. S. (1995): Global interdecadal and century-scale climate oscillations during the last five centuries. Nature **378**, 266–270.

Manoharan, P. K., et al. (1996): Evidence for large-scale solar magnetic reconnection from radio and X-ray measurements. Astrophysical Journal (Letters) **468**, L73–L76.

Marconi, G. (1899): Wireless telegraphy. Proceedings of the Institution of Electrical Engineers **28**, 273.

Margulis, L., Lovelock, J. E. (1974): Biological modulation of the Earth's atmosphere. Icarus **21**, 471–489.

Mariska, J. T. (1986): The quiet solar transition region. Annual Review of Astronomy and Astrophysics **24**, 23–28.

Mariska, J. T.: The solar transition region. New York: Cambridge University Press 1992.

Mariska, J. T., Doschek, G. A., Bentley, R. D. (1993): Flare plasma dynamics observed with the Yohkoh Bragg crystal spectrometer I. Properties of the Ca XIX resonance line. Astrophysical Journal **419**, 418–425.

Mariska, J. T., Feldman, U., Doschek, G. A. (1978): Measurements of extreme-ultraviolet emission-line profiles near the solar limb. Astrophysical Journal **226**, 698–705.

Marsch, E.: Working group 3: Coronal hole structure and high speed solar wind. In: The corona and solar wind near minimum activity. Proceedings of the fifth SOHO workshop. ESA SP-404. Noordwijk, The Netherlands: ESA Publications Division 1997, pp. 135–140.

Marsch, E., Goertz, C. K., Richter, K. (1982): Wave heating and acceleration of solar wind ions by cyclotron resonance. Journal of Geophysical Research **87**, A7, 5030–5044.

Marsch, E., Tu, C.-Y. (1997): The effects of high-frequency Alfvén waves on coronal heating and solar wind acceleration. Astronomy and Astrophysics **319**, L17–L20.

Marsch, E., Tu, C.-Y. (1997): Solar wind and chromospheric network. Solar Physics **176**, 87–106.

Marsden, R. G. (Ed.): The high latitude heliosphere. New York: Kluwer Academic Publishers 1995.

Marsden, R. G., et al. (1987): ISEE 3 observations of low-energy proton bidirectional events and their relation to isolated interplanetary magnetic structures. Journal of Geophysical Research **92**, 11,009–11,019.

Marsden, R. G., et al. (1996): Ulysses at high heliographic latitudes: an introduction. Astronomy and Astrophysics **316**, 279–286.

Marsden, R. G., Smith, E. J. (1996): Ulysses: solar sojourner. Sky and Telescope **91**, 24–30 – March.

Marsh, K. A. (1978): Ephemeral region flares and the diffusion of the network. Solar Physics **59**, 105–113.

Marsh, K. A., Hurford, G. J. (1980): VLA maps of solar bursts at 15 and 23 GHz with arcsecond resolution. Astrophysical Journal (Letters) **240**, L111–L114.

Martens, P. C. H., Kuin, N. P. M. (1989): A circuit model for filament eruptions and two-ribbon flares. Solar Physics **122**, 263–302.

Martinson, D. G., et al. (1987): Age dating and the orbital theory of the ice ages: Development of a high-resolution 0 to 300,000-year chronostratigraphy. Quaternary Research **27**, 1–29.

Martyn, D. F. (1946): Temperature radiation from the quiet Sun in the radio spectrum. Nature **158**, 632–633.

Masuda, S.: Hard X-ray sources and the primary energy release site in solar flares. Ph. D. Thesis, University of Tokyo, the Yohkoh HXT group, National Astronomical Observatory, Mitaka, Tokyo, 1994.

Masuda, S., et al. (1994): A loop-top hard X-ray source in a compact solar flare as evidence for magnetic reconnection. Nature **371**, 495–497.

Masuda, S., et al. (1995): Hard X-ray sources and the primary energy-release site in solar flares. Publications of the Astronomical Society of Japan **47**, 677–689.

Maunder, E. W. (1890): Professor Spoerer's researches on sunspots. Monthly Notices of the Royal Astronomical Society **50**, 251–252.

Maunder, E. W. (1894): A prolonged sunspot minimum. Knowledge **17**, No. 106, 173–176.

Maunder, E. W. (1905): Magnetic disturbances, 1882 to 1903, as recorded at the royal observatory, greenwich, and their association with sun-spots. Monthly Notices of the Royal Astronomical Society **65**, 2–34.

Maunder, E. W. (1922): The prolonged sunspot minimum, 1645–1715. Journal of the British Astronomical Association **32**, 140–145.

Maxwell, A., Swarup, G. (1958): A new spectral characteristic in solar radio emission. Nature **181**, 36–38.

Mazur, J. E., et al. (1992): The energy spectra of solar flare hydrogen, helium, oxygen, and iron: Evidence for stochastic acceleration. Astrophysical Journal **401**, 398–410.

Mc Allister, A. H., Crooker, N. U.: Coronal mass ejections, corotating interaction regions, and geomagnetic storms. In: Coronal mass ejections. Geophysical monograph 99 (Eds. N. Crooker, J. A. Joselyn and J. Feynman). Washington, DC: American Geophysical Union 1997, pp. 279–290.

Mc Comas, D. J., et al. (1998): Ulysses' return to the slow solar wind. Geophysical Research Letters **25**, 1–4.

Mc Crea, W. H. (1929): The hydrogen chromosphere. Monthly Notices of the Royal Astronomical Society **89**, 483–497.

Mc Crea, W. H. (1956): Shock waves in steady radial motion under gravity. Astrophysical Journal **124**, 461–468.

Mc Hargue, L. R., Damon, P. E. (1991): The global beryllium 10 cycle. Reviews of Geophysics **29**, 141–158.

Mc Kenzie, J. F., Axford, W. I., Banaszkiewicz, M. (1997): The fast solar wind. Geophysical Research Letters **24**, No. 22, 2877–2880.

Mc Kenzie, J. F., Banaszkiewicz, M., Axford, W. I. (1995): Acceleration of the high speed solar wind. Astronomy and Astrophysics **303**, L45–L46.

Mc Kenzie, J. F., Bornatici, M. (1974): Effect of sound waves, Alfvén waves, and heat flow on interplanetary shock waves. Journal of Geophysical Research **79**, 4589–4594.

Mc Kibben, R. B., et al. (1996): Observations of galactic cosmic rays and the anomalous helium during Ulysses passage from the south to the north solar pole. Astronomy and Astrophysics **316**, 547–554.

Mc Lean, D. J.: Metrewave solar radio bursts. In: Solar Radiophysics (Ed. D. J. Mc Lean and N. R. Labrum). Cambridge, England: Cambridge University Press 1985, pp. 37–52.

Mc Lean, D. J., Labrum, N. R. (Eds.): Solar radiophysics. Cambridge, England: Cambridge University Press 1985.

Mc Lennan, J. C., Shrum, G. M. (1926): On the origin of the auroral green line 5,577 Å and other spectra associated with the aurora borealis. Proceedings of the Royal Society (London) **A108**, 501.

Meadows, A. J.: Early solar physics. Oxford, England: Pergamon Press 1970.

Meadows, A. J. (1975): A 100 years of controversy over sunspots and weathers. Nature **256**, 95–97.

Meadows, A. J., Kennedy, J. E. (1982): The origin of solar-terrestrial studies. Vistas in Astronomy **25**, 419–426.

Melrose, D. B. (1995): Current paths in the corona and energy release in solar flares. Astrophysical Journal **451**, 391–401.

Melrose, D. B. (1997): A solar flare model based on magnetic reconnection between current-carrying loops. Astrophysical Journal **486**, 521–533.

Meyer, P., Parker, E. N., Simpson, J. A. (1956): Solar cosmic rays of February, 1956 and their propagation through interplanetary space. Physical Review **104**, 768–783.

Meyer, P., Simpson, J. A. (1954): Changes in amplitude of the cosmic-ray 27-day intensity variation with solar activity. Physical Review **96**, 1085–1088.

Meyer, P., Simpson, J. A. (1955): Changes in the low-energy particle cutoff and primary spectrum of cosmic radiation. Physical Review **99**, 1517–1523.

Meyer, P., Simpson, J. A. (1957): Changes in the low-energy particle cutoff and primary spectrum of cosmic rays. Physical Review **106**, 568–571.

Meyer, P., Vogt, R. (1961): Electrons in the primary cosmic radiation. Physical Review Letters **6**, 193–196.

Meyer, P., Vogt, R. (1962): High-energy electrons of solar origin. Physical Review Letters **8**, 387–389.

Michel, F. C., Dessler, A. J. (1965): Physical significance of inhomogeneities in polar cap absorption events. Journal of Geophysical Research **70**, 4305–4311.

Milankovitch, M. M.: Théorie mathématique des phénomenes thermiques produits par la radiation solarie. Académie Yugoslave des Sciences et des Arts de Zagreb. Paris: Gauthier-Villars 1920.

Milankovitch, M. M.: Kanon der Erdbestrahlung une sei Eiszeitenproblem (Canon of insolation and the ice-age problem). Königliche Serbische Akademie, Beograd, Publication 132, Section of Mathematics and Natural Science 33, 1941. (English translation by the Israel Program for Scientific Translations and published for the U. S. Department of Commerce and the National Science Foundation, Jerusalem 1970).

Miller, J. A., et al. (1997): Critical issues for understanding particle acceleration in impulsive solar flares. Journal of Geophysical Research **102**, 14,631–14,659.

Millikan, R. A. (1926): High frequency rays of cosmic origin. Proceedings of the National Academy of Sciences **12**, 48–55.

Millikan, R. A., Cameron, G. H. (1926): High frequency rays of cosmic origin III. Measurements in snow-fed lakes at high altitudes. Physical Review **28**, 851–868.

Mitchell, J. F. B. (1989): The "greenhouse" effect and climate change. Reviews of Geophysics **27**, 115–139.

Molina, M. J., Rowland, F. S. (1974): Stratospheric sink for chlorofluoromethanes. chlorine atomic-atalyzed destruction of ozone. Nature **249**, 810–812.

Montgomery, M. D., et al. (1974): Solar wind electron temperature depressions following some interplanetary shock waves: Evidence for magnetic merging? Journal of Geophysical Research **79**, 3103–3123.

Moore, R., et al.: The thermal X-ray flare plasma. In: Solar flares. Skylab solar workshop II (Ed. P. A. Sturrock). Boulder, Colorado: Colorado Associated University Press 1980, pp. 341–409.

Moreton, G. E. (1960): Hα observations of flare-initiated disturbances with velocities $\approx$ 1,000 km/sec. Astronomical Journal **65**, 494–495.

Moreton, G. E. (1961): Fast-moving disturbances on the sun. Sky and Telescope **21**, 145–147.

Moreton, G. E. (1964): Hα shock wave and winking filaments with the flare of 20 September 1963. Astronomical Journal **69**, 145.

Moreton, G. E., Severny, A. B. (1968): Magnetic fields and flares in the region cmp 20 September 1963. Solar Physics **3**, 282–297.

Morrison, P. (1954): Solar-connected variations of the cosmic rays. Physical Review **95**, 646.

Morrison, P. (1958): On gamma-ray astronomy. Nuovo Cimento **7**, 858–865.

Moses, D. et al. (1997): EIT observations of the extreme ultraviolet sun. Solar Physics **175**, 571–599. Reprinted in: The first results from SOHO (Eds. B. Fleck and Z. Svestka). Boston: Kluwer Academic Publishers 1997, pp. 571–599.

Mullen, E. G., et al. (1991): A double-peaked inner radiation belt: Cause and effect as seen on CRRES. IEEE Transactions on Nuclear Science **38**, 1713–1717.

Munro, R. H., et al. (1979): The association of coronal mass ejection transients with other forms of solar activity. Solar Physics **61**, 201–215.

Munro, R. H., Jackson, B. V. (1977): Physical properties of a polar coronal hole from 2 to 5 solar radii. Astrophysical Journal **213**, 874–886.

Murphy, R. J., Dermer, C. D., Ramaty, R. (1987): High-energy processes in solar flares. Astrophysical Journal Supplement **63**, 721–748.

Murphy, R. J., et al. (1985): Solar flare gamma-ray line spectroscopy. Proceedings 19th International Cosmic Ray Conference (La Jolla) **4**, 253–256.

Murphy, R. J., et al. (1991): Solar abundances from gamma-ray spectroscopy: Comparisons with energetic particle, photospheric, and coronal abundances. Astrophysical Journal **371**, 793–803.

Murphy, R. J., Ramaty, R. (1984): Solar flare neutrons and gamma rays. Advances in Space Research **4**, No. 7, 127–136.

Nakamura, M. et al. (1998): Reconnection event at the dayside magnetopause on January 10, 1997. Geophysical Research Letters **25**, 2529–2532.

National Research Council: Solar influences on global change Washington, DC: National Academy Press 1994.

Neftel, A., et al. (1982): Ice core sample measurements give atmospheric CO_2 content during the past 40,000 years. Nature **295**, 220–223.

Neftel, A., et al. (1985): Evidence from polar ice cores for the increase in atmospheric CO_2 in the past two centuries. Nature **315**, 45–47.

Neftel, A., Oeschger, H., Suess, H. E. (1981): Secular non-random variations of cosmogenic carbon-14 in the terrestrial atmosphere. Earth and Planetary Science Letters **56**, 127–147.

Nesme-Ribes, E., Baliunas, S. L., Sokoloff, D. (1996): The stellar dynamo. Scientific American **275**, 46–52 – August.

Nesme-Ribes, E., et al. (1993): Solar dynamics and its impact on solar irradiance and the terrestrial climate. Journal of Geophysical Research **98**, 18,923–18,935.

Nesme-Ribes, E., Mangeney, A. (1992): On a plausible physical mechanism linking the Maunder Minimum to the Little Ice Age. Radiocarbon **34**, No. 2, 263–270.

Nesme-Ribes, E., Sokoloff, D., Sadourny, R.: Solar rotation, irradiance changes and climate. In: The sun as a variable star (Ed. J. Pap, H. Hudson and S. Solanki). New York: Cambridge University Press 1994, pp. 244–251.

Ness, N. F. (1965): The earth's magnetic tail. Journal of Geophysical Research **70**, 2989–3005.

Ness, N. F. (1968): Observed properties of the interplanetary plasma. Annual Review of Astronomy and Astrophysics **6**, 79–114.

Ness, N. F. (1996): Pioneering the swinging 1960s into the 1970s and 1980s. Journal of Geophysical Research **101**, No. A5, 10,497–10,509.

Ness, N. F., Hundhausen, A. J., Bame, S. J. (1971): Observations of the interplanetary medium: Vela 3 and Imp 3, 1965–1967. Journal of Geophysical Research **76**, 6643–6660.

Ness, N. F., Scearce, C. S., Seek, J. B. (1964): Initial results of the Imp 1 magnetic field experiment. Journal of Geophysical Research **69**, 3531–3569.

Ness, N. F., Wilcox, J. M. (1964): Solar origin of the interplanetary magnetic field. Physical Review Letters **13**, 461–464.

Ness, N. F., Wilcox, J. M. (1965): Sector structure of the quiet interplanetary magnetic field. Science **148**, 1592–1594.

Ness, N. F., Wilcox, J. M. (1966): Extension of the photospheric magnetic field into interplanetary space. Astrophysical Journal **143**, 23–31.

Neugebauer, M. (1981): Observations of solar-wind helium. Fundamentals of Cosmic Physics **7**, 131–199.

Neugebauer, M., Snyder, C. W. (1962): The mission of mariner II – preliminary observations. Solar plasma experiment. Science **138**, 1095–1096.

Neugebauer, M., Snyder, C. W. (1966): Mariner 2 observations of the solar wind. Journal of Geophysical Research **71**, 4469–4484.

Neugebauer, M., Snyder, C. W. (1967): Mariner 2 observations of the solar wind 2. relation of plasma properties to the magnetic field. Journal of Geophysical Research **72**, 1823–1828.

Neupert, W. M. (1968): Comparison of solar X-ray line emission with microwave emission during flares. Astrophysical Journal (Letters) **153**, L59–L64.

Neupert, W. M., et al. (1967): Observation of the solar flare X-ray emission line spectrum of iron from 1.3 to 20 Å. Astrophysical Journal (Letters) **149**, L79–L83

Neupert, W. M., Pizzo, V. (1974): Solar coronal holes as sources of recurrent geomagnetic disturbances. Journal of Geophysical Research **79**, 3701–3709.

Newell, N. E., et al. (1989): Global marine temperature variation and the solar magnetic cycle. Geophysical Research Letters **16**, 311–314.

Newell, P. T., Meng, C.-I., Wing, S. (1998): Relation to solar activity of intense aurorae in sunlight and darkness. Nature **393**, 342–345.

Newkirk, G. Jr. (Ed.): Coronal disturbances: Proceedings of IAU symposium no. 57. Boston: D. Reidel 1974.

Newman, M. J., Rood, R. T. (1977): Implications of solar evolution for the Earth's early atmosphere. Science **198**, 1035–1037.

Newton, H. W. (1930): An active region of the sun on 1930 august 12. Monthly Notices of the Royal Astronomical Society **90**, 820–825.

Newton, H. W. (1932): The 27-day period in terrestrial magnetic disturbances. Observatory **55**, 256–261.

Newton, H. W. (1935): Note on two allied types of chromospheric eruptions. Monthly Notices of the Royal Astronomical Society **95**, 650–665.

Newton, H. W. (1942): Characteristic radial motions of Hα absorption markings seen with bright eruptions on the sun's disc. Monthly Notices of the Royal Astronomical Society **102**, 2–10.

Newton, H. W. (1943): Solar flares and magnetic storms. Monthly Notices of the Royal Astronomical Society **103**, 244–257.

Newton, H. W., Nunn, M. L. (1951): The Sun's rotation derived from sunspots 1934–1944 and additional results. Monthly Notices of the Royal Astronomical Society **111**, 413–421.

Ney, E. P. (1959): Cosmic radiation and the weather. Nature **183**, 451–452.

Noci, G., et al.: The quiescent corona and slow solar wind. In: The corona and solar wind near minimum activity Proceedings of the fifth SOHO workshop. ESA SP-404. Noordwijk, The Netherlands: ESA Publications Division 1997, pp. 75–84.

Noci, G., Kohl, J. L., Withbroe, G. L. (1987): Solar wind diagnostics from Doppler-enhanced scattering. Astrophysical Journal **315**, 706–715.

Nolte, J. T., et al. (1976).: Coronal holes as sources of solar wind. Solar Physics **46**, 303–322.

November, L. J., Koutchmy, S. (1996): White-light coronal dark threads and density fine structure. Astrophysical Journal **466**, 512–528.

Noyes, R. W. (1971): Ultraviolet studies of the solar atmosphere. Annual Review of Astronomy and Astrophysics **9**, 209–236.

Noyes, R. W., Baliunas, S. L., Guinan, E. F.: What can other stars tell us about the sun? In: The solar interior and atmosphere (Eds. A. N. Cox, W. C. Livingston and M. S. Matthews) Tucson, Arizona: University of Arizona Press 1991, pp. 1161–1186.

Noyes, R. W., et al. (1984): Rotation, convection, and magnetic activity in lower main sequence stars. Astrophysical Journal **279**, 763–777.

Noyes, R. W., Leighton, R. B. (1963): Velocity fields in the solar atmosphere II. The oscillation field. Astrophysical Journal **138**, 631–647.

Noyes, R. W., Weiss, N. O., Vaughan, A. H. (1984): The relation between stellar rotation rate and activity cycle periods. Astrophysical Journal **287**, 769–773.

Ofman, L., Davila, J. M., Shimizu, T. (1996): Signatures of global mode Alfvén resonance heating in coronal loops. Astrophysical Journal (Letters) **459**, L39–L42.

Oranje, B. J. (1983): The Ca II emission from the Sun as a star II. The plage emission profile. Astronomy and Astrophysics **124**, 43–49.

Orrall, F. Q., Rottman, G. J., Klimchuk, J. A. (1983): Outflow from the sun's polar corona. Astrophysical Journal (Letters) **266**, L65–L68.

Oster, L., Sofia, S., Schatten, K. (1982): Solar irradiance variations due to active regions. Astrophysical Journal **256**, 768–773.

Osterbrock, D. E. (1961): The heating of the solar chromosphere, plages, and corona by magnetohydrodynamic waves. Astrophysical Journal **134**, 347–388.

Owen, T., Cess, R. D., Ramanathan, V. (1979): Enhanced CO_2 greenhouse to compensate for reduced solar luminosity on early earth. Nature **277**, 640–642.

Paillard, D. (1998): The timing of Pleistocene glaciations from a simple multiple-state climate model. Nature **391**, 378–381.

Pallavicini, R., Serio, S., Vaiana, G. S. (1977): A survey of solar X-ray limb flare images: the relation between their structure in the corona and other physical parameters. Astrophysical Journal **216**, 108–122.

Pallé, P. L., Régulo, C., Roca-Cortés, T. (1989): Solar cycle induced variations of the low *l* solar acoustic spectrum. Astronomy and Astrophysics **224**, 253–258.

Parker, E. N. (1955): Dynamics of the interplanetary gas and magnetic fields. Astrophysical Journal **125**, 668–676.

Parker, E. N. (1957): Acceleration of cosmic rays in solar flares. Physical Review **107**, 830–836.

Parker, E. N. (1958): Dynamical instability in an anisotropic ionized gas of low density. Physical Review **109**, 1874–1876.

Parker, E. N. (1958): Interaction of the solar wind with the geomagnetic field. The Physics of Fluids **1**, 171–187.

Parker, E. N. (1958): Cosmic-ray modulation by solar wind. Physical Review **110**, 1445–1449. Reproduced in Hillas (1972).

Parker, E. N. (1958): Dynamics of the interplanetary gas and magnetic fields. Astrophysical Journal **128**, 664–676.

Parker, E. N. (1959): Extension of the solar corona into interplanetary space. Journal of Geophysical Research **64**, 1675–1681.

Parker, E. N. (1960): The hydrodynamic theory of solar corpuscular radiation and stellar winds. Astrophysical Journal **132**, 821–866.

Parker, E. N. (1961): Sudden expansion of the corona following a large solar flare and the attendant magnetic field and cosmic-ray effects. Astrophysical Journal **133**, 1014–1033.

Parker, E. N. (1963): The solar flare phenomenon and the theory of reconnection and annihilation of magnetic fields. Astrophysical Journal Supplement **8**, No. 77, 177–211.

Parker, E. N. (1964): Dynamical properties of stellar coronas and stellar winds I. Integration of the momentum equation. Astrophysical Journal **139**, 72–92.

Parker, E. N. (1972): Topological dissipation and the small-scale fields in turbulent gases. Astrophysical Journal **174**, 499–510.

Parker, E. N. (1979): Sunspots and the physics of magnetic flux tubes. I. The general nature of the sunspot. Astrophysical Journal **230**, 905–913.

Parker, E. N. (1983): Magnetic fields in the cosmos. Scientific American **249**, 44–65 – August.

Parker, E. N. (1983): Magnetic neutral sheets in evolving fields II. Formation of the solar corona. Astrophysical Journal **264**, 642–647.

Parker, E. N. (1988): Nanoflares and the solar X-ray corona. Astrophysical Journal **330**, 474–479.

Parker, E. N. (1991): Heating solar coronal holes. Astrophysical Journal **372**, 719–727.

Parker, E. N. (1997): Reflections on macrophysics and the sun. Solar Physics **176**, 219–247.

Parkinson, J. H., Morrison, L. V., Stephenson, F. V. (1980): The constancy of the solar diameter over the past 250 years. Nature **288**, 548–551.

Parnell, C. E., Jupp, P. E. (2000): Statistical analysis of the energy distribution of nanoflares in the quiet sun. Astrophysical Journal **529**, 554–569.

Parnell, C. E., Priest, E. R., Golub, L. (1994): The three-dimensional structures of X-ray bright points. Solar Physics **151**, 57–74.

Pawsey, J. L. (1946): Observation of million degree thermal radiation from the sun at a wave-length of 1.5 meters. Nature **158**, 633–634.

Payne-Scott, R., Little, A. G. (1952): The position and movement on the solar disk of sources of radiation at a frequency of 97 Mc/s. III – Outbursts. Australian Journal of Scientific Research **A5**, 32–46.

Payne-Scott, R., Yabsley, D. E., Bolton, J. G. (1947): Relative times of arrival of bursts of solar noise on different radio frequencies. Nature **160**, 256–257.

Pecker, J.-C., Runcorn, S. K., Eds. (1990): The earth's climate and variability of the sun over recent millennia: Geophysical, astronomical and archaeological aspects. Philosophical Transactions of the Royal Society (London) **A330**, 395–687. New York: Cambridge University Press 1990.

Peterson, L. E. (1963): The 0.5-MeV gamma-ray and low-energy gamma-ray spectrum to 6 grams per square centimeter over minneapolis. Journal of Geophysical Research **68**, 979–987.

Peterson, L. E., Winckler, J. R. (1959): Gamma ray burst from a solar flare. Journal of Geophysical Research **64**, 697–707.

Petit, J. R., et al. (1999): Climate and atmospheric history of the past 420,000 years from the Vostok ice core, Antarctica. Nature **399**, 429–436.

Petschek, H. E.: Magnetic field annihilation. In: AAS-NASA symposium on the physics of solar flares NASA SP-50 (Ed. W. Hess). Washington, DC: National Aeronautics and Space Administration 1964, pp. 425–439.

Petschek, H. E., Thorne, R. M. (1967): The existence of intermediate waves in neutral sheets. Astrophysical Journal **147**, 1157–1163.

Phillips, J. L., et al. (1995): Ulysses solar wind plasma observations at high southerly latitudes. Science **268**, 1030–1033.

Phillips, J. L., et. al. (1995): Ulysses solar wind plasma observations from pole to pole. Geophysical Research Letters **22**, No. 23, 3301–3304.

Phillips, J. L., et al. (1995): Sources of shocks and compressions in the high-latitude solar wind: Ulysses. Geophysical Research Letters **22**, No. 23, 3305–3308.

Phillips, K. J. H. (1991): Spectroscopy of high-temperature solar flare plasmas. Philosophical Transactions of the Royal Society (London) **A336**, 461–470.

Pick, M., Van Den Oord, G. H. J. (1990): Observations of beam propagation. Solar Physics **130**, 83–99.

Piddington, J. H. (1958): Interplanetary magnetic field and its control of cosmic ray variations. Physical Review **112**, 589–596.

Pneuman, G. W., Kopp, R. A. (1971): Gas-magnetic field interactions in the solar corona. Solar Physics **18**, 258–270.

Poincaré, H. (1896): Remarques sur une experience de M. Birkeland. Comptes Rendus de l'Academie des Sciences **123**, 530–533.

Poland, A. I., et al. (1981): Coronal transients near sunspot maximum. Solar Physics **69**, 169–175.

Pollack, H., Huang, S., Shen, P. Y. (1998): Climate change revealed by subsurface temperatures: A global perspective. Science **282**, 279–281.

Pomerantz, M. A., Duggal, S. P. (1973): Record-breaking cosmic ray storm stemming from solar activity in august 1972. Nature **241**, 331–333.

Porter, J. G., Dere, K. P. (1991): The magnetic network location of explosive events observed in the solar transition region. Astrophysical Journal **370**, 775–778.

Porter, J. G., et al. (1987): Microflares in the solar magnetic network. Astrophysical Journal **323**, 380–390.

Porter, J. G., Fontenla, J. M., Simnett, G. M. (1995): Simultaneous ultraviolet and X-ray observations of solar microflares. Astrophysical Journal **438**, 472–479.

Porter, J. G., Moore, R. L.: Coronal heating by microflares. In: Solar and stellar coronal structure and dynamics (Ed. R. C. Altrock). Sunspot, New Mexico: National Solar Observatory, Sacramento Peak 1988, pp. 125–129.

Porter, J. G., Toomre, J., Gebbie, K. B. (1984): Frequent ultraviolet brightenings observed in a solar active regions with solar maximum mission. Astrophysical Journal **283**, 879–886.

Porter, L. J., Klimchuk, J. A. (1995): Soft X-ray loops and coronal heating. Astrophysical Journal **454**, 499–511.

Porter, L. J., Klimchuk, J. A., Sturrock, P. A. (1994): The possible role of mhd waves in heating the solar corona. Astrophysical Journal **435**, 482–501.

Porter, L. J., Klimchuk, J. A., Sturrock, P. A. (1994): The possible role of high-frequency waves in heating solar coronal loops. Astrophysical Journal **435**, 502–514.

Priem, H. N. A. (1997): CO_2 and climate: A geologist's view. Space Science Reviews **81**, 173–198.

Priest, E. R. (1978): The structure of coronal loops. Solar Physics **58**, 57–87.

Priest, E. R.: Solar magnetohydrodynamics. Boston: D. Reidel 1982.

Priest, E. R. (1991): The magnetohydrodynamics of energy release in solar flares. Philosophical Transactions of the Royal Society (London) **A336**, 363–380.

Priest, E. R. (1996): Coronal heating by magnetic reconnection. Astrophysics and Space Science **237**, 49–73.

Priest, E. R. (1999): How is the solar corona heated? Solar and stellar activity: similarities and differences. asp conference series **158**, 321–333.

Priest, E. R, Foley, C. R., Heyvaerts, J., Arber, T. D., Culhane, J. L., Acton, L. W. (1998): Nature of the heating mechanism for the diffuse solar corona. Nature **393**, 545–547.

Priest, E. R., Forbes, T. G. (1986): New models for fast steady-state magnetic reconnection. Journal of Geophysical Research **91**, 5579–5588.

Priest, E. R., Forbes, T. G. (1990): Magnetic field evolution during prominence eruptions and two-ribbon flares. Solar Physics **126**, 319–350.

Priest, E. R., Parnell, C. E., Martin, S. F. (1994): A converging flux model of an X-ray bright point and an associated canceling magnetic feature. Astrophysical Journal **427**, 459–474.

Pudovkin, M. I., Veretenenko, S. V. (1996): Variations of the cosmic rays as one of the possible links between the solar activity and the lower atmosphere. Advances in Space Research **17**, No. 11, 161–164.

Purcell, J. D., Tousey, R., Watanabe, K. (1949): Observations at high altitudes of extreme ultraviolet and X-rays from the Sun. Physical Review **76**, 165–166.

Quinn, T. J., Fröhlich, C. (1999): Accurate radiometers should measure the output of the sun. Nature **401**, 841.

Radick, R. R., Lockwood, G. W., Baliunas, S. L. (1990): Stellar activity and brightness variations: A glimpse at the sun's history. Science **247**, 39–44.

Raisbeck, G. M., et al. (1981): Cosmogenic ^{10}Be concentrations in Antarctic ice during the past 30,000 years. Nature **292**, 825–826.

Raisbeck, G. M., et al. (1985): Evidence for an increase in cosmogenic ^{10}Be during a geomagnetic reversal. Nature **315**, 315–317.

Raisbeck, G. M., et al. (1987): Evidence for two intervals of enhanced ^{10}Be deposition in Antarctic ice during the last glacial period. Nature **326**, 273–277.

Raisbeck, G. M., et al. (1990): ^{10}Be and ^{2}H in polar ice cores as a probe of the solar variability's influence on climate. Philosophical Transactions of the Royal Society (London) **A300**, 463–470.

Raisbeck, G. M., Yiou, F.: ^{10}Be as a proxy indicator of variations in solar activity and geomagnetic field intensity during the last 10,000 years. In: Secular, solar and geomagnetic variations in the last 10,000 years (Eds. F. R. Stephenson and W. Wolfendale). Drodrecht, Holland: Kluwer 1988, pp. 287–296.

Ramanathan, V. (1988): The greenhouse theory of climate change. Science **240**, 293–299.

Ramanathan, V., et al. (1989): Cloud-radiative forcing and climate: Results from the earth radiation budget experiment. Science **243**, 57–62.
Ramaty, R. (1969): Gyrosynchrotron emission and absorption in a magnetoactive plasma. Astrophysical Journal **158**, 753–770.
Ramaty, R., et al. (1993): Acceleration in solar flares: interacting particles versus interplanetary particles. Advances in Space Research **13**, No. 9, 275–284.
Ramaty, R., et al. (1994): Gamma-ray and millimeter-wave emissions from the 1991 June x-class flares. Astrophysical Journal **436**, 941–949.
Ramaty, R., Kozlovsky, B., Lingenfelter, R. E. (1979): Nuclear gamma rays from energetic particle interactions. Astrophysical Journal Supplement **40**, 487–526.
Ramaty, R., Lingenfelter, R. E. (1966): Galactic cosmic-ray electrons. Journal of Geophysical Research **71**, 3687–3703.
Ramaty, R., Lingenfelter, R. E. (1979): γ-ray line astronomy. Nature **278**, 127–132.
Ramaty, R., Lingenfelter, R. E. (1983): Gamma-ray line astronomy. Space Science Reviews **36**, 305–317.
Ramaty, R., Mandzhavidze, N.: Theoretical models for high-energy solar flare emissions. In: High energy solar phenomena – a new era of spacecraft measurements (Eds. J. M. Ryan and W. T. Vestrand). New York: American Institute of Physics 1994, pp. 26–44.
Ramaty, R., Murphy, R. J. (1987): Nuclear processes and accelerated particles in solar flares. Space Science Reviews **45**, 213–268.
Ramaty, R., Petrosian, V. (1972): Free-free absorption of gyrosynchrotron radiation in solar microwave bursts. Astrophysical Journal **178**, 241–249.
Ramsay, W. (1901): The inert constituents of the atmosphere. Nature **65**, 161–164.
Ramsey, H. E., Smith, S. F. (1966): Flare-initiated filament oscillations. Astronomical Journal **71**, 197–199.
Rao, U. R. (1972): Solar modulation of galactic cosmic radiation. Space Science Reviews **12**, 719–809.
Raymond, J. C., et al. (1997): Composition of coronal streamers from the SOHO ultraviolet coronagraph spectrometer. Solar Physics **175**, 645–655. Reprinted in The first results from SOHO (Eds. B. Fleck and Z. Svestka). Boston: Kluwer Academic Publishers 1997, pp. 645–665.
Raynaud, D., et al. (1993): The ice record of greenhouse gases. Science **259**, 926–934.
Reames, D. V. (1990): Energetic particles from impulsive solar flares. Astrophysical Journal Supplement Series **73**, 235–251.
Reames, D. V. (1993): Non-thermal particles in the interplanetary medium. Advances in Space Research **13**, 331–339.
Reames, D. V. (1995): Solar energetic particles: A paradigm shift. Reviews of Geophysics Supplement **33**, 585–589.
Reames, D. V.: Energetic particles and the structure of coronal mass ejections. In: Coronal mass ejections. geophysical monograph 99 (Eds. N. Crooker, J. A. Joselyn, J. Feynman). Washington, DC: American Geophysical Union 1997, pp. 217–226.
Reames, D. V., Richardson, I. G., Wenzel, K.-P. (1992): Energy spectra of ions from impulsive solar flares. Astrophysical Journal **387**, 715–725.
Reedy, R. C., Arnold, J. R. (1972): Interaction of solar and galactic cosmic-ray particles with the moon. Journal of Geophysical Research **77**, 537–555.
Reedy, R. C., Arnold, J. R., Lal, D. (1983): Cosmic-ray record in solar system matter. Science **219**, 127–135.
Reid, G. C. (1976): Influence of ancient solar-proton events on the evolution of life. Nature **259**, 177–179.
Reid, G. C. (1987): Influence of solar variability on global sea surface temperatures. Nature **329**, 142–143.
Reid, G. C. (1991): Solar total irradiance variations and the global sea surface temperature record. Journal of Geophysical Research **96**, 2835–2844.

Reid, G. C. (1991): Solar irradiance variations and global ocean temperature. Journal of Geomagnetism and Geoelectricity **43**, 795–801.

Reid, G. C., Leinbach, H. (1959): Low-energy cosmic-ray events associated with solar flares. Journal of Geophysical Research **64**, 1801–1805.

Reiner, M. J. et al. (1998): On the origin of radio emissions associated with the January 6–11, 1997, CME. Geophysical Research Letters **25**, 2493–2496.

Reiner, M. J., Fainberg, J., Stone, R. G. (1995): Large-scale interplanetary magnetic field configuration revealed by solar radio bursts. Science **270**, 461–464.

Revelle, R., Suess, H. E. (1957): Carbon dioxide exchange between atmosphere and ocean and the question of an increase in atmospheric carbon dioxide during the past decades. Tellus **9**, 18–27.

Rhodes, E. J. Jr., et al. (1997): Measurements of frequencies of solar oscillation for the MDI medium-*l* program. Solar Physics **175**, 287–310. Reprinted in: The first results from SOHO (Eds. B. Fleck and Z. Svestka). Boston: Kluwer Academic Publishers 1997, pp. 287–310.

Rhodes, E. J. Jr., Ulrich, R. K., Simon, G. W. (1977): Observations of nonradial p-mode oscillations on the Sun. Astrophysical Journal **218**, 901–919.

Ribes, E. (1990): Astronomical determinations of the solar variability. Philosophical Transactions of the Royal Society (London) **A330**, 487–497.

Ribes, J. C., Nesme-Ribes, E. (1990): The solar sunspot cycle in the Maunder minimum AD 1645 to AD 1715. Astronomy and Astrophysics **276**, 549–563.

Richardson, I. G., Cane, H. V. (1995): Regions of abnormally low proton temperature in the solar wind (1965–1991) and their association with ejecta. Journal of Geophysical Research **100**, 23,397–23,412.

Richardson, R. S. (1939): Intensity changes in bright chromospheric disturbances. Astrophysical Journal **90**, 368–377.

Richardson, R. S. (1944): Solar flares versus bright chromospheric eruptions: A question of terminology. Publications of the Astronomical Society of the Pacific **56**, 156–158.

Richardson, R. S. (1951): Characteristics of solar flares. Astrophysical Journal **114**, 356–366.

Rickett, B. J., Coles, W. A.: Solar cycle changes in the high latitude solar wind. In: Study of the solar cycle from space. NASA conference publication 2098. Washington, DC: National Aeronautics and Space Administration 1980. pp. 233–243.

Rickett, B. J., Coles, W. A.: Solar cycle evolution of the solar wind in the three dimensions. In: Solar wind five. NASA conference publication CP-2280. Washington, DC: NASA 1982, pp. 315–321.

Rickett, B. J., Coles, W. A. (1991): Evolution of the solar wind structure over a solar cycle: Interplanetary scintillation velocity measurements compared with coronal observations. Journal of Geophysical Research **96**, A2, 1717–1736.

Roelof, E. C.: Coronal structure and the solar wind. In: Solar wind three (Ed. C. T. Russell). Los Angeles, California: Institute of Geophysics and Planetary Physics UCLA, 1974, pp. 98–131.

Rosenbauer, H. R., et al. (1977): A survey of initial results of the Helios plasma experiment. Journal of Geophysics **42**, 561–580.

Rosenberg, H. (1976): Solar radio observations and interpretations. Philosophical Transactions of the Royal Society (London) **A281**, 461–471.

Rosenberg, R. L. (1970): Unified theory of the interplanetary magnetic field. Solar Physics **15**, 72–78.

Rosenberg, R. L., Coleman, P. J. Jr. (1969): Heliographic latitude dependence of the dominant polarity of the interplanetary magnetic field. Journal of Geophysical Research **74**, 5611–5622.

Rosner, R., Tucker, W. H., Vaiana, G. S. (1978): Dynamics of the quiescent solar corona. Astrophysical Journal **220**, 643–665.

Rossi, B. (1991): The interplanetary plasma. Annual Review of Astronomy and Astrophysics **29**, 1–8.

Rostoker, G., Fälthammar, C.-G. (1967): Relationship between changes in the interplanetary magnetic field and variations in the magnetic field at the earth's surface. Journal of Geophysical Research **72**, 5853–5863.

Rottman, G. J. (1981): Rocket measurements of the solar spectral irradiance during solar minimum. Journal of Geophysical Research **86**, 6697–6705.

Rottman, G. J., Orrall, F. Q., Klimchuk, J. A. (1982): Measurements of outflow from the base of solar coronal holes. Astrophysical Journal **260**, 326–337.

Russell, C. T., Ed. (1995): The global geospace mission. Space Science Reviews **71**, 1–878.

Russell, C. T., Ed. (1997): Results of the IASTP program. Advances in Space Research **20**, 523–1107.

Russell, C. T., Mc Pherron, R. L. (1973): Semiannual variation of geomagnetic activity. Journal of Geophysical Research **78**, 92–108.

Rust, D. M. (1982): Solar flares, proton showers, and the space shuttle. Science **216**, 939–946.

Rust, D. M. (1983): Coronal disturbances and their terrestrial effects. Space Science Reviews **34**, 21–36.

Rust, D. M., Hildner, E. (1978): Expansion of an X-ray coronal arch into the outer corona. Solar Physics **48**, 381–387.

Rust, D, M., Kumar, A. (1996): Evidence for helically kinked magnetic flux ropes in solar eruptions. Astrophysical Journal (Letters) **464**, L199–L202.

Rust, D. M., Nakagawa, Y., Neupert, W. M. (1975): EUV emission, filament activation and magnetic fields in a slow-rise flare. Solar Physics **41**, 397–414.

Rust, D. M., Svestka, Z. (1979): Slowly moving disturbances in the X-ray corona. Solar Physics **63**, 279–295.

Rust, D. M., Webb, D. F. (1977): Soft X-ray observations of large-scale active region brightenings. Solar Physics **54**, 403–417.

Ruzmaikin, A. A., et al. (1996): Spectral properties of solar convection and diffusion. Astrophysical Journal **471**, 1022–1029.

Ryan, J., et al.: Neutron and gamma-ray measurements of the solar flare of 1991 June 9. In: High-energy solar phenomena – A new era of spacecraft measurements. AIP Conference Proceedings 294 (Eds. J. M. Ryan and W. T. Vestrand). New York: American Institute of Physics 1994.

Saba, J. L. R., Strong, K. T. (1991): Nonthermal broadening. Astrophysical Journal **375**, 789–799.

Sabine, E. (1852): Letter to John Herschel 16 March 1852. Herschel Letters No. 15,235. (Royal Society). Quoted by A. J. Meadows and J. E. Kennedy in: The origin of solar-terrestrial studies. Vistas in Astronomy **25**, 419–426 (1982).

Sabine, E. (1852): On periodical laws discoverable in the mean effects of the larger magnetic disturbances. Philosophical Transactions of the Royal Society (London) **142**, 103–124.

Sackmann, I.-J., Boothroyd, A. I., Kraemer, K. E. (1993): Our sun III. Present and future. Astrophysical Journal **418**, 457–468.

Sagan, C., Chyba, C. (1997): The early faint sun paradox: Organic shielding of ultraviolet-labile greenhouse gases. Science **276**, 1217–1221.

Sagan, C., Mullen, G. (1972): Earth and mars: Evolution of atmospheres and surface temperatures. Science **177**, 52–56.

Sagdeev, R. Z., Kennel, C. F. (1991): Collisionless shock waves. Scientific American **264**, 106–113 – April.

Saito, T. (1975): Two-hemisphere model of the three-dimensional magnetic structure of the interplanetary space. Science Reports of the Tohoku University, Series 5, **26**, 37–54.

Sakai, J.-I., De Jager, C. (1996): Solar flares and collisions between current-carrying loops. Space Science Reviews **77**, 1–192.

Sakao, T.: Characteristics of solar flare hard X-ray sources as revealed with the hard X-ray telescope aboard the Yohkoh satellite. Ph. D. Thesis, University of Tokyo, the Yohkoh HXT group, National Astronomical Observatory, Mitaka, Tokyo, 1994.

Sakurai, T. (1991): Observations from the Hinotori mission. Philosophical Transactions of the Royal Society (London) **A336**, 339–347.

Sakurai, T., Spangler, S. R. (1994): The study of coronal plasma structures and fluctuations with Faraday rotation measurements. Astrophysical Journal **434**, 773–785.

Sandbaek, O., Leer, E. (1995): Coronal heating and solar wind energy balance. Astrophysical Journal **454**, 486–498.

Sandbaek, O., Leer, E., Hansteen, V. H. (1994): On the relation between coronal heating, flux tube divergence, and the solar wind proton flux and flow speed. Astrophysical Journal **436**, 390–399.

Schatten, K. H. (1988): A model for solar constant secular changes. Geophysical Research Letters **15**, 121–124.

Schatten, K. H., et al. (1985): The importance of improved facular observations in understanding solar constant variations. Astrophysical Journal **294**, 689–696.

Schatzman, E. (1949): The heating of the solar corona and chromosphere. Annales d'Astrophysique **12**, 203–218.

Schein, M., Jesse, W. P., Wollan, E. O. (1941): The nature of the primary cosmic radiation and the origin of the mesotron. Physical Review **59**, 615.

Scherb, F. (1964): Velocity distributions of the interplanetary plasma detected by Explorer 10. Space Research **4**, 797–818.

Schlesinger, M. E., Ramankutty, N. (1992): Implications for global warming of intercycle solar irradiance variations. Nature **360**, 330–333.

Schmidt, A. (1924): Das erdmagnetische Aussenfeld. Zeitschrift Geophysikalische **1**, 3–13.

Schmidt, W. K. H., et al. (1980): On temperature and speed of He^{++} and O^{6+} ions in the solar wind. Geophysical Research Letters **7**, 697–700.

Schou, J., et al. (1997): Determination of the sun's seismic radius from the SOHO Michelson Doppler Imager. Astrophysical Journal (Letters) **489**, L197–L200.

Schove, D. J. (1955): The sunspot cycle 649 BC to 2000 AD. Journal of Geophysical Research **60**, 127–145.

Schrijver, C. J.: Working group 6: Magnetic fields, coronal structure and phenomena. In: The corona and solar wind near minimum activity. Proceedings of the fifth SOHO workshop ESA SP-404. Noordwijk, The Netherlands: ESA Publications Division 1997, pp. 149–153.

Schrijver, C. J., et al. (1989): Relations between the photospheric magnetic field and the emission from the outer atmospheres of cool stars. I. The solar Ca II K line core emission. Astrophysical Journal **337**, 964–976.

Schrijver, C. J., et al. (1997): Sustaining the quiet photospheric network: The balance of flux emergence, fragmentation, merging, and cancellation. Astrophysical Journal **487**, 424–436.

Schrijver, C. J., et al.: The dynamic quiet solar corona: 4 days of joint observing with MDI and EIT. In: The corona and solar wind near minimum activity. Proceedings of the fifth SOHO workshop.. ESA SP-404. Noordwijk, The Netherlands: ESA Publications 1997, pp. 669–674.

Schrijver, C. J., et al. (1998): Large-scale coronal heating by the small-scale magnetic field of the sun. Nature **394**, 152–154.

Schrijver, C. J., Zwaan, C.: Solar and stellar magnetic activity. New York: Cambridge University Press 2000.

Schröder, W. (1994): Behavior of auroras during the Spörer minimum (1450–1550). Annales Geophysicae **12**, 808–809.

Schröder, W. (1997): Some aspects of the earlier history of solar-terrestrial physics. Planetary and Space Science **45**, 395–400.

Schultz, M. (1973): Interplanetary sector structure and the heliomagnetic equator. Astrophysics and Space Science **24**, 371–383.

Schuster, A. (1911): The origin of magnetic storms. Proceedings of the Physical Society (London) **A85**, 61.

Schwabe, S. H. (1844): Sonnen-Beobachtungem im Jahre 1843. Astronomische Nachrichten **21**, No. 495, 233–236. Reprinted in Kosmos (Ed. A. Von Humboldt). English translation : Solar observations during 1843 in Early solar physics (Ed. A. J. Meadows). Oxford, England: Pergamon Press 1970, pp. 95–98.

Schwarzschild, M. (1948): On noise arising from the solar granulation. Astrophysical Journal **107**, 1–5.

Schwarzschild, M.: Structure and evolution of the stars. Princeton, New Jersey: Princeton University Press 1958, p. 207.

Schwenn, R. (1981): Solar wind and its interactions with the magnetosphere: measured parameters. Advances in Space Research **1**, 3–17.

Schwenn, R.: Large-scale structure of the interplanetary medium. In: Physics of the inner heliosphere I. (Eds. R. Schwenn and E. Marsch). New York: Springer-Verlag 1990, pp. 99–181.

Schwenn, R., Marsch, E. (Eds.): Physics of the inner heliosphere 1. large-scale phenomena, 2 particles, waves and turbulence. New York: Springer-Verlag 1990.

Severny, A. B. (1958): The appearance of flares in neutral points of the solar magnetic field and the pinch-effect. Crimean Astrophysical Observatory, Bulletin. **20**, 22–51.

Shackleton, N. J. (1977): The oxygen isotope stratigraphic record of the late Pleistocene. Philosophical Transactions of the Royal Society (London) **B280**, 169–182.

Share, G. H., Murphy, R. J. (1997): Intensity and directionality of flare-accelerated α-particles at the sun. Astrophysical Journal **485**, 409–418.

Share, G. H., Murphy, R. J., Ryan, J. (1997): Solar and stellar gamma ray observations with COMPTON. In: Proceedings of the fourth Compton symposium (Eds. C. D. Dermer, M. S. Strickman and J. D. Kurfess). New York: American Institute of Physics 1997, pp. 17–36.

Shea, M. A., Smart, D. F. (1990): A summary of major solar proton events. Solar Physics **127**, 297–320.

Sheeley, N. R., Jr., et al. (1975): Coronal changes associated with a disappearing filament. Solar Physics **45**, 377–392.

Sheeley, N. R. Jr., et al. (1977): A pictorial comparison of interplanetary magnetic field polarity, solar wind speed, and geomagnetic disturbances index during the sunspot cycle. Solar Physics **52**, 485–495.

Sheeley, N. R. Jr., et al. (1983): Associations between coronal mass ejections and soft X-ray events. Astrophysical Journal **272**, 349–354.

Sheeley, N. R. Jr., et al. (1984): Associations between coronal mass ejections and metric type II bursts. Astrophysical Journal **279**, 839–847.

Sheeley, N. R. Jr., et al. (1985): Coronal mass ejections and interplanetary shocks. Journal of Geophysical Research **90**, 163–175.

Sheeley, N. R. Jr., et al. (1997): Measurements of flow speeds in the corona between 2 and 30 solar radii. Astrophysical Journal **484**, 472–478.

Sheeley, N. R. Jr., Harvey, J. W., Feldman, W. C. (1976): Coronal holes, solar wind streams, and recurrent geomagnetic disturbances: 1973–1976. Solar Physics **49**, 271–278.

Sheeley, N. R. Jr., Wang, Y.-M. (1991): Magnetic field configurations associated with fast solar wind. Solar Physics **131**, 165–186.

Sheeley, N. R. Jr., Wang, Y.-M., Phillips, J. L.: Near-Sun magnetic fields and the solar wind. In: Cosmic winds and the heliosphere (Eds. J. R. Jokipii, C. P. Sonett and M. S. Giampapa). Tucson, Arizona: University of Arizona Press 1997, pp. 459–483.

Shevgaonkar, R. V., Kundu, M. R. (1984): Three-dimensional structures of two solar active regions from VLA observations at 2, 6 and 20 centimeters wavelength. Astrophysical Journal **283**, 413–420.

Shevgaonkar, R. K., Kundu, M. R. (1985): Dual frequency observations of solar microwave bursts using the VLA. Astrophysical Journal **292**, 733–751.

Shibata, K. (1996): New observational facts about solar flares from Yohkoh studies – Evidence of magnetic reconnection and a unified model of flares. Advances in Space Research **17**, No. 4/5, 9–18.

Shibata, K.: Rapidly time variable phenomena: Jets, explosive events, and flares. In: The corona and solar wind near minimum activity. Proceedings of the fifth SOHO workshop. ESA SP-404. Noordwijk, The Netherlands: ESA Publications Division 1997. pp. 103–112.

Shibata, K., et al. (1995): Hot plasma ejections associated with compact-loop solar flares. Astrophysical Journal (Letters) **451**, L83–L85.

Shimizu, T. (1995): Energetics and occurrence rate of active-region transient brightenings and implications for the heating of the active-region corona. Publications of the Astronomical Society of Japan **47**, 251–263.

Shimizu, T. (Ed.): Yohkoh views the sun – the first five years. Tokyo: The Institute of Space and Astronautical Science, National Astronomical Observatory, Yohkoh Group 1996.

Shimizu, T., Tsuneta, S. (1997): Deep survey of solar nanoflares with Yohkoh. Astrophysical Journal **486**, 1045–1057.

Shklovskii, I. S., Moroz, V. I., Kurt, V. G. (1960): The nature of the Earth's third radiation belt. Soviet Astronomy AJ **4**, 871–873.

Silverman, S. M. (1992): Secular variation of the aurora for the past 500 years. Reviews of Geophysics **30**, No. 4, 333–351.

Simnett, G. M. (1973): Relativistic electrons in space. Space Research **13**, 745–762.

Simnett, G. M. (1991): Energetic particle production in flares. Philosophical Transactions of the Royal Society (London) **A336**, 439–450.

Simon, G. W. (1967): Observations of horizontal motions in solar granulation: their relation to supergranulation. Zeitschrift für Astrophysick **65**, 345–363.

Simon, G. W., Leighton, R. B. (1964): Velocity fields in the solar atmosphere. III. Large-scale motions, the chromospheric network, and magnetic fields. Astrophysical Journal **140**, 1120–1147.

Simon, T., Herbig, G., Boesgaard, A. M. (1985): The evolution of chromospheric activity and the spin-down of solar-type stars. Astrophysical Journal **293**, 551–574.

Simpson, J. A. (1954): Cosmic-radiation intensity-time variations and their origin III. The origin of 27-day variations. Physical Review **94**, 426–440.

Simpson, J. A. (1983): Elemental and isotopic composition of the galactic cosmic rays. Annual Review of Nuclear Particle Science **33**, 323–381.

Simpson, J. A., et al. (1995): Cosmic ray and solar particle investigations over the south polar regions of the sun. Science **268**, 1019–1023.

Simpson, J. A., et al. (1995): The latitude gradients of galactic cosmic ray and anomalous helium fluxes measured on Ulysses from the sun's south polar region to the equator. Geophysical Research Letters **22**, No. 23, 3337–3340.

Singer, S. F. (1957): A new model of magnetic storms and aurorae. EOS **38**, 175–190.

Siscoe, G. L. (1980): Evidence in the auroral record for secular solar variations. Review of Geophysics and Space Physics **18**, 647–658.

Skumanich, A., et al. (1984): The sun as a star: Three-component analysis of chromospheric variability in the calcium K line. Astrophysical Journal **282**, 776–783.
Slottje, C. (1978): Millisecond microwave spikes in a solar flare. Nature **275**, 520–521.
Smith, E. J. (1962): A comparison of Explorer 6 and Explorer 10 magnetometer data. Journal of Geophysical Research **67**, 2045–2049.
Smith, E. J., Balogh, A. (1995): Ulysses observations of the radial magnetic field. Geophysical Research Letters **22**, No. 23, 3317–3320.
Smith, E. J., et al. (1995): Ulysses observations of Alfvén waves in the southern and northern solar hemispheres. Geophysical Research Letters **22**, No. 23, 3381–3384.
Smith, E. J., Marsden, R. G. (1995): Ulysses observations from pole-to-pole: an introduction. Geophysical Research Letters **22**, No. 23, 3297–3000.
Smith, E. J., Marsden, R. G. (1998): The Ulysses mission, Scientific American **278**, 74–79 – January.
Smith, E. J., Marsden, R. G., Page, D. E. (1995): Ulysses above the sun's south pole – an introduction. Science **268**, 1005–1006.
Smith, E. J., Sonett, C. P., Dungey, J. W. (1964): Satellite observation of the geomagnetic field during magnetic storms. Journal of Geophysical Research **69**, 2669–2688.
Smith, E. J., Tsurutani, B. T., Rosenberg, R. L. (1978): Observations of the interplanetary sector structure up to heliographic latitudes of 16 degrees by Pioneer 11. Journal of Geophysical Research **83**, 717–724.
Smith, E. J., Wolfe, J. H. (1976): Observations of interaction regions and co-rotating shocks between one and five AU: Pioneers 10 and 11. Geophysical Research Letters **3**, 137–140.
Snodgrass, H. B. (1985): Solar torsional oscillations: a net pattern with wavenumber 2 as artifact. Astrophysical Journal **291**, 339–343.
Snodgrass, H. B., Howard, R. (1985): Torsional oscillations of the sun. Science **228**, 945–952.
Snyder, C. W., Neugebauer, M. (1964): Interplanetary solar-wind measurements by mariner II. Space Research **4**, 89–113.
Snyder, C. W., Neugebauer, M., Rao, U. R. (1963): The solar wind velocity and its correlation with cosmic-ray variations and with solar and geomagnetic activity. Journal of Geophysical Research **68**, 6361–6370.
Soderblom, D. R. (1985): A survey of chromospheric emission and rotation among solar-type stars in the solar neighborhood. Astronomical Journal **90**, 2103–2115.
Soderblom, D. R., Baliunas, S. L.: The sun among the stars: What stars indicate about solar variability. In: Secular, solar and geomagnetic variations in the last 10,000 years (Eds. F. R. Stephenson and A. W. Wolfendale). Dordrecht, Holland: Kluwer 1988, pp. 25–48.
Solanki, S. K., Fligge, M. (1998): Solar irradiance since 1874 revisted. Geophysical Research Letters **25**, 341–344.
Solanki, S. K., Fligge, M. (1999): A reconstruction of total solar irradiance since 1700. Geophysical Research Letters **26**, 2465–2468.
Somov, B. V., Kosugi, T. (1997): Collisonless reconnection and high-energy particle acceleration in solar flares. Astrophysical Journal **485**, 859–868.
Sonett, C. P. (1984): Very long solar periods and the radiocarbon record. Reviews of Geophysics and Space Physics **22**, No. 3, 239–254.
Sonett, C. P., Colburn, D. S., Davis, L., Jr., Smith, E. J., Colman, P. J., Jr. (1964): Evidence for a collision-free magnetohydrodynamic shock in interplanetary space. Physical Review Letters **13**, 153–156.
Sonett, C. P., et al. (1960): Current systems in the vestigial geomagnetic field: Explorer 6. Physical Review Letters **4**, 161–163.
Sonett, C. P., Giampapa, M. S., Matthews, M. S. (Eds.): The sun in time. Tucson, Arizona: University of Arizona Press 1991.

Sonett, C. P., Suess, H. E. (1984): Correlation of bristlecone pine ring widths with atmospheric ^{14}C variations: A climate-sun relation. Nature **307**, 141–143.

Soon, W. H., Posmentier, E. S., Baliunas, S. L. (1996): Inference of solar irradiance variability from terrestrial temperature changes, 1880–1993: An astrophysical application of the sun-climate connection. Astrophysical Journal **472**, 891–902.

Southworth, G. C. (1945): Microwave radiation from the sun. Journal of the Franklin Institute **239**, 285–297.

Spiegel, E. A., Weiss, N. O. (1980): Magnetic activity and variations in solar luminosity. Nature **287**, 616–617.

Spörer, G. F. W.: Beobachtungen der Sonnenflecken zu Anclam. Leipzig, 1874–1876.

Spörer, G. F. W. (1887): Üeber die periodicität der Sonnenflecken seit dem Jahre 1618. Vierteljahrsschr Astronomische Gesellschaft (Leipzig) **22**, 323–329.

Spörer, G. F. W.. (1889): Üeber die periodicität der Sonnenflecken seit dem Jahre 1618. R. Leopold-Caroline Acad. Aston. Halle **53**, 283–324.

Spörer, G. F. W. (1899): Sur les différences que présentent l'Hémisphere nord el l'Hémisphere sud du Soleil. Bulletin Astronomique **6**, 60.

Spreybroeck, L. P. Van, Krieger, A. S., VAIANA, G. S. (1970): X-ray photographs of the Sun on March 7, 1970. Nature **227**, 818–822.

Spruit, H. C. (1982): Effect of spots on a star's radius and luminosity. Astronomy and Astrophysics **108**, 348–355.

Spruit, H. C.: Influence of magnetic activity on the solar luminosity and radius. In: Solar radiative output variations (Ed. P. V. Foukal). Cambridge, Massachusetts: Cambridge Research and Instrumentation 1988, pp. 254–288.

Stauffer, B., et al. (1998): Atmospheric CO_2 concentration and millennial-scale climate change during the last glacial period. Nature **392**, 59–61.

Steig, E. L., et al. (1996): Large amplitude solar modulation cycles of ^{10}Be in Antarctica: implications for atmospheric mixing processes and interpretation of the ice core record. Geophysical Research Letters **23**, 523–526.

Sterling, A. C., Hudson, H. S. (1997): Yohkoh SXT observations of X-ray "dimming" associated with a halo coronal mass ejection. Astrophysical Journal (Letters) **491**, L55–L58.

Stern, D. P. (1989): A brief history of magnetospheric physics before the spaceflight era. Reviews of Geophysics **27**, 103–114.

Stix, M. (1981): Theory of the solar cycle. Solar Physics **74**, 79–101.

Størmer, C. (1907): Sur les trajectories des corpuscles, electrisés dans l'espace sous l'action du magnétisme terrestre avec l'application aux aurores boréales. Archives des sciences physiques et naturelles (Geneva) **24**, 5, 113, 221, 317; **32**, 117–123, 190–219, 277–314, 415–436, 505–509 (1911); **33**, 51–69, 113–150 (1912).

Størmer, C. (1917): Corpuscular theory of the aurora borealis. Journal of Geophysical Research **22**, 23–34, 97–112.

Størmer, C. (1930): Periodische electronenbahnen im fielde lines elementarmagneton und ihre awendung auf eschenhagens elementarwellen des erdmagnetismus. Astrophysics **1**, 237.

Størmer, C.: The polar aurora. Oxford, England: The Clarendon Press 1955.

Strömgren, B. (1932): The opacity of stellar matter and the hydrogen content of the stars. Zeitschrift für Astrophysik **4**, 118–152.

Strong, K. T. (1991): Observations from the Solar Maximum Mission. Philosophical Transactions of the Royal Society (London) **A336**, 327–337.

Strong, K. T., et al. (1984): A multiwavelength study of a double impulsive flare. Solar Physics **91**, 325–344.

Strong, K. T., et al. (1992): Observations of the variability of coronal bright points by the soft X-ray telescope on Yohkoh. Publications of the Astronomical Society of Japan **44**, L161-L166.

Strong, K. T. et al. (Eds.): The many faces of the sun. A summary of the results from NASA's Solar Maximum Mission. New York: Springer Verlag 1999.
Stuiver, M. (1961): Variations in radiocarbon concentration and sunspot activity. Journal of Geophysical Research **66**, 273–276.
Stuiver, M. (1980): Solar variability and climatic change during the current millennium. Nature **286**, 868–871.
Stuiver, M., Braziunas, T. F. (1989): Atmospheric ^{14}C and century-scale solar oscillations. Nature **338**, 405–408.
Stuiver, M., Braziunas, T. F. (1993): Sun, ocean, climate and atmospheric 14 co2: an evaluation of causal and spectral relationships. Holocene **3**(4), 289–305.
Stuiver, M., Quay, P. D. (1980): Changes in atmospheric carbon-14 attributed to a variable sun. Science **207**, 11–19.
Sturrock, P. A. (1966): Model of the high-energy phase of solar flares. Nature **211**, 695–697.
Sturrock, P. A.: A model of solar flares. In: Structure and development of solar active regions. international astronomical union symposium No. 35 (Ed. K. O. Kiepenheuer). Dordrecht, Holland: D. Reidel Publishing Co. 1968, pp. 471–477.
Sturrock, P. A. (Ed.): Solar flares: A monograph from Skylab solar workshop II. Boulder, Colorado: Colorado Associated University Press 1980.
Sturrock, P. A. (1989): The role of eruption in solar flares. Solar Physics **121**, 387–397.
Sturock, P. A., Hartle, R. E. (1966): Two-fluid model of the solar wind. Physical Review Letters **16**, 628–631.
Sturrock, P. A., Wheatland, M. S., Acton, L. W. (1996): Yohkoh soft X-ray telescope images of the diffuse solar corona. Astrophysical Journal (Letters) **461**, L115–L117.
Suess, H. E. (1955): Radiocarbon concentration in modern wood. Science **122**, 415–417.
Suess, H. E. (1965): Secular variations of the cosmic-ray produced carbon 14 in the atmosphere and their interpretations. Journal of Geophysical Research **70**, 5937–5952.
Suess, H. E. (1968): Climate changes, solar activity, and cosmic-ray production rate of natural radiocarbon. Meteorology Monograph **8**, 146–150.
Suess, H. E. (1973): Natural radiocarbon. Endeavor **32**, 34–38.
Suess, H. E. (1980): Radiocarbon geophysics. Endeavor **4**, 113–117.
Suess, H. E., Linick, T. W. (1990): The ^{14}C record in bristlecone pine wood of the past 8000 years based on the dendrochronology of the late C. W. Ferguson. Philosophical Transactions of the Royal Society (London) **A330**, 403–412.
Suess, S. T. (1990): The heliopause. Reviews of Geophysics **28**, 97–115.
Suess, S. T., et al. (1996): Latitudinal dependence of the radial IMF component – interplanetary imprint. Astronomy and Astrophysics **316**, 304–312.
Sullivan, W. T. III. (Ed.): The early years of radio astronomy. New York: Cambridge University Press 1984.
Svalgaard, L., et al. (1975): The sun's sector structure. Solar Physics **45**, 83–91.
Svalgaard, L., Wilcox, J. M. (1976): Structure of the extended solar magnetic field and the sunspot cycle variation in cosmic ray intensity. Nature **262**, 766–768.
Svalgaard, L., Wilcox, J. M. (1978): A view of solar magnetic fields, the solar corona, and the solar wind in three dimensions. Annual Review of Astronomy and Astrophysics **16**, 429–443.
Svalgaard, L., Wilcox, J. M., Duvall, T. L. (1974): A model combining the polar and the sector structured solar magnetic fields. Solar Physics **37**, 157–172.
Svensmark, H., Friis-Christensen, E. (1997): Variation of cosmic ray flux and global cloud coverage – a missing link in solar-climate relationships. Journal of Atmospheric and Solar-Terrestrial Physics **59**, 1225–1232.
Svestka, Z.: Solar flares. Norwell, Massachusetts: Kluwer 1976.
Svestka, Z. (1995): On "the solar flare myth" postulated by Gosling. Solar Physics **160**, 153–156.

Svestka, Z., Cliver, E. W.: History and basic characteristics of eruptive flares. In: Eruptive solar flares. Proceedings of international astronomical union colloquium No. 133 (Eds. Z. Svestka, B. V. Jackson, and M. E. Machado). New York: Springer-Verlag 1992, pp. 1–14.

Svestka, Z., et al. (1982): Observations of a post-flare radio burst in X-rays. Solar Physics **75**, 305–329.

Svestka, Z. et al. (1987): Multi-thermal observations of newly formed loops in a dynamic flare. Solar Physics **108**, 237–250.

Sweet, P. A.: The neutral point theory of solar flares. In: Electromagnetic phenomena in cosmical physics. international astronomical union symposium no. 6 (Ed. B. Lehnert). Cambridge, England: Cambridge at the University Press 1958, pp. 123–134.

Sweet, P. A. (1969): Mechanisms of solar flares. Annual Review of Astronomy and Astrophysics **7**, 149–176.

Syrovatskii, S. I. (1981): Pinch sheets and reconnection in astrophysics. Annual Review of Astronomy and Astrophysics **19**, 163–229.

Takakura, T. (1961): Acceleration of electrons in the solar atmosphere and type IV radio outbursts. Publications of the Astronomical Society of Japan **13**, 166–172.

Takakura, T. (1967): Theory of solar bursts. Solar Physics **1**, 304–353.

Takakura, T. (1995): Imaging spectra of hard X-rays from the foot points of impulsive loop flares. Publications of the Astronomical Society of Japan **47**, 355–364.

Takakura, T., et al. (1993): Time variation of the hard X-ray image during the early phase of solar impulsive bursts. Publications of the Astronomical Society of Japan **45**, 737–753.

Takakura, T., et al. (1995): Imaging spectra of hard X-rays from the footpoints of solar impulsive loop flares. Publications of the Astronomical Society of Japan **47**, 355–364.

Takakura, T., Kai, K. (1966): Energy distribution of electrons producing microwave impulsive bursts and X-ray bursts from the sun. Publications of the Astronomical Society of Japan **18**, 57–76.

Tanaka, K. (1987): Impact of X-ray observations from the Hinotori satellite on solar flare research. Publications of the Astronomical Society of Japan **39**, 1–45.

Tanaka, K., et al. (1982): High-resolution solar flare X-ray spectra obtained with rotating spectrometers on the Hinotori satellite. Astrophysical Journal (Letters) **254**, L59–L63.

Tandberg-Hanssen, E., Emslie, A. G.: The physics of solar flares. New York: Cambridge University Press 1988.

Tandon, J. N., Das, M. K. (1982): The effect of a magnetic field on solar luminosity. Astrophysical Journal **260**, 338–341.

Tett, S. F. B., et al. (1999): Causes of twentieth-century temperature change near the earth's surface. Nature **399**, 569–572.

Thomas, B. T., Smith, E. J. (1980): The Parker spiral configuration of the interplanetary magnetic field between 1 and 8.5 AU. Journal of Geophysical Research **85**, 6861–6867.

Thomas, B. T., Smith, E. J. (1981): The structure and dynamics of the heliospheric current sheet. Journal of Geophysical Research **86**, 11,105–11,110.

Thompson, M. J., et al. (1996): Differential rotation and dynamics of the solar interior. Science **272**, 1300–1305.

Thomsen, M. F. et al. (1998): The magnetospheric response to the CME passage of January 10–11, 1997, as seen at geosynchronous orbit. Geophysical Research Letters **25**, 2545–2548.

Thomson, W. (Baron Kelvin): Presidential address to the Royal Society on November 30, 1892. In: Popular lectures and addresses by Sir William Thomson Baron Kelvin. Volume II. Geology and general physics. London: Macmillan and Company 1894, 508–529.

Timothy, A. F., Krieger, A. S., Vaiana, G. S. (1975): The structure and evolution of coronal holes. Solar Physics **42**, 135–156.

Tinsley, B. A. (1988): The solar cycle and the QBO influences on the latitude of storm tracks in the North Atlantic. Geophysical Research Letters **15**, 409–412.

Tinsley, B. A. (1994): Solar wind mechanism suggested for weather and climate change. EOS Transactions of the American Geophysical Union **75**, No. 32, 369–376.

Tomczyk, S., Schou, J., Thompson, M. J. (1995): Measurement of the rotation rate in the deep solar interior. Astrophysical Journal (Letters) **448**, L57–L60.

Torsti, J. et al. (1998): Energetic (~ 1 to 50 MeV) protons associated with earth-directed coronal mass ejections. Geophysical Research Letters **25**, 2525–2528.

Tousey, R. (1963): The extreme ultraviolet spectrum of the sun. Space Science Review **2**, 3–69.

Tousey, R. (1967): Some results of twenty years of extreme ultraviolet solar research. Astrophysical Journal **149**, 239–252.

Tousey, R. (1973): The solar corona. Space Research **13**, 713–730.

Tousey, R. (1976): Eruptive prominences recorded by the X u.v. spectroheliograph on Skylab. Philosophical Transactions of the Royal Society (London) **A281**, 359–364.

Tousey, R., et al. (1946): The solar ultraviolet spectrum from a V-2 rocket. Astronomical Journal **52**, 158–159.

Tousey, R., et al. (1973): A preliminary study of the extreme ultraviolet spectroheliograms from Skylab. Solar Physics **33**, 265–280.

Trattner, K. J., et al. (1996): Ulysses COSPIN/LET: latitudinal gradients of anomalous cosmic ray O, N and Ne. Astronomy and Astrophysics **316**, 519–527.

Tsuneta, S. (1995): Particle acceleration and magnetic reconnection in solar flares. Publications of the Astronomical Society of Japan **47**, 691–697.

Tsuneta, S. (1996): Interacting active regions in the solar corona. Astrophysical Journal (Letters) **456**, L63–L65.

Tsuneta, S. (1996): Structure and dynamics of magnetic reconnection in a solar flare. Astrophysical Journal **456**, 840–849.

Tsuneta, S., et al. (1983): Vertical structure of hard X-ray flare. Solar Physics **86**, 313–321.

Tsuneta, S., et al. (1992): Observation of a solar flare at the limb with the Yohkoh soft X-ray telescope. Publications of the Astronomical Society of Japan **44**, L63–L69.

Tsuneta, S., et al. (1992): Global restructuring of the coronal magnetic fields observed with the Yohkoh Soft X-ray Telescope. Publications of the Astronomical Society of Japan **44**, L211–L214.

Tsurutani, B. T. et al. (1990): Interplanetary Alfvén waves and auroral (substorm) activity: IMP 8. Journal of Geophysical Research **95**, 2241–2252.

Tsurutani, B. T., et al. (1994): The relationship between interplanetary discontinuities and Alfvén waves: Ulysses observations. Geophysical Research Letters **21**, No. 21, 2267–2270.

Tsurutani, B. T. et al. (1995): Interplanetary origin of geomagnetic activity in the declining phase of the solar cycle. Journal of Geophysical Research **100**, 21,717–21,733.

Tsurutani, B. T., et al. (1995): Large amplitude IMF fluctuations in corotating interaction regions: Ulysses at midlatitudes. Geophysical Research Letters **22**, No. 23, 3397–3400.

Tsurutani, B. T. et al. (1996): Interplanetary discontinuities and Alfvén waves at high heliographic latitudes: Ulysses. Journal of Geophysical Research **101**, 11,027–11,038.

Tsurutani, B. T., Gonzalez, W. D. (1987): The cause of high-intensity, long-duration continuous AE activity (HILDCAAS): interplanetary Alfvén wave trains. Planetary and Space Science **35**, 405–412.

Tsurutani, B. T., Gonzalez, W. D.: The interplanetary causes of magnetic storms: A review. In: Magnetic storms: Geophysical monograph 98. Washington, DC: American Geophysical Union 1997. pp. 77–89.

Tsurutani, B. T., Gonzalez, W. D., Kamide, Y., Arballo, J. K. (Eds.). Magnetic storms. Geophysics monograph 98. Washington, DC: American Geophysical Union 1997.

Turck-Chieze, S, et al. (1988): Revisiting the solar model. Astrophysical Journal **335**, 415–424.

Turck-Chieze, S., et al. (1997): First results of the solar core from GOLF acoustic modes. Solar Physics **175**, 247–265. Reprinted in: The first results from SOHO (Eds. B. Fleck and Z. Svestka). Boston: Kluwer Academic Publishers 1997, pp. 247–265.

Tyndall, J. (1861): On the absorption and radiation of heat by gases and vapors, and on the physical connection of radiation, absorption, and conduction. Philosophical Magazine and Journal of Science **22A**, 276–277.

Tzedakis, P. C., et al. (1997): Comparison of terrestrial and marine records of changing climate of the last 500,000 years. Earth and Planetary Science Letters **150**, 171–176.

Uchida, Y. (1963): An effect of the magnetic field in the shock wave heating theory of the solar corona. Publications of the Astronomical Society of Japan **15**, 376–399.

Uchida, Y. (1974): Behavior of flare-produced coronal mhd wavefront and the occurrence of type II radio bursts. Solar Physics **39**, 431–449.

Uchida, Y., Altschuler, M. D., Newkirk, G. Jr. (1973): Flare-produced coronal mhd-fast-mode wavefronts and Moreton's wave phenomenon. Solar Physics **28**, 495–516.

Uchida, Y., Canfield, R. C., Watanabe, T., Hiei, E. (Eds.): Flare physics in solar activity maximum 22. New York: Springer-Verlag 1991.

Uchida, Y., et al. (1992): Continual expansion of the active-region corona observed by the Yohkoh soft X-ray telescope. Publications of the Astronomical Society of Japan **44**, L155–L160.

Uchida, Y., et al. (Eds.): X-ray solar physics from Yohkoh. Tokyo: University Academy Press 1994.

Uchida, Y., Kosugi, T., Hudson, H. S. (Eds.): Magnetodynamic phenomena in the solar atmosphere – prototypes of stellar magnetic activity. IAU colloquium No. 153. Boston: Kluwer Academic Publishers 1996.

Ulmschneider, P., Priest, E. R., Rosner, R. (Eds.): Mechanisms of chromospheric and coronal heating. New York: Springer-Verlag 1991.

Ulrich, R. K. (1970): The five-minute oscillations on the solar surface. Astrophysical Journal **162**, 993–1002.

Ulrich, R. K. (1975): Solar neutrinos and variations in the solar luminosity. Science **190**, 619–624.

Ulrich, R. K., Bertello, L. (1995): Solar-cycle dependence of the sun's apparent radius in the neutral iron spectral line at 525 nm. Nature **377**, 214–215.

Ulrich, R. K., Rhodes, E. J. Jr. (1977): The sensitivity of nonradial p mode eigenfrequencies to solar envelope structure. Astrophysical Journal **218**, 521–529.

Underwood, J. H., et al. (1976): Preliminary results from S-056 X-ray telescope experiment aboard the Skylab-Apollo Telescope Mount. Progress in Astronautics and Aeronautics **48**, 179–195.

Unsöld, A. (1928): Über die Struktur der Fraunhoferschen Linien und die quantitative Spektralanalyse der Sonnenatmosphäre. Zeitschrift fur Physik **46**, 765.

Vaiana, G. S., et al. (1973): X-ray observations of characteristic structures and time variations from the solar corona: Preliminary results from Skylab. Astrophysical Journal (Letters) **185**, L47–L51.

Vaiana, G. S., Krieger, A. S., Timothy, A. F. (1973): Identification and analysis of structures in the corona from X-ray photography. Solar Physics **32**, 81–116.

Vaiana, G. S., Reidy, W. P., Zehnpfennig, T., Van Speybroeck, L., Giacconi, R. (1968): X-ray structures of the sun during the importance 1n flare of 8 June 1968. Science **161**, 564–567.

Vaiana, G. S., Rosner, R. (1978): Recent advances in coronal physics. Annual Review of Astronomy and Astrophysics **16**, 393–428.

Van Allen, J. A. (1975): Interplanetary particles and fields. Scientific American **233**, 160–162 – September.

Van Allen, J. A., Fennell, J. F., Ness, N. F. (1971): Asymmetric access of energetic solar protons to the earth's north and south polar caps. Journal of Geophysical Research **76**, 4262–4275.

Van Allen, J. A., Krimigis, S. M. (1965): Impulsive emission of $\approx 40\,\mathrm{keV}$ electrons from the sun. Journal of Geophysical Research **70**, 5737–5751.

Van Allen, J. A., Mc Ilwain, C. E., Ludwig, G. H. (1959): Radiation observations with satellite 1958ε. Journal of Geophysical Research **64**, 271–286. Reproduced in: A source book in astronomy and astrophysics 1900–1975 (Eds. K. R. Lang and O. Gingerich). Cambridge, Massachusetts: Harvard University Press 1979, pp. 149–151.

Van Allen, J. A., Ness, N. F. (1969): Particle shadowing by the moon. Journal of Geophysical Research **74**, 91–93.

Van Ballegooijen, A. A., Martens, P. C. H. (1989): Formation and eruption of solar prominences. Astrophysical Journal **343**, 971–984.

Van De Hulst, H. C. (1947): Zodiacal light in the solar corona. Astrophysical Journal **105**, 471–488.

Van Driel-Gesztelyi, L., et al. (1996): X-ray bright point flares due to magnetic reconnection. Solar Physics **163**, 145–170.

Van Loon, H., Labitzke, K. (1988): Association between the 11-year solar cycle, the QBO and the atmosphere, Part II. Surface and 700 mb on the northern hemisphere in winter. Journal of Climate **1**, 905–920.

Van Loon, H., Labitzke, K. (1990): Association between the 11-year solar cycle and the atmosphere. Part IV. The stratosphere, not grouped by the phase of the QBO. Journal of Climate **3**, 827–837.

Van Speybroeck, L. P., Krieger, A. S., Vaiana, G. S. (1970): X-ray photographs of the sun on March 7, 1970. Nature **227**, 818–822.

Vaughan, A. H., Preston, G. W. (1980): A survey of chromospheric Ca II H and K emission in field stars of the solar neighborhood. Publications of the Astronomical Society of the Pacific **92**, 385–391.

Vegard, L. (1913): On spectra of the aurora borealis. Physikalishe Zeitschrift **14**, 677.

Vestrand, W. T. (1991): High-energy flare observations from the Solar Maximum Mission. Philosophical Transactions of the Royal Astronomical Society **A336**, 349–362.

Vial, J. C., Bocchialini, K., Boumier, P. (Eds.): Space solar physics. Theoretical and observational issues in the context of the SOHO mission. Lecture notes in physics no. 507. Heidelberg: Springer Verlag 1999.

Völk, H. J. (1975): Cosmic ray propagation in interplanetary space. Review of Geophysics and Space Physics **13**, 547–566.

Von Steiger, R., et al. (1992): Variable carbon and oxygen abundances in the solar wind as observed in earth's magnetosheath by AMPTE/CCE. Astrophysical Journal **389**, 791–799.

Wagner, W. J. (1984): Coronal mass ejections. Annual Review of Astronomy and Astrophysics **22**, 267–289.

Waldmeier, M. (1938): Chromosphärische Eruptionen I. Zeitschrift für Astrophysik **16**, 276–290.

Waldmeier, M. (1940): Chromosphärische Eruptionen II. Zeitschrift für Astrophysik **20**, 46–66.

Waldmeier, M. (1951): Spektralphotometrische Klassifikation der Protuberanzen. Zeitschrift für Astrophysik **28**, 208–218.

Waldmeier, M.: Die Sonnenkorona I, II. Basel, Switzerland: Birkhäuser 1951, 1957.

Waldmeier, M. The sunspot activity in the years 1610–1960. Zurich: Schulthess 1961.

Wallerstein, G. (1988): Mixing in stars. Science **240**, 1743–1750.
Wang, Y.-M. (1994): Polar plumes and the solar wind. Astrophysical Journal (Letters) **435**, L153–L156.
Wang, Y.-M. (1994): Two types of slow solar wind. Astrophysical Journal (Letters) **437**, L67–L70.
Wang, Y.-M. (1998): Network activity and the evaporative formation of polar plumes. Astrophysical Journal (Letters) **501**, L145–L150.
Wang, Y.-M., et al. (1997): The green line corona and its relation to the photospheric magnetic field. Astrophysical Journal **485**, 419–429.
Wang, Y.-M., et al. (1997): Solar wind stream interactions and the wind speed-expansion factor relationship. Astrophysical Journal (Letters) **488**, L51–L54.
Wang, Y.-M., et al. (1998): Origin of streamer material in the outer corona. Astrophysical Journal (Letters) **498**, L165–L168.
Wang, Y.-M., et al. (1998): Observations of correlated white-light and extreme ultraviolet jets from polar coronal holes. Astrophysical Journal **508**, 899–907.
Wang, Y.-M., et al. (1998): Coronagraph observations of inflows during high solar activity. Geophysical Research Letters **26**, 1203–1206.
Wang, Y.-M., et al. (1999): Streamer disconnection events observed with the LASCO coronagraph. Geophysical Research Letters **26**, 1349–1352.
Wang, Y.-M., Hawley, S. H., Sheeley, N. R. Jr. (1996): The magnetic nature of coronal holes. Science **271**, 464–469.
Wang, Y.-M., Nash, A. G., Sheeley, N. R. Jr. (1989): Magnetic flux transport on the sun. Science **245**, 712–718.
Wang, Y.-M., Sheeley, N. R. Jr. (1990): Solar wind speed and coronal flux-tube expansion. Astrophysical Journal **355**, 726–732.
Wang, Y.-M., Sheeley, N. R. Jr. (1990): Magnetic flux transport and the sunspot-cycle evolution of coronal holes and their wind streams. Astrophysical Journal **365**, 372–386.
Wang, Y.-M., Sheeley, N. R. Jr. (1991): Why fast solar wind originates from slowly expanding coronal flux tubes. Astrophysical Journal (Letters) **372**, L45–L48.
Wang, Y.-M., Sheeley, N. R. Jr. (1997): The high-latitude solar wind near sunspot maximum. Geophysical Research Letters **24**, No. 24, 3141–3144.
Wang, Y.-M., Sheeley, N. R. Jr., Nash, A. G. (1990): Latitudinal distribution of solar-wind speed from magnetic observations of the sun. Nature **347**, 439–444.
Wang, Y.-M., Sheeley, N. R. Jr., Nash, A. G. (1991): A new solar cycle model including meridional circulation. Astrophysical Journal **383**, 431–442.
Warren, H. P., et al. (1997): Doppler shifts and nonthermal broadening in the quiet solar transition region: O VI. Astrophysical Journal (Letters) **484**, L91–L94.
Webb, D. F.: The solar sources of coronal mass ejections. In: Eruptive solar flares (Eds. Z. Svestka, B. V. Jackson, M. E. Machado). Berlin: Springer-Verlag 1992, pp. 234–247.
Webb, D. F. (1995): Coronal mass ejections: The key to major interplanetary and geomagnetic disturbances. Reviews of Geophysics, Supplement **33**, 577–583.
Webb, D. F. et al. (1998): The solar origin of the January 1997 coronal mass ejection, magnetic cloud and geomagnetic storm. Geophysical Research Letters **25**, 2469–2472.
Webb, D. F., Howard, R. A. (1994): The solar cycle variation of coronal mass ejections and the solar wind mass flux. Journal of Geophysical Research **99**, 4201–4220.
Webb, D. F., Hundhausen, A. J. (1987): Activity associated with the solar origin of coronal mass ejections. Solar Physics **108**, 383–401.
Webb, D. F., Krieger, A. S., Rust, D. M. (1976): Coronal X-ray enhancements associated with Hα filament disappearances. Solar Physics **48**, 159–186.
Weber, E. J., Davis, L. Jr. (1967): The angular momentum of the solar wind. Astrophysical Journal **148**, 217–227.
Weiss, J. E., Weiss, N. O. (1979): Andrew Marvell and the Maunder minimum. Quarterly Journal of the Royal Astronomical Society **20**, 115–118.

Weiss, N. O. (1990): Periodicity and aperiodicity in solar magnetic activity. Philosophical Transactions of the Royal Society (London) **A330**, 617–625.
Wheatland, M. S., Sturrock, P. A., Acton, L. W. (1997): Coronal heating and the vertical temperature structure of the quiet corona. Astrophysical Journal **482**, 510–518.
White, O. R. (Ed.): The solar output and its variation. Boulder, Colorado: Colorado Associated University Press 1977.
White, O. R., Livingston, W. C. (1981): Solar luminosity variation III. Calcium K variation from solar minimum to maximum in cycle 21. Astrophysical Journal **249**, 798–816.
White, O. R., Livingston, W. C., Wallace, L. (1987): Variability of chromospheric and photospheric lines in solar cycle 21. Journal of Geophysical Research **92**, 823–827.
White, W. B., et al. (1997): Response of global upper ocean temperature to changing solar irradiance. Journal of Geophysical Research **102**, 3255–3266.
Wigley, T. M. L. (1976): Spectral analysis: Astronomical theory of climatic change. Nature **264**, 629–631.
Wigley, T. M. L., Kelly, P. M. (1990): Holocene climatic change, ^{14}C wiggles and variations in solar irradiance. Philosophical Transactions of the Royal Society (London) **A330**, 547–560.
Wigley, T. M. L., Raper, S. C. B. (1990): Climatic change due to solar irradiance changes. Geophysical Research Letters **17**, 2169–2172.
Wilcox, J. M. (1968): The interplanetary magnetic field, solar origin and terrestrial effects. Space Science Reviews **8**, 258–328.
Wilcox, J. M., Ness, N. F. (1965): Quasi-stationary corotating structure in the interplanetary medium. Journal of Geophysical Research **70**, 5793–5805.
Wild, J. P. (1950): Observations of the spectrum of high-intensity solar radiation at meter wavelengths. II – Outbursts, III – Isolated Bursts. Australian Journal of Scientific Research **A3**, 399–408, 541–557.
Wild, J. P.: Fast phenomena in the solar corona. In: The solar corona. Proceedings of IAU symposium no. 16 (Ed. J. W. EVANS). New York: Academic Press 1963, pp. 115–127.
Wild, J. P., Mc Cready, L. L. (1950): Observations of the spectrum of high-intensity solar radiation at meter wavelengths. I – The apparatus and spectral types. Australian Journal of Scientific Research **A3**, 387–398.
Wild, J. P., Murray, J. D., Rowe, W. C. (1953): Evidence of harmonics in the spectrum of a solar radio outburst. Nature **172**, 533–534.
Wild, J. P., Roberts, J. A., Murray, J. D. (1954): Radio evidence of the ejection of very fast particles from the sun. Nature **173**, 532–534.
Wild, J. P., Sheridan, K. V., Neylan, A. A. (1959): An investigation of the speed of the solar disturbances responsible for type III radio bursts. Australian Journal of Physics **12**, 369–398.
Wild, J. P., Smerd, S. F. (1972): Radio bursts from the solar corona. Annual Review of Astronomy and Astrophysics **10**, 159–196.
Wild, J. P., Smerd, S. F., Weiss, A. A. (1963): Solar bursts. Annual Review of Astronomy and Astrophysics **1**, 291–366.
Wilhelm, K., et al. (1998): The solar corona above polar coronal holes as seen by SUMER on SOHO. Astrophysical Journal **500**, 1023–1038.
Willson, R. C. (1982): Solar irradiance variations and solar activity. Journal of Geophysical Research **87**, 4319–4324.
Willson, R. C. (1984): Measurements of solar total irradiance and its variability. Space Science Reviews **38**, 203–242.
Willson, R. C. (1991): The Sun's luminosity over a complete solar cycle. Nature **351**, 42–44.
Willson, R. C. (1997): Total solar irradiance trend during solar cycles 21 and 22. Science **277**, 1963–1965.

Willson, R. C., et al. (1986): Long-term downward trend in total solar irradiance. Science **234**, 1114–1117.

Willson, R. C, Gulkis, S., Janssen, M., Hudson, H. S., Chapman, G. A. (1981): Observations of solar irradiance variability. Science **211**, 700–702.

Willson, R. C., Hudson, H. S. (1988): Solar luminosity variations in solar cycle 21. Nature **332**, 810–812.

Willson, R. C., Hudson, H. S. (1991): The Sun's luminosity over a complete solar cycle. Nature **351**, 42–44.

Willson, R. F., Lang, K. R. (1984): Very large array observations of solar active regions IV. Structure and evolution of radio bursts from 20 centimeter loops. Astrophysical Journal **279**, 427–437.

Willson, R. F., Lang, K. R., Gary, D. E. (1993): Particle acceleration and flare triggering in large-scale magnetic loops joining widely separated active regions. Astrophysical Journal **418**, 490–495.

Wilson, O. C. (1966): Stellar chromospheres. Science **151**, 1487–1498.

Wilson, O. C. (1966): Stellar convection zones, chromospheres, and rotation. Astrophysical Journal **144**, 695–708.

Wilson, O. C. (1978): Chromospheric variations in main-sequence stars. Astrophysical Journal **226**, 379–396.

Wilson, O. C., Woolley, R. (1970): Calcium emission intensities as indicators of stellar age. Monthly Notices of the Royal Astronomical Society **148**, 463–475.

Winograd, I. J., et al. (1988): A 250,000-year climatic record from great basin vein calcite: Implications for Milankovitch theory. Science **242**, 1275–1280.

Withbroe, G. L.: Activity and outer atmosphere of the sun. In: Activity and outer atmospheres of the sun and stars, 11th advanced course of the swiss national academy of sciences. Sauverny, Switzerland: Observatoire de Geneve 1981.

Withbroe, G. L. (1988): The temperature structure, mass, and energy flow in the corona and inner solar wind. Astrophysical Journal **325**, 442–467.

Withbroe, G. L. (1989): The solar wind mass flux. Astrophysical Journal (Letters) **337**, L49–L52.

Withbroe, G. L. et al. (1982): Probing the solar wind acceleration region using spectroscopic techniques. Space Science Reviews **33**, 17–52.

Withbroe, G. L., Feldman, W. C., Ahluwalia, H. S.: The solar wind and its coronal origins. In: Solar interior and atmosphere (Eds. A. N. Cox, W. C. Livingston, and M. S. Matthews). Tucson: University of Arizona Press 1991, pp. 1087–1106.

Withbroe, G. L., Noyes, R. W. (1977): Mass and energy flow in the solar chromosphere and corona. Annual Review of Astronomy and Astrophysics **15**, 363–387.

Woch, J., et al. (1997): SWICS/Ulysses observations: The three-dimensional structure of the heliosphere in the declining/minimum phase of the solar cycle. Geophysical Research Letters **24**, No. 22, 2885–2888.

Wolff, C. L. (1972): Free oscillations of the sun and their possible stimulation by solar flares. Astrophysical Journal **176**, 833–842.

Wolff, C. L. (1972): The five-minute oscillations as nonradial pulsations of the entire sun. Astrophysical Journal (Letters) **177**, L87–L92.

Wolfson, R. (1983): The active solar corona. Scientific American **248**, 104–119 – February.

Wollaston, W. H. (1802): A method of examining refractive and dispersive power by prismatic reflection. Philosophical Transactions of the Royal Society (London) **92**, 365–380.

Woo, R., et al. (1995): Fine-scale filamentary structure in coronal streamers. Astrophysical Journal (Letters) **449**, L91–L94.

Woo, R., Habbal, S. R. (1997): Extension of coronal structure into interplanetary space. Geophysical Research Letters **24**, No. 10, 1159–1162.

Woodard, M. F., Hudson, H. (1983): Frequencies, amplitudes and line widths of solar oscillations from total solar irradiance observations. Nature **305**, 589–593.

Woodard, M. F., Hudson, H. (1983): Solar oscillations observed in the total irradiance. Solar Physics **82**, 67–73.

Woodard, M. F., Noyes, R. C. (1985): Change of the solar oscillation eigenfrequencies with the solar cycle. Nature **318**, 449–450.

Wu, S. T., et al.: Flare energetics. In: Energetic phenomena on the sun (Eds. M. R. Kundu, B. Woodgate, E. J. Schmahl). Boston: Kluwer Academic Publishers 1989, pp. 377–492.

Yiou, F., et al. (1985): ^{10}Be in ice at Vostok Antarctica during the last climatic cycle. Nature **316**, 616–617.

Yokoyama, T., Shibata, K. (1995): Magnetic reconnection as the origin of X-ray jets and Hα surges on the sun. Nature **375**, 42–44.

Yoshida, T., Tsuneta, S. (1996): Temperature structure of solar active regions. Astrophysical Journal **459**, 342–346.

Yoshimori, M. (1989): Observational studies of gamma-rays and neutrons from solar flares. Space Science Reviews **51**, 85–115.

Yoshimori, M., et al. (1983): Gamma-ray observations from Hinotori. Solar Physics **86**, 375–382.

Young, C. A. (1869): On a new method of observing contacts at the sun's limb, and other spectroscopic observations during the recent eclipse. American Journal of Sciences and Arts **48**, 370–378. Reproduced in: Early solar physics (Ed. A. J. Meadows). Oxford, England: Pergamon Press 1970, pp. 125–134.

Young, C. A.: The sun. New York: Appleton 1896.

Zank, G. P., Gaisser, T. K. (Eds.): Particle acceleration in cosmic plasmas. New York: American Institute of Physics 1992.

Zhang, G., Burlaga, L. F. (1988): Magnetic clouds, geomagnetic disturbances, and cosmic ray decreases. Journal of Geophysical Research **93**, 2511–2518.

Zhang, Q., et al. (1994): A method of determining possible brightness variations of the sun in past centuries from observations of solar-type stars. Astrophysical Journal (Letters) **427**, L111–L114.

Zirin, H., Moore, R., Walters, J. (1976): Proceedings of the workshop: The solar constant and the earth's atmosphere. Solar Physics **46**, 377–409.

Zirker, J. B.: Coronal holes – an overview. In: Coronal holes and high speed wind streams (Ed. J. B. Zirker). Boulder, Colorado: Colorado Associated University Press, 1977, pp. 1–26.

Zirker, J. B. (Ed.): Coronal holes and high speed wind streams. Boulder, Colorado: Colorado Associated University Press 1977.

Zirker, J. B. (1993): Coronal heating. Solar Physics **148**, 43–60.

Author Index

Acton, Loren W. 140, 197
Adhémar, Joseph Alphose 262
Akasofu, Syun-Ichi 265
Alfvén, Hannes 120, 152, 206
Anderson, Carl D. 33, 55
Ando, Hiroyasu 93
Antia, H. M. 93
Antonucci, Ester 209
Appleton, Edward Victor 206, 236, 263
Arnoldy, Roger L. 207
Aschwanden, Markus J. 118, 122, 173, 177, 209, 210
Athay, R. Grant 121
Avrett, Eugene 106
Axford, William Ian 121, 126, 129, 143, 152, 154, 265

Babcock, Harold D. 56, 151
Babcock, Horace W. 56, 151
Bagenal, Frances 227
Bahcall, John N. 94
Baliunas, Sallie L. 250, 267
Balogh, Andre 145, 147, 153
Barnett, M. A. F. 263
Bartels, Julius 218
Bartoe, John-David F. 115, 121
Basu, Sarbani 94
Beer, Juerg 244, 250, 268
Belcher, John W. 120, 129, 152
Benz, Arnold O. 117, 122, 164, 173, 209
Berger, André 256, 266
Bethe, Hans A. 61
Biermann, Ludwig F. 25, 56, 96, 120, 124, 151, 264
Bigelow, Frank H. 41, 54, 150
Birkeland, Kristian 19, 54, 150, 263
Blackett, Patrick M. S. 33
Bogart, Richard S. 94
Bohlin, J. David 108, 120
Boischot, André 160, 207
Bolton, John G. 158, 206
Bondi, Herman 150
Boriakoff, Valentin 243

Bos, Randall J. 72
Bradley, Raymond E. 244, 250, 268
Bradt, H. L. 55
Braun, Douglas C. 94
Breit, Gregory 263
Bridge, Herbert A. 56
Brown, Timothy M. 93
Brueckner, Guenter E. 115, 121
Bruner, Elmo C., Jr. 121
Bumba, V. 206
Bunsen, Robert 21, 54
Burlaga, Leonard F. 203
Burnight, T. R. 55, 150, 206
Burton, Rande K. 217, 265

Cahill, Larry J. 216, 265
Cane, Hilary V. 210, 266
Canfield, Richard C. 230, 231, 268
Carmichael, Hugh 194, 198
Carrington, Richard C. 54, 156, 205, 262
Cayan, Daniel R. 268
Celsius, Anders 215, 262
Changery, Michael 244
Chapman, Gary A. 266
Chapman, Sydney 19, 27, 55, 56, 151, 216, 263
Chappellaz, J. 266
Chenette, David L. 219
Chitre, Shashikumar M. 93
Christensen, Eigil Friis 267
Christensen-Dalsgaard, Jørgen 93, 94
Chupp, Edward L. 182, 208, 209
Chyba, Christopher 266
Claverie, Andre 93
Clay, Jacob 32, 55
Cliver, Edward W. 218, 243
Cocconi, Giuseppe 204
Colburn, David S. 216, 265
Coleman, Paul J. 265
Coles, William A. 126, 153
Compton, Arthur H. 32, 55
Cox, Arthur N. 72
Craig, I. J. D. 104

Critchfield, Charles L. 61
Croll, James 263
Crooker, Nancy U. 218
Crowley, Thomas J. 245, 268

Damon, Paul E. 245, 251, 268
Davis, Leverett, Jr. 56, 120, 151, 152, 265
Davis, Raymond, Jr. 69
Debrunner, Hermann 209
De Jager, Kees 207
Dellinger, J. Howard 206, 264
Demarque, Pierre 94
Denisse, Jean-Francoise 160, 207
Dere, Kenneth P. 115, 121
Dettinger, Michael D. 268
Deubner, Franz-Ludwig 93
Dietrich, William F. 208
Dobson, Helen W. 206
Doppler, Christiaan 76
Doschek, George A. 108, 120, 209
Dowdy, James F., Jr. 153
Dröge, Franz 208
Duijveman, André 209
Dungey, James W. 193, 206, 216, 265
Duvall, Thomas L., Jr. 90, 93, 94
Dziembowski, Wojciech A. 93

Eddy, John A. 251, 266
Edlén, Bengt 23, 55, 120, 150
Einstein, Albert 61
Ellison, Mervyn Archdall 206
Elsasser, Walter M. 212
Elsworth, Y. P. 94
Emiliani, Cesare 264
Enome, Shinzo 190
Espenak, Fred 22

Fairfield, Donald H. 216, 265
Falconer, David A. 100, 102, 115, 121
Feldman, Uri 108, 120
Ferraro, Vincent C. A. 19, 55, 216, 263
Feynman, Joan 243
Fisk, Lennard A. 147
Fitzgerald, George Francis 216, 263
Forbes, Terry G. 113, 121, 188, 210
Forbush, Scott E. 42, 56, 151, 200, 206, 207, 264
Forrest, Baldwin 18
Forrest, David J. 209, 210
Fossat, Eric 93
Foukal, Peter V. 267
Frank, Louis A. 266
Fraunhofer, Joseph, von 20, 54
Frazier, Edward N. 93
Freier, Phyllis 55
Friedman, Herbert 45, 55, 56, 151, 206, 264
Friis-Christensen, Eigil 242, 243, 268
Fröhlich, Claus 241, 268

Gabriel, Alan H. 108, 110, 120, 152, 153
Gaizauskas, Victor 105
Galilei, Galileo 53, 246, 262
Gauss, Carl Friedrich 54, 212, 262
Geiss, Johannes 138, 152
Genthon, C. 266
Gilbert, William 53, 212, 262
Giles, Peter M. 94
Ginzburg, Vitaly L. 24, 55, 120, 150
Giovanelli, Ronald G. 120, 193, 206, 216, 265
Gleissberg, Wolfgang 251, 264
Gloeckler, George 138
Gokhale, M. H. 93
Gold, Thomas 102, 120, 192, 195, 203, 204, 207, 214, 264
Goldreich, Peter 93, 94
Golub, Leon 113, 120, 121
Gonzalez, Walter D. 221, 267
Goode, Philip R. 93
Gosling, John Thomas 208, 265, 267
Gough, Douglas O. 93, 94, 264
Graham, George 53, 215, 262
Greaves, William, M. H. 263
Grec, Gérard 93
Gringauz, Konstantin I. 28, 56, 124, 151
Grotrian, Walter 23, 55, 120, 150
Guenther, D. B. 94
Gulkis, Samuel 266

Habbal, Shadia R. 129
Haisch, Bernhard 267
Hale, George Ellery 35, 38, 55, 106, 150, 156, 205, 206, 215, 263, 264
Halley, Edmund 18, 53, 262
Hanaoka, Yoichiro 210
Hansen, Richard T. 188, 208
Harkness, William 23, 54, 150
Harrison, Richard A. 115, 121
Hartle, R. E. 129, 152
Hartz, T. R. 207
Harvey, John W. 93, 94
Haselgrove, C. B. 264
Hassler, Donald M. 134, 136, 154
Hathaway, David 37

Hays, James D. 256, 266
Heaviside, Oliver 263
Herlofson, Nicolai 206
Herschel, John 215
Hess, Victor Franz 31, 55
Hey, J. Stanley 206
Hickey, John R. 267
Hildner, Ernest 188
Hines, Colin O. 265
Hirayama, Tadashi 198, 208
Hirshberg, Joan 208, 216, 265
Hodgson, Richard 54, 156, 205, 262
Holzer, Thomas E. 129, 152
Howard, Russell A. 187, 209, 210, 266
Howe, Rachel 94
Hoyle, Fred 102, 120, 150, 192, 195, 207, 264
Hoyng, Peter 209
Hoyt, Douglas V. 267
Hsieh, K. C. 208
Hudson, Hugh S. 93, 114, 121, 230, 266, 267, 268
Hughes, Malcolm K. 244, 268
Humason, Milton L. 206
Humboldt, Alexander von 54, 215, 262
Hundhausen, Arthur J. 52, 209
Hurford, Gordon J. 173, 209

Imbrie, John 256, 266
Inhester, Bernd 121
Innes, Davina 113, 114, 121
Isaak, G. R. 93, 94

Janssen, Jules 54
Janssen, Michael 266
Jastrow, Robert L. 250, 267
Jefferies, Stuart Mark 94
Jockers, Klaus 129, 152
Johnson, Thomas H. 55
Jokipii, J. Randy 146, 153
Joselyn, Jo Ann 266
Joy, Alfred H. 206
Jupp, P. E. 117, 122

Kahler, Steven W. 267
Kakinuma, Takakiyo 126, 153
Kale, D. M. 93
Kanbach, Gottfried 210
Kane, Sharad R. 207
Kano, Ryouhei 103, 104
Keeley, Douglas A. 93
Kennelly, Arthur E. 263
Kim, Kwang-Yul 245, 268
Kirchhoff, Gustav 21, 54
Klimchuk, James A. 50, 57, 104, 153
Kohl, John L. 140, 145, 154
Kojima, Masayoshi 126, 153
Kolhörster, Werner 32
Kopp, Roger A. 124, 129, 152, 198, 208
Kosovichev, Alexander G. 84, 90, 92, 94
Kóta, Joseph 146, 153
Koutchmy, Serge 137
Krieger, Allen S. 49, 57, 120, 152
Krimigis, Stramatios M. 202
Krucker, Säm 117, 122
Kumar, Pawan 94
Kundu, Mukul R. 173, 207, 210

Labitzke, Karin 239, 267
LaBonte, Barry J. 94
Lang, Kenneth R. 173, 209
Lassen, Knud 242, 267
Lean, Judith 237, 244, 245, 250, 267, 268
Leer, Egil 129
Leibacher, John W. 93
Leighton, Robert B. 73, 93
Libbrecht, Kenneth G. 94
Lin, Robert P. 209
Lindemann, Frederick Alexander 19, 55, 150, 203, 205, 216, 263
Lindsey, Charles 94
Linford, Gary A. 101
Livingston, William C. 39
Lockwood, G. W. 250, 267
Lockyer, Norman 23, 54
Lodge, Oliver 216, 263
Loomis, Elias 54
Lorius, C. 266
Lovelock, James E. 261, 265
Lyot, Bernard 51

Machado, Marcos E. 210
Mairan, Jean Jacques d'Ortous, de 18, 54, 262
Mann, Michael E. 244, 246, 268
Marconi, Guglielmo 235, 263
Margulis, Lynn 261, 265
Mariska, John T. 120
Marsch, Eckart 143, 154
Marsh, Kenneth A. 173, 209
Martin, Sara F. 121
Martyn, David F. 23, 55, 120, 150
Masuda, Satoshi 198, 210
Maunder, Edward Walter 218, 246, 263
Maxwell, Alan 207
Mc Allister, Alan 133

Mc Clymont, A. N. 104
Mc Crea, William H. 55, 150
Mc Intosh, Patrick S. 266
Mc Kenzie, David E. 230, 268
Mc Kenzie, James F. 143, 154
McLeod, C. P. 93, 94
Mc Pherron, Robert L. 217, 265
Meyer, Peter 56, 151, 203
Milankovitch, Milutin 254, 263
Millikan, Robert A. 32, 55
Moore, Ronald L. 100, 102, 115, 121, 153
Moreton, Gail E. 160, 207
Morrison, Philip 203, 207, 208, 264
Morrow, Cherilynn A. 93
Mullen, Edward G. 224
Mullen, George 260, 265

Ness, Norman F. 42, 57, 151, 265
Neugebauer, Marcia 29, 56, 124, 151, 265
Neupert, Werner M. 167, 207
New, R. 94
Newman, Michael J. 266
Newton, Harold W. 206, 263, 264
Ney, Edward P. 242, 264
Noci, Giancarlo 140, 154
Numazawa, Shigemi 137

Orrall, Frank Q. 50, 57, 153
Osaki, Yoji 93

Pallavicini, Roberto 208
Pallé, Pere Lluis 94
Parker, Eugene N. 27, 42, 56, 102, 114, 121, 124, 127, 128, 151, 264
Parnell, Clare E. 114, 117, 121, 122
Pawsey, Joseph L. 23, 55, 120, 150
Payne-Scott, Ruby 158, 206
Peristykh, Alexei N. 245, 251, 268
Peters, Bernard 55
Peterson, Laurence E. 166, 207
Petit, J. R. 266
Petschek, Harry E. 113, 120, 194, 207
Phillips, John L. 129, 153
Pneuman, Gerald W. 124, 129, 152, 198, 208
Pomerantz, Martin A. 93, 94
Porter, Jason G. 115, 121
Porter, Lisa J. 104
Powell, Cecil F. 33
Priest, Eric R. 104, 113, 114, 119, 121, 122, 188, 210

Quay, Paul D. 266

Rabin, Douglas M. 153
Radick, Richard R. 250, 267
Ramsay, William 23, 54
Raynaud, D. 266
Reames, Donald V. 210
Reeves, Hubert 152
Régulo, Clara 94
Reid, George C. 242, 267
Reiner, Michael J. 202
Rhodes, Edward J., Jr. 93
Richardson, Robert S. 206
Rickett, Barney J. 126, 153
Rimmele, Thomas R. 71
Roca-Cortés, T. 93, 94
Roelof, Edmond C. 49, 57, 152
Rood, Robert T. 266
Rosenbauer, Helmuth R. 153
Rosenberg, Ronald L. 57, 153
Rosner, Robert 104, 121
Rottman, Gary J. 50, 57, 153
Russell, Christopher T. 217, 265
Rust, David M. 188

Sabine, Edward 54, 205, 215, 262
Sagan, Carl 260, 265, 266
Sakao, Taro 210
Schatzman, Evry 96, 120
Scherrer, Philip H. 88, 90, 94
Schmidt, Wolfgang K. H. 153
Schou, Jesper 94
Schrijver, Carolus J. 110, 118, 121
Schuster, Arthur 19, 55, 263
Schwabe, Samuel Heinrich 36, 54, 215, 262
Schwarzschild, Martin 96, 120, 264
Serio, Salvatore 208
Severny, A. B. 206
Shackleton, Nicholas J. 256, 264, 266
Sheeley, Neil R., Jr. 132, 141, 154, 210, 266
Shevgaonkar, Raghunath K. 173, 210
Shibata, Kazunari 112, 113, 121, 210
Shimizu, Toshifumi 121
Shing, Lawrence 101
Silk, J. Kevin 120
Simpson, John A. 56, 151, 153, 208
Slater, Gregory L. 101
Slottje, Cornelius 208
Smith, Edward J. 43, 57, 120, 145, 147, 152, 153, 218, 265, 266
Snyder, Conway W. 29, 56, 124, 151, 265

Sonett, Charles P. 203, 207, 218, 251, 265, 266
Spörer, Gustav Friedrich Wilhelm 246, 263
Stein, Robert F. 93
Sterling, Alphonse C. 230, 268
Strömgren, Bengt 55
Størmer, Carl 223, 226
Strong, Keith T. 168, 267
Stuiver, Minze 265, 266
Sturrock, Peter A. 129, 152, 194, 198, 207
Suess, Hans E. 251, 265
Svalgaard, Lief 152
Svensmark, Henrik 243, 268
Svestka, Zdenek 209
Swarup, Govind 207
Sweet, Peter A. 194

Takakura, Tatsuo 209
Tanaka, Katsuo 209
Tett, Simon F. B. 245, 268
Thompson, Michael J. 94
Thomson, William 215, 263
Timothy, Adrienne F. 49, 57, 120, 152
Title, Alan M. 111, 118, 121
Tomczyk, Steven 94
Tousey, Richard 45, 55, 151, 208
Tsuneta, Saku 103, 104, 121, 197, 210
Tsurutani, Bruce T. 57, 145, 153, 221, 267
Tu, Chuan-Yi 154
Tucker, Wallace H. 104, 121
Tuve, M. A. 263
Tyndall, John 247

Uchida, Yutaka 133, 134, 153, 208
Ulrich, Roger K. 93, 264
Underwood, J. H. 104
Unsöld, Albrecht 21, 55

Vaiana, Giuseppe S. 57, 104, 120, 121, 152, 207, 208
Van Allen, James A. 28, 56, 202, 223, 225, 264
van der Raay, H. B. 93
Van Loon, Harold 239, 267
Vogt, Rochus 56, 203

Waldmeier, Max 48, 56, 151, 206
Wang, Haimin 157
Wang, Yi-Ming 132, 154
Weber, Edmund J. 129, 152
White, Oran R. 121, 266
White, Warren B. 242, 268
Wilcox, John M. 43, 57, 151, 152
Wild, John Paul 158, 206
Wilhelm, Klaus 121, 154
Willson, Richard C. 266, 267
Willson, Robert F. 173, 209
Wilson, Charles T. R. 32
Wilson, Olin C. 250, 265
Winckler, John Randolph 166, 207
Withbroe, George L. 140
Woch, Joachim 130
Wolfe, John H. 218, 266
Wolff, Charles L. 93
Wollaston, William Hyde 20, 54
Woo, Richard 129
Woodard, Martin F. 93, 94

Yabsley, D. E. 158, 206
Yoshimori, Masato 209
Young, Charles A. 23, 54, 150

Zeeman, Pieter 35, 38
Zhang, Qizhou 250, 268
Zharkova, Valentina V. 92, 94

Subject Index

absorption lines 20, 54
absorption, polar cap 201
abundance, solar wind helium 152
accelerated electrons 165
acceleration
– electron 209
– fast solar wind 140, 143
– particle 207
– slow solar wind 140, 141
acceleration process, solar flares 164
acceleration site, coronal 210
accretion of interstellar matter by a star 150
Active Cavity Radiometer Irradiance Monitor (ACRIM) 93, 266, 267
active region on the back side 94
active regions 36, 40, 48, 91, 99, 102, 132, 231
active-region coronal loops 102
activity cycle
– maximum 41
– minimum 41
air temperature 242
– Antarctica 258
Alfvén waves 98, 99, 120, 145, 152–154, 270
Alouette-I satellite 158
alpha particles 32
annihilation
– electron–positron 179, 180
anomalous cosmic rays 147
Antarctica 256
Antarctica, air temperature 258
Antarctica, temperature record 256
Apollo Telescope Mount 167
arcade formation 190
areas, unipolar 151
astronomical cycles 254
astronomical unit 62
asymmetry, solar wind 129
atmosphere
– pressure 236
– temperature 234, 236
– upper, temperature 234, 237
atmospheric carbon dioxide 266
atmospheric drag 224
aurora 53, 54, 218, 219, 262
– australis 18, 219
– borealis 18, 219
– electrons 220
– geomagnetic activity 19
– northern and southern lights 18
aurora oval 219, 220, 266
aurora rays 262
azimuthal number 79

base of the convective zone 94
BCS (Yohkoh) 13
beams, electron 159, 160, 163, 209
belt, equatorial streamer 124, 153
beryllium 252
^{10}Be 252
bi-directional jets 121
bipolar pairs 206
– sunspots 206
bipolar regions 151
bipolar sunspots 36
blinkers 121
blobs 132, 141
bow shock 214, 265
bremsstrahlung 168, 173, 174
– non-thermal 171, 174
– non-thermal hard X-ray 178
– power 174
– thermal 174, 178
bright points 57
brightenings, chromosphere 206
bursts
– centimeter 159
– decimeter 164
– decimetric type III 173
– millisecond 159
– type I 159
– type II 159
– type III 158, 159, 163, 207
– type IV 159, 160, 163
– U-type 159

calcium, H and K line 240
capture, neutron 180
carbon dioxide 247, 258
– atmospheric 248, 266
carbon 14, 252
– 200-year periodicity 265
– atmospheric, variable Sun 266
carbon nuclei, excited 179
CDS (SOHO) 6
CELIAS (SOHO) 6
centimeter bursts 159
central temperature 64
CFC, chlorofluorocarbon 239
CGRO (Compton Gamma Ray Observatory) 177, 182, 201, 210
Cherenkov radiation 70
chlorofluorocarbon 239
chromosphere 106, 156, 240
chromosphere brightenings 206
chromosphere ribbons 194
chromospheric evaporation 170, 173, 178, 209
chromospheric flares 156
climate 256, 258
– Earth 243
climate change 263
cloud cover 264, 268
clouds, magnetic 159, 204, 207
coherent radiation 166
comet ion tails 56, 151, 264
comet tails 25, 27
communication satellites 222
components, solar wind 129, 130
composition, fast wind 127, 139
composition, slow wind 127, 139
Compton Gamma Ray Observatory *see* CGRO
convection 68
convection zone 67, 69, 93
– boundary 81
convection zone depth 94
corona 1, 21, 23, 24, 47, 137, 154, 160, 198, 199
– coronium 23
– depletions 188
– green emission line 54
– hard X-ray source 210
– heating mechanisms 119
– heating of the 152
– hot, static 151
– interaction of magnetic loops 210
– magnetized loops 110
– million-degree solar 150
– plasma oscillations 160
– polar 144
– solar 150, 208
– static, isothermal 128
– X-rays 99
coronagraph 50, 51, 182, 184
– Large Angle and Spectrometric 132
– space-borne 184
coronal acceleration site 210
coronal dimming 188
coronal electron density 162
coronal emission lines 55, 120, 150
coronal energy loss 96
coronal forbidden emission lines 24
coronal hard X-ray source 199
coronal heating 99, 103
coronal heating mechanisms 103
coronal heating models 103
coronal heating problem 107
coronal hole temperature 140
coronal holes 47–50, 56, 57, 124, 130, 131, 134, 135, 140, 143, 151, 152, 154
coronal loops 40, 47, 48, 57, 102, 104, 121, 133, 152
– magnetic interaction 210
– scaling law 102, 103, 121
– sheared 199
– twisted 199
coronal Lyman alpha line 152
coronal mass ejections 51, 52, 182, 183–186, 191, 197, 203, 204, 208–210, 215, 217, 222
– halo 232
– energy 186
– erupting prominences 188
– mass 186
– mass flux 186
– satellites 224
– size 191
– time delay 186
– twisted soft X-ray structures 230
coronal streamer 129, 140
coronal streamer temperature 140
Co-rotating Interacting Regions 218, 221, 266
corpuscles, solar 151
cosmic ray electrons 32, 56
cosmic ray flux 268
cosmic ray proton 33, 55
cosmic rays 31, 32, 42, 55, 146, 147, 151, 153, 207, 226, 242, 249, 252, 264
– average fluxes 33
– block incoming 153

– diffusion 146
– electrons 56
– energy 35
– galactic 201
– protons 55
– solar 201
COSPIN (Ulysses) 10
COSTEP (SOHO) 6
current sheet 43, 98, 122, 125, 198
cusp geometry 197, 198, 210
cusp-shaped loop structures 198
cycle of magnetic activity 51, 191, 207, 234, 235, 237, 242, 254, 270, 271
cyclotron resonance theory 143

decay, meson 210
decimeter bursts 164
decimetric type III bursts 173
Deep Space Network (NASA) 8, 10, 14
density, proton 127
density, fast wind 127
density, slow wind 127
depletions, sudden 208
deuterium formation 179
differential rotation 83, 93
diffuse corona, heating mechanism 104
diffusion of cosmic rays 146
dimming, coronal 188
dipolar magnetic field 56, 151
dipolar magnetic model 152
dipole field, Earth 213
discharge, neutral-point 193
disk, visible 178
Doppler effect 76
Doppler shift 76, 107
double hard X-ray flare 176, 177, 209
downflowing material 108
DUST (Ulysses) 10
dynamic spectra 158
dynamo, Earth's magnetic field 212

Earth
– climate 243
– core 262
– dipole field 213
– long, cold spell on 250
– magnetic field 212, 213
– – variations 215
– magnetosphere 214
– mean distance from the Sun 62
– orbit 254
– – shape 263
– rotational axis 263

Earth Radiation Budget (ERB) 267
Earth's magnetic field
– dynamo 212
east-west effect 55
eclipse, total solar 21
edge, solar system 148, 150
effective temperature 64
eighth Orbiting Solar Observatory (OSO 8) 96
eighty-year cycle 264
EIT (SOHO) 6, 118, 135, 136, 233
ejection, plasma 205
electrical power grids 228
electrical power outages 225
electromagnetic waves 45
electron acceleration 178, 209
– low corona 209
electron beams 160, 163, 173, 209
electron density
– atmosphere 231
– coronal 162
electron neutrino 66, 67
electron–positron annihilation 179–181
electron temperature, fast wind 127
electron temperature, slow wind 127
electron volts 33
electrons 129, 158, 165, 200
– accelerated 165
– beams of 159
– energetic 201
– flare-associated 201
– flaring 158
– high energy 158, 182
– high-speed 163, 164
– interplanetary 201, 202
– non-thermal 171, 172, 178
11-year magnetic activity cycle 51, 191, 215, 234, 235, 237, 242, 254, 265, 270, 271
emission, gamma-ray 179
emission lines 21, 107, 150
– solar 170
– solar, below 200.0 nanometers 170
energetic charged particles, Sun 264
energetic electrons 201
energetic interplanetary electrons 202
energetic ions 209
energetic particles 204
energy
– kinetic 185, 186
– magnetic 192–194, 207
– solar flares 192
– thermal 142

energy flux, solar wind 30
energy of coronal mass ejections 186
energy release, flare 210
energy-generating core 63
EPAC/GAS (Ulysses) 10
equatorial slow-speed wind 130
equatorial steamer belt 153
ERB (Earth Radiation Budget) 267
ERNE (SOHO) 6
erupting filament 190
erupting prominences and coronal mass ejections 188
eruptive prominences 188–190
escape velocity 27, 61
evaporation, chromospheric 170, 173, 178, 209
events
– gradual 205
– impulsive 204, 205
– solar energetic particle 210
excitation, proton 180
excited carbon nuclei 179
excited heavier nuclei 179
excited nitrogen nuclei 179
excited nuclei 178
excited oxygen nuclei 179
expanding coronal loops 132, 133, 153
expanding loops 134
experiment
– GALLEX 65
– HOMESTAKE 65
– KAMIOKANDE 65
– SAGE 65
Experimenters' Operations Facility 5
Explorer 1 56, 223, 225
Explorer 10 56
Explorer 3 225
Explorer 12 216
extreme ultraviolet 106
Extreme-ultraviolet Imaging Telescope *see* EIT (SOHO)

faculae 240, 242, 267
faint-young-Sun paradox 260
fast solar wind 29, 50, 124, 125, 127, 130, 131, 134–136, 140, 143, 144, 153
– composition 127
– density 127
– electron temperature 127
– helium temperature 127
– helium to proton abundance 127
– proton density 127
– proton flux 127
– proton speed 127
– proton temperature 127
– source 127
– temperature 127
fast stream
– composition differences 139
– temperature differences 139
fields
– dipolar magnetic 151
– interplanetary magnetic 151, 216
– large-scale solar magnetic 150
– magnetic 1, 150, 193, 204, 205
– solar magnetic 151
filaments 106, 188
– erupting 190
– disruptions 266
five-minute oscillations 73, 93
flare energy release 210
flare ribbons 157
– hydrogen alpha 206
flare-associated electrons 201
flare-associated neutrons 201
flares
– chromospheric 156
– double hard X-ray 176
– gradual 190
– Hα 156
– hard X-ray 175
– impulsive 199
– long-duration 198
– loop-top impulsive hard X-ray 199
– soft X-ray 167, 207, 208
– solar 2, 156, 166, 168, 169, 173, 178, 179, 181, 182, 192, 202, 205, 206, 208, 210, 216, 262, 264
– white light 156, 176, 177
– X-ray 166
flaring electrons 158
flaring soft X-ray loops, temperatures 167
flow
– high-speed 135
– solar-wind 152
fluctuations 252
flux
– proton 127
– solar wind 30, 31
footpoints 210
forbidden emission lines 23
Forbush effect 42
formation
– arcade 190
– deuterium 179

free magnetic energy 196
– release 195
frequency, plasma 158, 161, 162
frequency of occurrence, coronal mass ejections 185

Gaia hypothesis 261, 265
galactic cosmic rays 201
GALLEX 70
gallium experiment 70
gamma ray lines 179–181
– positron annihilation 181
– solar 208
gamma rays 13, 178, 179, 181, 210
gamma-ray emission 179
gamma-ray radiation 207
gas, magnetized clouds 207, 264
gas pressure 40, 41
General Theory of Relativity 86, 93
geomagnetic storms 54, 150, 205, 215, 217, 262–264
– 27-day interval 263
– non-recurrent 216, 263
– recurrent 218
– sudden commencement 218
GEOTAIL satellite 230
giant star 262
Gleissberg cycle 251
Global Oscillations Network Group 77
global surface temperatures, sunspot cycle 243
global temperature variations 244, 246
global warming 248, 268
GOLF (SOHO) 6, 80
GONG 77
gradual events 205
gradual flares 190
gradual phase 169, 175
gradual soft X-ray flares 203
GRB (Ulysses) 10
green emission line 23
greenhouse effect 244, 247, 260, 266
– "natural" 247
– "unnatural" 247
greenhouse gases 245, 247, 256–258
– man-made 268
Greenland 256
GWE (Ulysses) 10
gyration radius 34

halo coronal mass ejections 187, 209, 232
halo mass ejections 187
halo orbit 3
Hα 156
Hα flares 156
hard X-ray bremsstrahlung, non-thermal 178
hard X-ray flare, loop-top impulsive 199
hard X-ray imaging instruments 172
hard X-ray phase, impulsive 173
hard X-ray solar flares 175, 207
hard X-ray sources 198
– corona 210
hard X-rays 12, 166, 171, 172, 209, 210
heat, magnetic waves 154
heated waves 144
heating coronal loops 104
heating models 102
heating of the corona 120, 152
heavier nuclei, excited 179
heavy particles 143
heliopause 149
Helios 1 57, 126, 127
Helios 1 and 2 spacecraft 152
Helios 2 126, 127
helioseismic holography 91
helioseismology 73, 93
heliosphere 8, 56, 148, 149, 151
– shape and content 148
– size 150
helium 54, 153, 208
helium nuclei 32
helium temperature, fast wind 127
helium temperature, slow wind 127
helium to proton abundance, fast wind 127
helium to proton abundance, slow wind 127
helmet streamer 132, 137, 187, 188
high-energy electrons 158, 182
high-energy jets 121
high-speed component, solar wind 125, 127, 130, 132, 153, 154, 187
high-speed electrons 163, 164
high-speed flows 135
high-speed solar wind 29, 50, 124, 125, 131, 134, 136, 140, 153
high-speed wind streams 56, 57, 151, 152, 265
high-speed, solar-wind outflow velocity 154
Hinotori 170, 172, 208
– hard X-ray imaging instruments 172
HISCALE (Ulysses) 10

holes, coronal 124, 130, 131, 134, 135, 140, 143, 151, 152, 154
holes, polar coronal 153
Holocene period 253
Homestake experiment 70
Homestake Gold Mine 69
hot, static corona 151
HXT (Yohkoh) 13
hydrazine 4
hydrogen 55, 140
– emission line 106
– Hα 156
– Hα flares 156
hydrogen alpha 21, 156, 206
hydrogen alpha flare ribbons 206
hydrogen nuclei 32

ice age 253, 258, 262–264, 266
– astronomical theory 254
ice age temperatures 257
ice core 256, 266
ideal gas law 128
IMP (Interplanetary Monitoring Platform) 152, 203
IMP-1 (Interplanetary Monitoring Platform 1) 42, 213, 265
impulsive and gradual events 204
impulsive events 204, 205
impulsive flares 199
impulsive hard X-ray phase 173
impulsive particle events 205
impulsive phase of solar flares 169, 171, 175
impulsive radio bursts 207
impulsive solar flares 171
Institute of Space and Astronautical Science (ISAS) 12
instruments, hard X-ray imaging 172
Interacting Regions, Co-rotating 218, 221
interaction of magnetic loops, corona 210
interglacial 253
internal flows 86
internal rotation 83, 84
International Solar-Terrestrial Physics (ISTP) program 5, 228, 229
interplanetary electrons 201, 202
– energetic 202
interplanetary magnetic clouds 203
interplanetary magnetic field 27, 42, 43, 56, 57, 151, 216
– spiral pattern 202
interplanetary medium 204
Interplanetary Monitoring Platform 1 (IMP-1) 42, 213, 265
Interplanetary Monitoring Platforms (IMPs) 152, 203
interplanetary plasma clouds 203
interplanetary proton and electron events 203
interplanetary protons 200
interplanetary scintillation 126, 153
interplanetary scintillations 153
interplanetary shock wave 204
interplanetary shocks 203, 207, 210, 265, 266
interplanetary space, magnetic sectors 57
interstellar pressure 148
ion tails, comet 264
ionized calcium
– H lines 250
– K lines 250
ionosphere 55, 151, 162, 220, 235–237, 263, 264
– layer 238
– – D 238
– – E 238
– – F 238
ionosphere disturbances, sudden 206, 264
ions
– energetic 209
– oxygen 140, 154
– pick-up 148
– preferentially accelerate heavier 154
irradiance
– oscillations 93
– solar 249
irradiance decrease, sunspots 242
irradiance increase, faculae or plage 242
isotope, radioactive 252

Kamioka 70
Kepler's third law 62
Key events
– coronal heating 120
– discovery of space 53
– explosive solar activity 205
– helioseismology 93
– solar-terrestrial interactions 262
– solar wind 150
kilo electron volts 167
kinetic energy 185, 186
kinetic temperature 142

Lagrangian point 3
land air temperature 267
land temperatures 243
Large Angle Spectrometric COronagraph *see* LASCO (SOHO)
large-scale coronal heating 122
large-scale solar magnetic fields 150
largest mass, ejected 185
LASCO (SOHO) 6, 132, 154, 184, 187, 229
latitude increase, cosmic rays 32
law, ideal gas 128
layer, ozone 239
lines
– coronal emission 150
– coronal Lyman alpha 152
– emission 150
– gamma ray 181, 182
– hydrogen alpha 21, 106, 156, 206
– magnetic neutral 157, 177
Little Ice Age 246, 251
l-nu diagram 79, 80
long-duration flares 198
long-lasting soft X-ray flare 210
loop apex, flare energy 209
loop heating 102
loop structures, cusp-shaped 198
loop-top impulsive hard X-ray flare 199
loops
– coronal 133, 152
– expanding 134
– expanding coronal 133, 153
– magnetic 132
– scaling law 104
low corona, electron acceleration 209
low-energy flares 117
low luminosity, Sun 250
low solar activity 251
Lunik 2 56

magnetic activity 37
– long-term 265
– maximum 41
– solar-type stars 268
– stars 250, 267
magnetic activity cycle 51, 191, 207, 234, 237, 242, 254, 271
magnetic carpet 118, 119
magnetic clouds 159, 204, 207
– interplanetary 203
magnetic connections, low corona 119
magnetic diffusion 97
magnetic energy 102, 120, 192–194, 207
– free 196
magnetic field 193
– Earth 212, 213
– – variations 215
– interplanetary 216
magnetic field reconnection 207
magnetic field strength 31, 38, 40
magnetic fields 1, 120, 150, 204, 205
– direction 36
– non-potential 193, 194
magnetic fields in sunspots 150, 205
magnetic interaction of coronal loops 210
magnetic loops 101, 102, 132
– tops 178
magnetic network 109, 110, 115, 120, 134, 136, 153, 154, 242
magnetic neutral line 157, 177, 206
magnetic neutral point 206
magnetic pressure 40, 41
magnetic reconnection 98, 111–113, 119–122, 188, 192, 194–199, 208, 210, 216, 265
– location 197
– low corona 113
– site 199
magnetic sectors 43, 124
magnetic storms 53, 55, 215, 262, 263, 266
magnetic tail 265
magnetic waves 97, 143, 146
magnetic waves heat 154
magnetism 262
– interplanetary 151
magnetized clouds of gas 207, 264
magnetized plasma clouds 203
magnetogram 39, 41
magnetograph 39
magnetohydrodynamic waves 120, 208
magnetohydrodynamics 97
magnetopause 214
magnetosphere 207, 214
– Earth 214
magnetosphere substorm 265
magnetotail 214, 217
Mariner 2 56, 124, 207, 265
Mariner 4 202
Mariner 5 99
mass, Sun 62
mass ejections
– coronal 182, 183–186, 191, 197, 203, 204, 208–210

– coronal and erupting prominences 188
– coronal, mass of 186
– coronal, mass flux 186
– coronal, size 191
– halo coronal 187, 209
mass loss rate, solar wind 30
Maunder Minimum 246, 250, 251, 263
MDI/SOI (SOHO) 6, 82, 87, 118
medium, interplanetary 204
Mercury's orbital motion 86, 93
meridional surface flow 89
meson decay 210
mesons 33, 179, 210
methane 247, 258
– atmospheric 266
Michelson Doppler Imager (MDI, SOTTO) 75, 88, 91
microflares 115, 117, 122
microwave spikes 208
Milankovitch cycles 254
million-degree solar corona 27, 55, 97, 120, 150
millisecond bursts 159
model
– dipolar magnetic 152
– two-component (electrons and protons) 152
modern satellites 222
momentum flux 30
Moreton waves 207

nanoflares 115, 117, 119, 121, 122
NASA's Deep Space Network 8, 10
NASA's Orbiting Solar Observatory (OSO) 167
National Climatic Data Center (NCDC) 244
National Oceanic and Atmospheric Administration (NOAO) 244
Naval Research Laboratory 55, 56, 115, 150
network 135
– magnetic 134, 136, 153, 154
Neupert effect 167, 172
neutral current sheet 125, 152
neutral line 198
– magnetic 206
neutral point 193, 194
neutral-point discharge 193
neutron 180
neutron capture 180
neutrons
– flare-associated 201
– relativistic 182
– solar 209
Nimbus 7 satellite 267
nitrogen nuclei, excited 179
Nobeyama Radioheliograph 190
noise storms 159
non-potential magnetic fields 193, 194
non-recurrent geomagnetic storms 216, 263, 267
non-thermal bremsstrahlung 171, 174
non-thermal electrons 171, 172, 178
non-thermal hard X-ray bremsstrahlung 178
non-thermal motions 99
non-thermal radiation 158
non-thermal radio sources 210
non-thermal transition region 120
northern lights 219
nuclear energy 63
nuclear fusion 63
nuclear reactions 61, 63, 178, 179
nuclei, excited 178
number of sunspots 151, 207

occurrence frequency, coronal mass ejections 185
opacity 68
open magnetic field lines 49, 50
orbit 9
– Earth 254
– halo 3
Orbiting Solar Observatory 51
Orbiting Solar Observatory (OSO) 3 170, 207
Orbiting Solar Observatory (OSO) 4 170
Orbiting Solar Observatory (OSO) 5 170
Orbiting Solar Observatory (OSO) 6 170
Orbiting Solar Observatory (OSO) 7 182, 184, 208
oscillations, plasma 160
OSO *see* Orbiting Solar Observatory
outflow, solar-wind 154
outflow velocity, solar wind 50, 51, 153, 154
oxygen ion 140, 154
oxygen nuclei, excited 179
ozone layer 237, 239
ozone-destroying chemicals 239

pairs, bipolar 206
Parker spiral 42, 125
particles
– acceleration 207

– density, solar wind 31
– energetic 204
– heavy 143
– kinetic energy 35
– speeds 35
– thermal energy, solar wind 31
– thermal energy density, solar wind 31
permitted lines 107
phases
– gradual 169, 175
– impulsive 169, 175
– impulsive hard X-ray 173
– precursor 169
– thermal 173
physical properties,
coronal mass ejections 185
pick-up ions 148
Pioneer 11 57
plage 240, 242
plasma 24, 55, 150, 160
plasma clouds 216, 263
– interplanetary 203
– magnetized 203
plasma ejection 205, 263
plasma frequency 158, 161, 162, 237
plasma oscillations 160
plumes, polar 134, 135, 154
point
– magnetic neutral 206
– stagnation 148
– X-ray bright 152
polar cap absorptions 201
polar corona 144
polar coronal holes 48, 57, 153
polar plumes 134, 135, 154
polar regions 146
POLAR satellite 229, 230
poles 130
positron 55
positron annihilation,
gamma ray lines 182
post-eruption cusp 232
post-flare arches, X-ray 209
power
– bremsstrahlung 174
– solar flares 120
power generation plants 228
power grids 228
precession 254
precipitating electrons 220
precursor phase, solar flare 169
pre-eruption twist 232
preferentially accelerate heavier ions 154
pressure
– gas 40, 128, 148
– interstellar 148
– magnetic 40
– solar wind 128, 148
prominence 188
– eruptive 188–190
properties, physical, coronal mass
ejections 185
Prosteyshiy Sputnik 56
proton 154
proton and electron events,
interplanetary 203
proton density, fast wind 127
proton density, slow wind 127
proton excitation 180
proton flux, fast wind 127
proton flux, slow wind 127
proton kinetic energy, solar wind 31
proton kinetic energy density,
solar wind 31
proton speed, fast wind 127
proton speed, slow wind 127
proton temperature, fast wind 127
proton temperature, slow wind 127
proton velocities, solar wind 141
proton–proton chain 65
protons
– cosmic rays 32
– energy, interplanetary 200
– flux, interplanetary 200
– interplanetary 200
– solar 201
– solar wind 129, 153
P78-1 170, 187, 209

quadrupole moment 86, 93

radial order 79
radiation
– coherent 166
– gamma-ray 207
– non-thermal 158
– soft X-ray 168, 178
– synchrotron 163, 164, 206
– thermal 168
radiation belts
– Van Allen 214, 223, 225
– – inner 226
radioactive isotope 252
Radio Astronomy Explorer (RAE-1) 158
radio bursts
– impulsive 207

– solar 157, 158, 163
– type II 163, 203, 206
– type III 161, 163, 201, 206
– type IV 207
– U-type 207
radiocarbon 249
radio communication 44, 236
radio noise 206
– solar 158
radio scintillations 126, 129
radio sources, non-thermal 210
radio spikes 165, 166
radio transmission 263
radius of gyration 34
RAE-1 (Radio Astronomy Explorer) 158
rays 137
– anomalous cosmic 147
– block incoming cosmic 153
– cosmic 146, 147, 151, 153, 207, 252
– diffusion of cosmic 146
reaction
– nuclear 178, 179
– spallation 179
reconnection
– magnetic 188, 192, 194–199, 208, 210
– magnetic field 207
recurrent geomagnetic storm 218, 263
redshift 76
regions
– active 132
– bipolar 151
– polar 146
– solar active 165
– Sun's polar 146
relativistic electrons, simultaneous acceleration with energetic ions 209
relativistic neutrons 182
release of free magnetic energy 195
ribbons, chromatosphere flare 157, 194, 207
rotation rate 84

SAGE 70
satellite
– Alouette-I 158
– Explorer 1 28
– GEOTAIL 230
– Helios 1 and 2 152
– Lunik 2 28
– Mariner 2 29
– Nimbus 7 267
– POLAR 229
– Prosteyshiy Sputnik 28
– P78-1 170
– SOHO 3, 5, 6, 75, 82, 107, 117, 119, 130, 132, 134, 148, 154, 184, 187, 230
– Solar Maximum Mission (SMM) 52, 93, 169, 172, 182, 184, 201, 208, 209, 224, 269, 270
– Sputnik 28
– Ulysses 8–11, 129, 145, 201
– Yohkoh 12–15, 101, 110, 117, 119, 121, 130, 168, 172, 188, 190, 196, 197
satellites
– communication 222
– deconstruction by coronal mass ejection 224
– deconstruction by solar flares 224
– modern 222
– weather 222
scaling law, coronal loops 102, 103, 121
SCE (Ulysses) 10
scintillation
– interplanetary 126, 153
– radio 126, 129
sea-surface temperature 242, 267, 268
– global 242
sector, magnetic 124
shape, spiral 151
sheared, twisted coronal loops 199
sheet
– current 125
– neutral current 125, 152
shock waves 159, 160, 163, 200, 203, 204, 218
shocks, interplanetary 203, 204, 207, 210
sidereal rotation period 83
sigmoid shape 231, 268
sigmoids 230
simultaneous acceleration, relativistic electrons and energetic ions 209
size, heliosphere 150
size, coronal mass ejections 191
Skylab 57, 102, 152, 167, 184, 208, 224
slow stream
– composition differences 139
– temperature differences 139
slow solar wind 125, 127, 132, 139, 153, 154, 187
– acceleration 139
– composition 127
– density 127
– electron temperature 127
– helium temperature 127

– helium to proton abundance 127
– proton density 127
– proton flux 127
– proton speed 127
– proton temperature 127
– source 127
– sporadic 132
– temperature 127
small-scale magnetic fields 118
SMM (Solar Maximum Mission) satellite 52, 93, 172, 182, 184, 169, 201, 208, 209, 224, 269, 270
SMM, instruments
– ACRIM (Active Cavity Radiometer Irradiance Monitor & Ultraviolet to Infrared) 171
– Active Cavity Radiometer Irradiance Monitor & Ultraviolet to Infrared (ACRIM) 171
– C/P (Coronagraph and Polarimeter) 171
– Coronagraph and Polarimeter (C/P) 171
– Gamma Ray Spectrometer (GRS) 171
– GRS (Gamma Ray Spectrometer) 171
– Hard X-ray Burst Spectrometer (HXRBS) 171
– Hard X-ray Imaging Spectrometer (HXIS) 171
– HXIS (Hard X-ray Imaging Spectrometer) 171
– HXRBS (Hard X-ray Burst Spectrometer) 171
– UV Spectrometer and Polarimeter (UVSP) 171
– UVSP (UV Spectrometer and Polarimeter) 171
– X-ray Polychromator (XRP, with BCS and FCS) 171
– XRP (X-ray Polychromator), with BCS and FCS 171
soft X-ray flares 167, 207, 208, 210
soft X-ray radiation 168, 178
Soft X-ray Telescope (SXT) 13, 101, 110, 121, 131, 153, 188, 190, 196, 197, 210
soft X-rays 12, 166–168, 173
SOHO 3, 5, 6, 75, 82, 107, 117, 119, 130, 132, 134, 148, 154, 184, 187, 230
– CDS 6
– CELIAS 6
– COSTEP 6
– EIT 6, 135, 233
– ERNE 6
– GOLF 6
– home page 8
– instruments 5, 6
– LASCO 6, 154, 229
– Lost in Space 3
– MDI/SOI 6, 86, 94, 109
– mission 8
– scientific goals 8
– SUMER 6, 134
– SWAN 6, 148
– UVCS 6, 154
– VIRGO 6
solar active regions 165
solar activity 227
– back side of the Sun 91
– 11-year cycle 267
– low 251
– maximum 37
– minimum 37
– temperature fluctuation 245
– warning time 227
SOlar and Heliospheric Observatory *see* SOHO
solar atmosphere 24
solar constant 240, 267
solar corona 150, 208
solar corpuscles 25, 56, 151, 264
solar cosmic rays 201
solar cycle
– 11-year magnetic activity cycle 51, 191, 234, 235, 268
– length 242
– modulation of cosmic rays 267
– X-ray view 235
Solar Data Analysis Center 15
solar emission lines 170
solar energetic particle events 210
solar energy 266
solar flares 2, 91, 156, 166, 168, 169, 173, 178, 179, 181, 182, 192, 202, 205, 206, 208, 210, 216, 227, 236, 262, 264
– hard X-ray 207
– impulsive 171
– impulsive phase 171
– magnetic energy 207
– model 177
– phases 175
– power 207
– satellites 224
– two-ribbon 208
– X-rays 206
solar gamma ray lines 208

solar irradiance 240, 249
– total variations 266
– total of Earth 242
– variation, reconstruction 244
– varies over time scales of days and years 241
solar latitude 8
solar magnetic activity 11-year cycle 51, 191, 234, 235, 268
solar magnetic fields 41, 56, 151
Solar Maximum Mission *see* SMM
solar meteorology 88
solar models 94
solar neutrino experiments 65, 69
solar neutrino problem 65, 81, 94
solar neutrons 209
solar oblateness 93
solar oscillations, dispersion law 93
solar protons 201
solar protons, kill an unprotected astronaut 225
solar radiation 245
solar radio bursts 157, 158, 163
solar radio noise 158
solar storm, early warning 228
solar system 148
– edge 150
solar-type stars 250
– increasing brightness 268
– magnetic activity 268
solar ultraviolet radiation 239
solar variability 243
solar wind 2, 11, 27, 29, 30, 43, 56, 124, 127, 129, 130, 148, 151, 152, 264, 265
– helium abundance 152
– high-speed 29, 124, 125, 127, 130, 131, 134–136, 140, 143, 144, 153
– outflow velocity 154
– parameters 31
– slow-speed 29, 125, 127, 132, 139, 153, 154, 187
– speeds 130, 153
– variation 243
solid-body rotation 83
Solwind 184
Solwind coronagraph 187
sound speed 81, 82
sound waves 72, 73, 83, 94, 120, 121
– periods 80
source,
– coronal hard X-ray 199
– double hard X-ray 177
– fast wind 127
– hard X-ray 198
– high-speed solar wind 136
– slow wind 127
South Atlantic Anomaly 223
southern lights 219
Soviet–American Gallium Experiment 70
Space Environment Center 225
Space Shuttle 224
space weather 2, 51, 222, 224
– astronauts 224
– communication systems 222
– electric power grids 222
– endanger humans 224
– high-altitude aircraft crews 224
– radio communications 224
– radio navigation systems 222
space-borne coronagraphs 184
spacecraft *see* satellite
spallation 178
spallation reaction 179
spatial asymmetry, solar wind 129
spectra, dynamic 158
spectrohelioscope 156
speed
– coronal mass ejection 185, 186
– solar wind 130, 153
– sound 93
spherical harmonic degree 79
spherical harmonics 79
spikes, microwave 208
spiral, Parker 125
spiral shape, interplanetary magnetic field 56, 57, 151
Spörer Minimum 251
stagnation point 148
stalks, streamer 132
Starfish 222
stars
– accretion of interstellar matter 150
– brightness 250
– magnetic activity 250, 267
– solar-type 247, 250
– – increasing brightness 268
– – magnetic activity 268
– Sun-like 265
Stefan-Boltzmann law 64
storms
– geomagnetic 150, 205, 215, 217, 262–264
– – 27-day interval 263
– – non-recurrent 216, 263, 267
– – recurrent 218
– – sudden commencement 218

– magnetic 215, 262, 263, 266
– noise 159
– solar, early warning 228
streamer 132
– belt 124, 188
– coronal 129, 140, 153
– helmet 132, 137, 187, 188
– stalks 132
streams
– high-speed 152
– high-speed wind 151
Sudbury Neutrino Observatory 67
sudden depletions 208
sudden ionosphere disturbances 206
Suess "wiggles" 251
SUMER (SOHO) 6, 134, 136, 154
Sun
– absolute luminosity 64
– activity 155–210
– age 60
– atmosphere 20
– brightening 259
– central temperature 64
– chemical ingredients 21, 60
– chromosphere 21
– chorona 1, 21, 23, 24, 47, 137, 154, 160, 199
– density 60
– distance 61
– Earth climate 242
– effective temperature 64
– energetic charged particles 264
– escape velocity 61
– faint brightness 265
– faint-young-Sun paradox 260
– giant star 262
– growing luminosity 262, 264, 267
– growing size 262
– low luminosity 250
– luminosity 60, 64
– mass 60–62
– mean distance 60
– polar regions 130–160
– pressure 60
– principal chemical constituents 60
– radiation 60
– radio emission 24
– radius 60, 61
– slow growth in luminous intensity 259
– solar constant 60
– Sun's radio emission 23
– temperature 60
– ultraviolet radiation 237
– variability 267
– variable atmospheric carbon 14 266
– variable brightness 60
– volume 60
sunlight arriving at Earth
– distribution 266
– intensity 266
Sun-like stars 249, 265
– brightness variations 246, 249, 250
sunquakes 91
sunspot activity
– fluctuations in the radiocarbon, or carbon 14 265
sunspot cycle 51, 191, 215, 234, 235, 237, 242, 254, 265, 270, 271
– global surface temperatures 243
sunspot minimum, prolonged 266
sunspots 53, 54, 89, 94, 206, 262, 265, 267
– bipolar pairs 55, 206
– 80-year cycle 264
– 11-year cycle 51, 191, 215, 234, 235, 237, 242, 254, 265, 270, 271
– irradiance decrease 242
– irradiance increase 242
– magnetic fields 55, 150, 205
– number of 151, 207
– polarity 36
supergranular convection cells 109, 110, 120
Super-Kamiokande 67
SWAN (SOHO) 6, 148
SWICS (Ulysses) 10
SWOOPS (Ulysses) 10
SXT (Soft X-ray Telescope) 13, 101, 110, 121, 131, 153, 188, 190, 196, 197, 210
synchrotron radiation 163, 164, 206
system, solar 148

tails, comet ion 151
tau neutrinos 67
Telescope, Soft X-ray (SXT) 131, 153
temperature
– corona 24
– coronal hole 140
– coronal streamer 140
– kinetic 142
– solar wind 31, 127
temperature differences of fast and slow streams 139
temperature, fast wind 127
temperature, slow wind 127

temperature variations, global 244
terminal velocity 138
terrestrial clouds 242
thermal bremsstrahlung 174, 178
thermal energy 142
thermal phase, solar flare 173
thermal radiation 168
time delay
 of coronal mass ejections 186
time-distance helioseismology 86, 94
tops of magnetic loops 178
– electron acceleration 178
transition region 106, 109, 115
– falling down 107, 120
Transition Region and Coronal Explorer (TRACE) 111, 112, 117
travel time, from Sun to Earth 227
tree rings 249
troposphere 239
tunneling 61
turbulent convection 94
two-component (electrons and protons), solar wind 129, 152
two-ribbon solar flares 208
type I bursts 159
type II bursts 159, 163, 203, 206
type III bursts 158, 159, 161, 163, 201, 206, 207
– decimetric 173
type IV bursts 159, 160, 163, 207

ultraviolet 106, 234
UltraViolet Coronagraph Spectrometer *see* UVCS (SOHO)
ultraviolet flares 115
ultraviolet radiation, solar 44, 237, 239
Ulysses 8–11, 129, 145, 201
– COSPIN 10
– DUST 10
– EPAC/GAS 10
– GRB 10
– GWE 10
– HISCALE 10
– instruments 9, 10
– SCE 10
– scientific objective 11
– SWICS 10
– SWOOPS 10
– URAP 10
– VHM/FGM 10
Ulysses home page 11
unipolar areas 151
URAP (Ulysses) 10
U-type bursts 159, 207
UVCS (SOHO) 6, 140, 141, 145, 154

Van Allen radiation belts 214, 223, 225
– inner 226
Variability of solar IRradiance and Gravity Oscillations *see* VIRGO (SOHO)
variable gusty slow wind 124
variations in the Sun's radiative output 244
varying solar wind 152, 243
velocity
– coronal mass ejection 185, 186
– outflow 154
– proton 141
– solar wind 31, 121, 130, 153, 154
– sound 81, 82
– terminal 138
– type II bursts 158–161
– type III bursts 158–161
velocity of light 45
Very Large Array (VLA) 164, 165, 209, 210
VHM/FGM (Ulysses) 10
VIRGO (SOHO) 6, 77, 80, 82
visible disk 178
VLA *see* Very Large Array
volcanoes 245

warning time, solar activity 227
water vapor 247
wavelength 46
waves
– Alfvén 145, 152–154
– magnetic 143, 146
– magnetohydrodynamic 208
– shock 159, 160, 163
– sound 72, 73, 83, 94, 120, 121
WBS (Yohkoh) 13
weather, space 2, 51, 222, 224
weather satellites 222
white light 205, 262
white-light flares 156, 176, 177
wind, solar 2, 27, 29, 30, 124, 127, 129, 130, 148, 151, 152, 264, 265
– composition, fast 127
– composition, slow 127
– density, fast 127
– density, slow 127
– electron temperature, fast 127
– electron temperature, slow 127
– equatorial, slow-speed 130

– fast 29, 124, 125, 127, 130, 131, 134–136, 140, 143, 144, 152, 153
– helium temperature, fast 127
– helium temperature, slow 127
– helium to proton abundance, fast 127
– helium to proton abundance, slow 127
– high-speed 29, 124, 125, 130, 131, 134–136, 140, 143, 144, 153
– middle atmosphere 267
– proton density, fast 127
– proton density, slow 127
– proton flux, fast 127
– proton flux, slow 127
– proton speed, fast 127
– proton speed, slow 127
– proton temperature, fast 127
– proton temperature, slow 127
– slow 125, 127, 132, 139, 153, 154, 187
– source, fast 127, 136
– source, slow 127
– streams, high-speed 265
– temperature, fast 127
– temperature, slow 127
– varying, slow-speed 124, 152
WIND satellite 229, 230

X-ray bright points 47, 50, 113, 120, 152
X-ray bright points by magnetic reconnection 121
X-ray corona 102, 106
X-ray flares 166
– gradual soft 203
X-ray images 102
X-ray jet 113
X-ray loops 101, 102
X-ray post-flare arches 209
X-ray wavelengths 12, 166
X-rays 55, 206
– hard 12, 166, 171, 172
– soft 12, 166–168, 173

Yohkoh 12–15, 101, 110, 117, 119, 121, 130, 168, 172, 188, 190, 196, 197, 210, 229
– BCS 13
– hard X-ray imaging instruments 172
– home page 15
– HXT 13
– instruments 13
– scientific objective 15
– SXT 13
– WBS 13

Zeeman effect 38, 39